T0320815

Markov Chains and Dependability Theory

Dependability metrics are omnipresent in every engineering field, from simple ones to more complex measures combining performance and dependability aspects of systems. This book presents the mathematical basis of the analysis of these metrics in the most used framework, Markov models, describing both basic results and specialized techniques.

The authors first present both discrete and continuous-time Markov chains before focusing on dependability measures, which necessitate the study of Markov chains on a subset of states representing different user satisfaction levels for the modeled system. Topics covered include Markovian state lumping, analysis of sojourns on subsets of states of Markov chains, analysis of most dependability metrics, fundamentals of performability analysis, and bounding and simulation techniques designed to evaluate dependability measures. The book is of interest to graduate students and researchers in all areas of engineering where the concepts of life-time, repair duration, availability, reliability, and risk are important.

Gerardo Rubino is a Senior Researcher at INRIA, France. His main research interests are in the quantitative analysis of computer and communication systems through the use of stochastic models and different analysis techniques.

Bruno Sericola is a Senior Researcher at INRIA, France. His main research activities are in computer and communication systems performance evaluation, dependability and performability analysis of fault-tolerant systems, and applied stochastic processes.

Markov Chains and Dependability Theory

GERARDO RUBINO and BRUNO SERICOLA

Inria Rennes – Bretagne Atlantique, France

CAMBRIDGE
UNIVERSITY PRESS

CAMBRIDGE
UNIVERSITY PRESS

Shaftesbury Road, Cambridge CB2 8EA, United Kingdom

One Liberty Plaza, 20th Floor, New York, NY 10006, USA

477 Williamstown Road, Port Melbourne, VIC 3207, Australia

314–321, 3rd Floor, Plot 3, Splendor Forum, Jasola District Centre, New Delhi – 110025, India

103 Penang Road, #05–06/07, Visioncrest Commercial, Singapore 238467

Cambridge University Press is part of Cambridge University Press & Assessment, a department of the University of Cambridge.

We share the University's mission to contribute to society through the pursuit of education, learning and research at the highest international levels of excellence.

www.cambridge.org
Information on this title: www.cambridge.org/9781107007574

© Cambridge University Press & Assessment 2014

First published 2014

A catalogue record for this publication is available from the British Library

Library of Congress Cataloging-in-Publication data
Rubino, Gerardo, 1955– author.
Markov chains and dependability theory / Gerardo Rubino and Bruno Sericola.
   pages cm
Includes bibliographical references and index.
ISBN 978-1-107-00757-4 (hardback)
1. Markov processes.   I. Sericola, Bruno, author.   II. Title.
QA274.7.R83   2014
519.2′33–dc23   2013048010

ISBN   978-1-107-00757-4   Hardback

Contents

1   Introduction

1.1   Preliminary words

From the theoretical point of view, Markov chains are a fundamental class of stochastic processes. They are the most widely used tools for solving problems in a large number of domains. They allow the modeling of all kinds of systems and their analysis allows many aspects of those systems to be quantified. We find them in many subareas of operations research, engineering, computer science, networking, physics, chemistry, biology, economics, finance, and social sciences. The success of Markov chains is essentially due to the simplicity of their use, to the large set of theoretical associated results available, that is, the high degree of understanding of the dynamics of these stochastic processes, and to the power of the available algorithms for the numerical evaluation of a large number of associated metrics.

In simple terms, the Markov property means that given the present state of the process, its past and future are independent. In other words, knowing the present state of the stochastic process, no information about the past can be used to predict the future. This means that the number of parameters that must be taken into account to represent the evolution of a system modeled by such a process can be reduced considerably. Actually, many random systems can be represented by a Markov chain, and certainly most of the ones used in practice. The price to pay for imposing the Markov property on a random system consists of cleverly defining the present of the system or equivalently its state space. This can be done by adding a sufficient amount of information about the past of the system into the definition of the states. The theory of Markov models is extremely rich, and it is completed by a large set of numerical procedures that allow the analysis in practice of all sorts of associated problems.

Markov chains are at the heart of the tools used to analyze many types of systems from the point of view of their dependability, that is, of their ability to behave as specified when they were built, when faced with the failure of their components. The reason why a system will not behave as specified can be, for instance, some fault in its design, or the failure of some of its components when faced with unpredicted changes in the system's environment [3]. The area where this type of phenomenon is analyzed is globally called *dependability*. The two main associated keywords are *failures* and *repairs*. Failure is the transition from a state where the system behaves as specified to a state where this is not true anymore. Repair is the name of the opposite transition. Markov chains play a central role in the quantitative analysis of the behavior of a system that faces failure occurrences and possibly the repair of failed components, or at least of part of them. This book develops a selected set of topics where different aspects of these mathematical objects are analyzed, having in mind mainly applications in the dependability analysis of multicomponent systems.

In this chapter, we first introduce some important dependability metrics, which also allow us to illustrate in simple terms some of the concepts that are used later. At the same time, small examples serve not only to present basic dependability concepts but also some of the Markovian topics that we consider in this book. Then, we highlight the central pattern that can be traced throughout the book, the fact that in almost all chapters, some aspect of the behavior of the chains in subsets of their state spaces is considered, from many different viewpoints. We finish this Introduction with a description of the different chapters that compose the book, while commenting on their relationships.

1.2   Dependability and performability models

In this section we introduce the main dependability metrics and their extensions to the concept of performability. At the same time, we use small Markov models that allow us to illustrate the type of problems this book is concerned with. This section also serves as an elementary refresher or training in Markov analysis techniques.

1.2.1   Basic dependability metrics

Let us start with a single-component system, that is, a system for which the analyst has no structural data, and let us assume that the system can not be repaired. At time 0, the system works, and at some random time T, the system's *lifetime*, a failure occurs and the system becomes forever failed. We obviously assume that T is finite and that it has a finite mean. The two most basic metrics defined in this context are the Mean Time To Failure, MTTF, which is the expectation of T, MTTF $= \mathbb{E}\{T\}$, and the *reliability at time t*, $R(t)$, defined by

$$R(t) = \mathbb{P}\{T > t\},$$

that is, the tail of the distribution of the random variable, T. Observe that we have

$$\mathbb{E}\{T\} = \text{MTTF} = \int_0^\infty R(t)\,dt.$$

The simplest case from our Markovian point of view is when T is an exponentially distributed random variable with rate λ. We then have MTTF $= 1/\lambda$ and $R(t) = e^{-\lambda t}$. Defining a stochastic process $X = \{X_t, t \in \mathbb{R}^+\}$ on the state space $S = \{1, 0\}$ as $X_t = 1$ when the system is working at time t, 0 otherwise, X is a continuous-time Markov chain whose dynamics is represented in Figure 1.1.

Let us assume now that the system (always seen as made of a single component) can be repaired. After a repair, it becomes operational again as it was at time 0. This behavior then cycles forever, alternating periods where the system works (called *operational* or

Figure 1.1   A single component with failure rate λ and $X_0 = 1$

up periods) and those where it is being repaired and thus does not provide any useful work (called *nonoperational* or *down* periods). Thus, after a first failure at some time F_1, the system becomes nonoperational until it is repaired at some time $R_1 \geq F_1$, then it works until the occurrence of a second failure at some time $F_2 \geq R_1$, etc. Let us call $U_1 = F_1$ the *length of the first up period*, $D_1 = R_1 - F_1$ the *length of the first down period*, $U_2 = F_2 - R_1$ *the length of the second up period*, etc. Let us consider now the main case for this framework, which occurs when the two sequences $(U_i)_{i \geq 1}$ and $(D_j)_{j \geq 1}$ are both i.i.d. and independent of each other (this is called an *alternating renewal process* in some contexts).

In this model, there is an infinite number of failures and repairs. By definition, the MTTF is the mean time until the first system's failure:

$$\text{MTTF} = \mathbb{E}\{U_1\},$$

and

$$R(t) = \mathbb{P}\{U_1 > t\}.$$

We may now consider other relevant metrics. First, the Mean Time To Repair, MTTR, is given by

$$\text{MTTR} = \mathbb{E}\{D_1\},$$

and the Mean Time Between Failures, MTBF, is given by MTBF = MTTF + MTTR. The reliability at time t measures the *continuity* of the service associated with the system, but one may also need to know if the system will be operational *at time t*. We define the *point availability at time t*, $PAV(t)$, as the probability that the system will be working at t.

Assume now that the U_i are exponentially distributed with rate λ and that the D_j are also exponentially distributed with rate μ. We then have MTTF = $1/\lambda$, MTTR = $1/\mu$, and $R(t) = e^{-\lambda t}$. If we define a stochastic process $X = \{X_t, t \in \mathbb{R}^+\}$ such that $X_t = 1$ if the system works at time t, and $X_t = 0$ otherwise, X is the continuous-time Markov chain whose associated graph is depicted in Figure 1.2. Let us denote $p_i(t) = \mathbb{P}\{X_t = i\}$, $i = 1, 0$. In other words, $(p_1(t), p_0(t))$ is the distribution of the random variable X_t, seen as a row vector (a convention that is followed throughout the book). Solving the Chapman–Kolmogorov differential equations in the $p_i(t)$ and adding the initial condition $X_0 = 1$, we get

$$PAV(t) = \mathbb{P}\{X_t = 1\} = p_1(t) = \frac{\mu}{\lambda + \mu} + \frac{\lambda}{\lambda + \mu} e^{-(\lambda + \mu)t}.$$

This example allows us to introduce the widely used *asymptotic availability* of the system, which we denote here by $PAV(\infty)$, defined as $PAV(\infty) = \lim_{t \to \infty} PAV(t)$. Taking

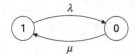

Figure 1.2   A single component with failure rate λ and repair rate μ; $X_0 = 1$

the limit in $p_1(t)$, we get $PAV(\infty) = \mu/(\lambda+\mu)$. Of course, if $\pi = (\pi_1, \pi_0)$ is the stationary distribution of X, we have $PAV(\infty) = \pi_1$. The stationary distribution, π, can be computed by solving the linear system of equilibrium equations of the chain: $\pi_1 \lambda = \pi_0 \mu$, $\pi_1 + \pi_0 = 1$.

1.2.2   More complex metrics

Let us now illustrate the fact that things can become more complex when dealing with more sophisticated metrics. Suppose we are interested in the behavior of the system in the interval $[0, t]$, and that we want to focus on how much time the system works in that interval. This is captured by the *interval availability on the interval* $[0, t]$, $IA(t)$, defined by the fraction of that interval during which the system works. Formally,

$$IA(t) = \frac{1}{t} \int_0^t 1_{\{X_s=1\}} ds.$$

Observe that $IA(t)$ is itself a random variable. We can be interested just in its mean, the *expected interval availability on* $[0, t]$. In the case of the previous two-state example, it is given by

$$\mathbb{E}\{IA(t)\} = \frac{1}{t} \int_0^t PAV(s)\, ds = \frac{\mu}{\lambda+\mu} + \frac{\lambda}{(\lambda+\mu)^2 t} \left(1 - e^{-(\lambda+\mu)t}\right).$$

If, at the other extreme, we want to evaluate the distribution of this random variable, things become more complex. First, see that $\mathbb{P}\{IA(t) = 1\} = e^{-\lambda t}$, that is, there is a *mass* at $t = 1$. Then, for instance in [5], building upon previous work by Takàks, it is proved that if $x < 1$,

$$\mathbb{P}\{IA(t) \leq x\} = 1 - e^{-\lambda xt}\left(1 + \sqrt{\lambda \mu xt} \int_0^{(1-x)t} \frac{e^{-\mu y}}{\sqrt{y}} I_1(2\sqrt{\lambda \mu xty})\, dy\right), \tag{1.1}$$

where I_1 is the modified Bessel function of the first kind defined, for $z \geq 0$, by

$$I_1(z) = \sum_{j \geq 0} \left(\frac{z}{2}\right)^{2j+1} \frac{1}{j!(1+j)!}.$$

In the well-known book by Gnedenko *et al.* [39], the following expression is proposed:

$$\mathbb{P}\{IA(t) \leq x\} = \sum_{n \geq 0} e^{-\mu(1-x)t} \frac{(\mu(1-x)t)^n}{n!} \sum_{k=n+1}^{\infty} e^{-\lambda xt} \frac{(\lambda xt)^k}{k!}. \tag{1.2}$$

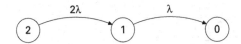

Figure 1.3 Two identical components in parallel with failure rate λ and no repair; $X_0 = 2$

Actually, there is an error in [39] on the starting index value of the embedded sum. The expression given here is the correct one.

In the book [85] by S. Ross, using the uniformization method (see Chapter 3 of this book if you are not familiar with this technique), the following expression is derived:

$$\mathbb{P}\{IA(t) \leq x\} = \sum_{n \geq 1} e^{-vt} \frac{(vt)^n}{n!} \sum_{k=1}^{n} \binom{n}{k-1} p^{n-k+1} q^{k-1} \sum_{i=k}^{n} \binom{n}{i} x^i (1-x)^{n-i}, \quad (1.3)$$

where $p = \lambda/(\lambda + \mu) = 1 - q$ and $v = \lambda + \mu$. In Chapter 6 this approach is followed for the analysis of the interval availability metric in the general case. This discussion illustrates that even for elementary stochastic models (here, a simple two-state Markov chain), the evaluation of a dependability metric can involve some effort.

The previous model is irreducible. Let us look at simple systems modeled by absorbing chains. Consider a computer system composed of two identical processors working in parallel. Assume that the behavior of the processors, with respect to failures, is independent of each other, and that the lifetime of each processor is exponentially distributed, with rate λ. When one of the processors fails, the system continues to work with only one unit. When this unit fails, the system is dead, that is, failed forever. If X_t is the number of processors working at time t, then $X = \{X_t, t \in \mathbb{R}^+\}$ is a continuous-time Markov chain on the state space $S = \{2, 1, 0\}$, with the dynamics shown in Figure 1.3. The system is considered operational at time t if $X_t \geq 1$.

There is no repair here. The MTTF of the system, the mean time to go from the initial state 2 to state 0, is the sum of the mean time spent in state 2 plus the mean time spent in state 1, that is,

$$\text{MTTF} = \frac{1}{2\lambda} + \frac{1}{\lambda} = \frac{3}{2\lambda}.$$

To evaluate the reliability at time t, which is given by

$$R(t) = \mathbb{P}\{U_1 > t\} = \mathbb{P}\{X_t \geq 1\},$$

we need the transient distribution of the model, the distribution $p(t)$ of the random variable X_t, that is, the row vector

$$p(t) = (p_2(t), \, p_1(t), \, p_0(t)).$$

After solving the Chapman–Kolmogorov differential equations satisfied by vector $p(t)$, we have

$$p_2(t) = e^{-2\lambda t}, \quad p_1(t) = 2e^{-\lambda t}(1 - e^{-\lambda t}), \quad p_0(t) = 1 - 2e^{-\lambda t} + e^{-2\lambda t}.$$

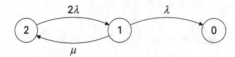

Figure 1.4   Two identical components in parallel, each with failure rate λ; a working component can repair a failed one, with repair rate μ; if both components are failed, the system is dead; $X_0 = 2$

We obtain

$$R(t) = p_2(t) + p_1(t) = 1 - p_0(t) = 2e^{-\lambda t} - e^{-2\lambda t}.$$

Observe that in this particular case, there is another elementary method to obtain the reliability function. Since the lifetime T of the system is the sum of two independent and exponentially distributed random variables (the sojourn times of X in states 2 and 1), the convolution of the two density functions of these sojourn times gives the density function of T. Integrating the latter, we obtain the cumulative distribution function of T:

$$\mathbb{P}\{T \le t\} = 1 - R(t) = \int_0^t \int_0^s 2\lambda e^{-2\lambda x} \lambda e^{-\lambda(s-x)} dx\, ds.$$

Now, suppose that when a processor fails, the remaining operational one can repair the failed unit, while doing its normal work at the same time. The repair takes a random amount of time, exponentially distributed with rate μ, and it is independent of the processors' lifetimes. If, while repairing the failed unit, the working one fails, then the system is dead since there is no operational unit able to perform a repair. With the same definition of X_t, we obtain the continuous-time Markov chain depicted in Figure 1.4.

The best way to evaluate the MTTF $= \mathbb{E}\{T\}$ is to define the conditional expectations, $x_i = \mathbb{E}\{T \mid X_0 = i\}$ for $i = 2, 1$, and write the equations

$$x_2 = \frac{1}{2\lambda} + x_1, \qquad x_1 = \frac{1}{\lambda + \mu} + \frac{\mu}{\lambda + \mu} x_2,$$

leading to

$$\text{MTTF} = x_2 = \frac{3\lambda + \mu}{2\lambda^2}.$$

For the reliability at time t, we must again solve for the distribution $p(t)$ of X_t. We obtain

$$p_2(t) = \frac{(\lambda - \mu + G)e^{-a_1 t} - (\lambda - \mu - G)e^{-a_2 t}}{2G},$$

$$p_1(t) = 2\lambda \frac{e^{-a_2 t} - e^{-a_1 t}}{G},$$

where

$$G = \sqrt{\lambda^2 + 6\lambda\mu + \mu^2},$$

and

$$a_1 = \frac{3\lambda + \mu + G}{2} > a_2 = \frac{3\lambda + \mu - G}{2} > 0.$$

This leads to

$$R(t) = \mathbb{P}\{U_1 > t\} = \mathbb{P}\{X_t \geq 1\} = p_2(t) + p_1(t) = \frac{a_1 e^{-a_2 t} - a_2 e^{-a_1 t}}{G}.$$

Let us include here a brief introduction to the concept of *quasi-stationary distribution*, used in Chapter 4. When the model is absorbing, as in the last example, the limiting distribution is useless: at ∞, the process will be in its absorbing state (with probability 1), meaning that $p(t) \rightarrow (0,0,1)$ as $t \rightarrow \infty$. But we can wonder if the conditional distribution of X_t knowing that the process is not absorbed at time t has a limit. When it exists, this limit is called the quasi-stationary distribution of process X. In the example, we have

$$\lim_{t \to \infty} \mathbb{P}\{X_t = 2 \,|\, X_t \neq 0\} = \lim_{t \to \infty} \frac{p_2(t)}{p_2(t) + p_1(t)} = \frac{G - \lambda + \mu}{3\lambda + \mu + G} = \frac{G - \lambda - \mu}{2\lambda},$$

and

$$\lim_{t \to \infty} \mathbb{P}\{X_t = 1 \,|\, X_t \neq 0\} = \lim_{t \to \infty} \frac{p_1(t)}{p_2(t) + p_1(t)} = \frac{4\lambda}{3\lambda + \mu + G} = \frac{3\lambda + \mu - G}{2\lambda}.$$

When the system is not repairable, the point availability and the reliability functions coincide ($PAV(t) = R(t)$ for all t). At the beginning of this chapter we used the elementary model given in Figure 1.2 where the system could be repaired, and thus $PAV(t) \neq R(t)$.

To conclude this section, let us consider the example given in Figure 1.5, and described in the figure's caption. Observe that the topology of the model (its Markovian graph) is the same as in the model of Figure 1.4 but the transition rates and the interpretation are different.

First of all, we have MTTF $= 1/\lambda$. The mean time until the system is dead, the mean absorption time of the Markov chain, can be computed as follows. If W is the absorption time, and if we denote $w_i = \mathbb{E}\{W \,|\, W_0 = i\}$, $i = 1, 0$, we have

$$w_1 = \frac{1}{\lambda} + w_0, \qquad w_0 = \frac{1}{\mu} + cw_1,$$

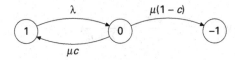

Figure 1.5   A single component with failure rate λ; there is a repair facility with repair rate μ; when being repaired, the system does not work; the repair can fail, and this happens with probability $1 - c$; the repair is successful with probability c, called the *coverage factor*, usually close to 1; if the repair fails, the system is dead; if the repair is successful, the system restarts as new

leading to

$$\mathbb{E}\{W\} = w_1 = \frac{1}{1-c}\left(\frac{1}{\lambda}+\frac{1}{\mu}\right).$$

The system can be repaired. This leads to $R(t) = e^{-\lambda t}$ and $PAV(t) = p_1(t)$. Solving the differential equations $p_1'(t) = -\lambda p_1(t) + \mu c p_0(t)$, $p_0'(t) = -\mu p_0(t) + \lambda p_1(t)$ with $p_1(0) = 1$, $p_0(0) = 0$, we obtain

$$PAV(t) = \frac{\mu - a_1}{H} e^{-a_1 t} + \frac{a_2 - \mu}{H} e^{-a_2 t},$$

where $H = \sqrt{(\lambda - \mu)^2 + 4(1-c)\lambda\mu}$, $a_1 = (\lambda + \mu - H)/2$ and $a_2 = (\lambda + \mu + H)/2$.

We are also interested in the total time, TO, during which the system is operational. Let us first compute its mean. Observing that the number of visits that the chain makes to state 1 (that is, the number of operational periods) is geometric, we have

$$\mathbb{E}\{TO\} = \sum_{n\geq 1} \frac{n}{\lambda}(1-c)c^{n-1} = \frac{1}{(1-c)\lambda}.$$

The distribution of TO is easy to derive using Laplace transforms. If $\widetilde{TO}$ denotes the Laplace transform of TO, we have

$$\widetilde{TO}(s) = \sum_{n\geq 1} \left(\frac{\lambda}{\lambda + s}\right)^n (1-c)c^{n-1} = \frac{(1-c)\lambda}{(1-c)\lambda + s},$$

that is, TO has the exponential distribution with rate $(1-c)\lambda$.

1.2.3   Performability

Consider the model of Figure 1.4 and assume that when the system works with only one processor, it generates r \$ per unit of time, while when there are two processors operational, the reward per unit of time is equal to αr, with $1 < \alpha < 2$. The reward is not equal to $2r$ because there is some capacity cost (some *overhead*) in being able to work with two parallel units at the same time. We can now look at the amount of money produced by the system until the end, T, of its lifetime. Let us call it R, and name it the *accumulated reward until absorption*. Looking for the expectation of R is easy. As for the evaluation of the MTTF, we use the conditional expectations, $y_i = \mathbb{E}\{R \mid X_0 = i\}$, $i = 2, 1$, which must satisfy

$$y_2 = \frac{\alpha r}{2\lambda} + y_1, \qquad y_1 = \frac{r}{\lambda + \mu} + \frac{\mu}{\lambda + \mu}y_2.$$

This leads to

$$\mathbb{E}\{R\} = y_2 = r\frac{2\lambda + \alpha(\lambda + \mu)}{2\lambda^2}.$$

This is an example of a performability metric: instead of looking at the system as either working or not, we now distinguish between two different levels when it works, since it does not produce the same reward when it works with two units or with only one.

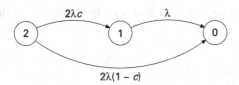

Figure 1.6   Two identical components in parallel; the failure rate is λ, the failure coverage factor is c; $X_0 = 2$

We can of course look for much more detailed information. To illustrate this and, at the same time, to provide more examples of Markovian models for the reader, we will use a variation of a previous example. We always have two identical processors working in parallel. When one of the two units fails, the system launches a procedure to try to continue working using the remaining one. This procedure works in general, but not always, and it takes very little time. We neglect this delay in the model, and we take into account the fact that the recovering procedure is not always successful by stating that when one of the units fails, the system continues to work with the other one with some fixed probability, c, and ends its life with probability $1 - c$. In the former case, when a new failure occurs, the remaining processor stops working and the whole system is dead. The parameter c is sometimes called the *coverage factor* in the dependability area, and its value is usually close to 1. Figure 1.6 shows the resulting absorbing model (to the previous assumptions we add the usual independence conditions on the events controlling the system's dynamics).

Let us first look at the previously considered metrics. If we just want to evaluate the MTTF of the system, we simply have

$$\text{MTTF} = \frac{1}{2\lambda} + c\frac{1}{\lambda} = \frac{1+2c}{2\lambda}.$$

The reader can check that this system behaves better than a single component from the point of view of the MTTF only if $c > 1/2$. The evaluation of the reliability at time t (or the point availability at time t: here, both metrics are identical) needs the transient distribution of the system. Solving the Chapman–Kolmogorov differential equations, we obtain

$$p_2(t) = e^{-2\lambda t}, \quad p_1(t) = 2ce^{-\lambda t}(1 - e^{-\lambda t}), \quad p_0(t) = 1 - 2ce^{-\lambda t} - (1 - 2c)e^{-2\lambda t}.$$

This gives

$$R(t) = PAV(t) = p_2(t) + p_1(t) = 2ce^{-\lambda t} + (1 - 2c)e^{-2\lambda t}.$$

Now, assume again that a reward of r \$ per unit of time is earned when the system works with one processor, and αr \$ per unit of time when it works with two processors, where $1 < \alpha < 2$. Denote the accumulated reward until absorption by R. The mean accumulated reward until absorption, $\mathbb{E}\{R\}$, is, using the same procedure as before,

$$\mathbb{E}\{R\} = \frac{\alpha r}{2\lambda} + c\frac{r}{\lambda} = r\frac{\alpha + 2c}{2\lambda}.$$

But what about the distribution of the random variable R? In Chapter 7, we show why, for any $x \geq 0$ and for $r = 1$,

$$\mathbb{P}\{R > x\} = \frac{2c}{2-\alpha}e^{-\lambda x} + \left(1 - \frac{2c}{2-\alpha}\right)e^{-2\lambda x/\alpha}.$$

Obtaining this distribution is much more involved. For instance, observe that this expression does not hold if $\alpha = 2$. In Chapter 7, the complete analysis of this metric is developed, for this small example, and of course in the general case.

1.2.4   Some general definitions in dependability

As we have seen in the previous small examples, we consider systems modeled by Markov chains in continuous time, where we can distinguish two main cases: either the model is irreducible (as in Figure 1.2), or absorbing (as in Figures 1.1, 1.3, 1.4, and 1.6). In all cases, we have a partition $\{U, D\}$ of the state space S; U is the set of *up* states (also called *operational* states), where the system works, and D is the set of *down* states (also called *nonoperational* states), where the system is failed and does not provide the service it was built for. For instance, in Figure 1.1, $U = \{1\}$ and $D = \{0\}$; in Figure 1.6, $U = \{2, 1\}$ and $D = \{0\}$.

We always have $X_0 \in U$ and, if a is an absorbing state, $a \in D$. Otherwise, the possible interest of the model in dependability is marginal. As we have already stated, transitions from U to D are called failures, and transitions from D to U are called repairs. Observe that this refers to the global system. For instance, if we look at Figure 1.4, the transition from state 1 to state 2 corresponds to the repair of a *component*, not of the whole system. We always have at least one failure in the model; we may have models without repairs (as in Figure 1.1 or in Figure 1.3).

With the previous assumptions, we always have at least one first sojourn of X on the set of up states, U. As when we described the example of Figure 1.2, the lengths in time of the successive sojourns of X in U are denoted by U_1, U_2, etc.; these sojourns are also called operational (or up) periods. The corresponding unoperational (or down, or nonoperational) periods (if any), have lengths denoted by D_1, D_2, etc. A first remark here is that these sequences of random variables need not be independent and identically distributed anymore, as they were in the model of Figure 1.2, neither do they need to be independent of each other. The analysis of these types of variables is the object of the whole of Chapter 5 in this book. To see an example where these sequences are not independent and identically distributed, just consider the one in Figure 1.4 but assume now that we add another repair facility that is activated when both processors are failed, that is, in state 0. This means that we add a transition from 0 to 1 with some rate, η. The new model is given in Figure 1.7.

We can observe that the first sojourn in $U = \{2, 1\}$ starts in state 2, while the remaining sojourns in U start in state 1. It is easy to check that, in distribution, we have $U_1 \neq U_2 = U_3 = \cdots$ (see next section where some details are given). Here, just observe that the mean sojourn times have already been computed on page 6, when the model in

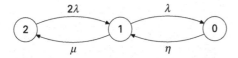

Figure 1.7 The model of Figure 1.4 with an additional repair facility that works when both processors are down; $X_0 = 2$

Figure 1.4 was discussed, and that they are different:

$$\mathbb{E}\{U_1\} = x_2 = \frac{3\lambda + \mu}{2\lambda^2},$$

and

$$\mathbb{E}\{U_k\} = x_1 = \frac{2\lambda + \mu}{2\lambda^2}, \quad \text{for } k \geq 2.$$

Coming back to the general concepts, for the two basic metrics MTTF and reliability at time t, the convention is to define them by MTTF $= \mathbb{E}\{U_1\}$ and $R(t) = \mathbb{P}\{U_1 > t\}$. This works in all cases, as there is always such a first up period.

The availability at time t is defined by $PAV(t) = \mathbb{P}\{X_t \in U\}$. The interval availability on $[0, t]$, $IA(t)$, is defined by the fraction of that interval during which the system works, i.e.

$$IA(t) = \frac{1}{t} \int_0^t 1_{\{X_s = 1\}} ds.$$

Recall that this is a random variable. The expected interval availability on $[0, t]$ is given by

$$\mathbb{E}\{IA(t)\} = \frac{1}{t} \int_0^t PAV(s) ds.$$

This closes the brief discussion about our basic model settings and the main dependability metrics that we deal with. Let us focus now on the behavior of a Markov model related to subsets of its state space.

1.3 Considering subsets of the state space

We move now to some issues related to the fact that we often have some interest in the behavior of the chain on subsets of its state space of our model, or on a partition of that state space. As we have just seen, this is typically the case when we consider basic dependability metrics such as the MTTF, the reliability, or the availability. Our objective is to introduce the reader to different specific topics developed in the book, and to some of the relationships between them.

1.3.1 Aggregation

Let us start with the general problem of state aggregation, or lumping. Consider the discrete-time homogeneous Markov chain $X = \{X_n, n \in \mathbb{N}\}$ whose dynamics is given

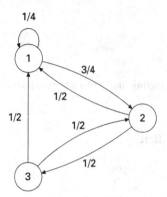

Figure 1.8   An example illustrating the concept of strong lumpability

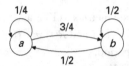

Figure 1.9   The aggregation of the chain depicted in Figure 1.8

in Figure 1.8. Assume for instance that $X_0 = 1$. Define a new process $Y = f(X)$ on the state space $\{a,b\}$, where function f defines a partition of $S = \{1,2,3\}$ in two classes, $f : \{1,2,3\} \mapsto \{a,b\}$, with $f(1) = a$, $f(2) = b$, and $f(3) = b$. That is, $Y_n = a \iff X_n = 1$ and $Y_n = b \iff X_n \in \{2,3\}$. It is not hard to see that Y is also a Markov chain; the reason is that if $Y_n = b$, the probability that $Y_{n+1} = a$ is equal to $1/2$, no matter which of the two states of $\{2,3\}$ X is in at time n. That is, given the state where Y is at step n, the future of Y is independent of the states visited by Y before time n. Observe that it does not matter where X starts (that is, the value of X_0) for the previous reasoning to hold. We say that X is *strongly lumpable* with respect to the partition $\{1\},\{2,3\}$ of its state space. The transition probability matrix of Y is obviously

$$\begin{pmatrix} \dfrac{1}{4} & \dfrac{3}{4} \\ \dfrac{1}{2} & \dfrac{1}{2} \end{pmatrix}.$$

Figure 1.9 shows the dynamics of Y (the same information given by the transition probability matrix of Y).

Let us now change the dynamics of X to the one given in Figure 1.10, keeping the same partition of the state space, that is, the same function, f. It is as easy as before to see that the new aggregated process Y is not Markov, no matter how the chain starts, that is, no matter the distribution of X_0: the sojourn time of Y in state b is exactly equal to 2, which is impossible with a Markov chain. To see it in a different way, let us give explicit cases where future and past events are dependent given the present of the chain.

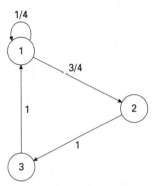

Figure 1.10 The aggregation of this chain is never Markov

For example,

$$\mathbb{P}\{Y_{n+1} = a \mid Y_n = b, Y_{n-1} = b\} = 1$$

but

$$\mathbb{P}\{Y_{n+1} = a \mid Y_n = b, Y_{n-1} = a\} = 0.$$

In the well-known book [50], Kemeny and Snell showed that a third type of behavior is possible: if X has for instance the dynamics given in Figure 1.11, then something curious happens. If $X_0 = 1$, then the aggregated Y is a Markov chain, while if $X_0 = 2$, it is not. This is called *weak lumpability*. It can be seen as an extension of the concept of strong lumpability, since the latter is a particular case of the former. What we have here is a situation where for some initial distributions of X, the aggregated process Y is Markovian (for instance, for the initial distribution $(1, 0, 0)$), while for others, as for the initial distribution $(0, 1, 0)$, it is not. Chapter 4 answers many questions associated with this type of model. These examples show that there are couples (chain, partition of the state space) where (i) the aggregated process is not Markov, (ii) the aggregated process is Markov for some initial distributions, and, as a particular case, (iii) the aggregated process is always Markov, that is, for any initial distribution. Chapter 4 deals with these ideas, and in particular it explores in depth the weak lumpability concept. At this point we recognize the work done by Kemeny and Snell in [50]: our contributions on the topic, which are the subject of Chapter 4, come from the analysis of the phenomenon presented in their book.

Why is all this of interest? The first obvious comment is that if our target is Y rather than X, the fact that Y is Markov is an important advantage, because of the previously mentioned benefits that come with the Markov property: a solid theory and a rich set of algorithmic procedures. Moreover, observe that Y is also *smaller* than X, even if in the previous three-state examples the difference in size between X and Y is minimal. Last, in most of the applications, especially in dependability models, there is a partition of the state space with respect to which the aggregation process Y has all the information

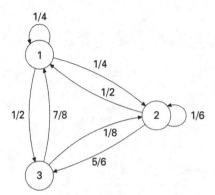

Figure 1.11   The aggregation of this chain is sometimes Markov (this example comes from [50])

we need. Let us come back to the situation where process X represents the behavior of some system subject to failures and repairs. The process lives in the state space S, which is partitioned into the two subsets, U (the set of up states) and D (the set of down states). If f is the function $f : S \longrightarrow \{u,d\}$, where if $x \in U, f(x) = u$ and if $x \in D, f(x) = d$, then process $Y = f(X)$ allows us to define most dependability metrics of the system. For instance, $PAV(t) = \mathbb{P}\{Y_t = u\}$.

1.3.2   Lumping and performance evaluation

Let us look first at the performance area. Assume that we are using a continuous-time Markov model $X = \{X_t, t \in \mathbb{R}\}$ for a performance evaluation of some system. In this area, most of the time we are interested in steady-state behavior. The state space of X is S and the distribution of the random variable X_t is $p(t) = (p_j(t))_{j \in S}$, where $p_j(t) = \mathbb{P}\{X_t = j\}$. To be more specific, assume that X_t is the state of an open queuing network at time t, composed of M nodes of the $./M/1$ type. We then have $S = \mathbb{N}^M$.

If the network is a Jackson one (Poisson arrivals and Bernoulli routing, plus the usual independence assumptions), then, in the stable case, we know explicitly the steady-state distribution $\pi = \lim_{t \to \infty} p(t)$, which is of the product-form type: if $\pi = (\pi_n)_{n \in S}$, then it is of the following form:

$$\pi_n = \prod_{i=1}^{M} (1 - \varrho_i)\varrho_i^{n_i}, \quad n = (n_1, \dots, n_M), \quad n_i = \text{\# of customers at node } i,$$

where $\varrho_i < 1$ is the load at station i. If we have some state-dependent form of routing between the nodes of the network, then in general π is unknown in closed form. If we need it, we must solve the linear equilibrium equations (in this example, a formidable task, since the state space is infinite), or try to bound it (see Chapter 10), or to simulate it.

Vector π however, contains too much information. We are typically interested in quantities such as the mean number of customers at, say, node 1. What we want to point out here is the fact that the mean number of customers at station 1 is the expectation of the asymptotic distribution of the stochastic process Y on the state space $\mathbb{N}$, where Y_t is the number of customers at station 1 at time t. Now, Y is the projection of X over its first coordinate, and it is also the aggregation of X with respect to the partition $(S_0, S_1, S_2, \ldots)$ of S, where S_k is the set of states of S in which there are k customers in node 1. Of course, in the Jackson case, we know explicitly this expectation (equal to $\varrho_1/(1 - \varrho_1)$).

This situation appears quite frequently: we can build a Markov chain on some large state space S, and we are interested in some metric built on the aggregation Y of X associated with some partition of S. Obviously, we would like to know whether Y is Markov or not, because if it is, all the theoretical results and algorithmic tools available for these stochastic processes are there to help us in the analysis of Y.

We can illustrate another concept discussed in Chapter 4 with a Jackson network. Using the results of that chapter, we can check that the aggregated process just discussed, the number of customers in node 1, is never Markov for any initial distribution of the Jackson model. In other words, we never have the strong lumpability property. For the weak lumpability one, the question is open.

Call $X = (N_1, \ldots, N_M)$ the Markov chain describing the configuration of the queuing network where N_i is the number of customers at node i and M is the number of nodes. The distribution of N_1 in equilibrium is obviously known: $\mathbb{P}\{N_1 = n\} = (1 - \varrho_1)\varrho_1^n$, where ϱ_1, the load at node 1, is equal to F_1/μ_1, with μ_1 the service rate of node 1 and F_1 the mean throughput of that node. The mean throughputs are obtained by solving the linear system of flow conservation equations. For the equilibrium regime to exist, we have the necessary and sufficient condition $F_i < \mu_i$ for each node i. Since the only way for N_1 to go from state $j \geq 1$ to $j - 1$ triggered by only one event is by a departure from that node, we immediately think of N_1 as the birth-and-death process of the $M/M/1$-type with arrival rate F_1 and departure rate μ_1. This is not in contradiction with the preceding remarks. This birth-and-death process, that can always be defined, is not in general the aggregation (or the projection) of the vector state X of the queuing network with respect to the appropriate partition of its state space: it is a Markov chain by construction, having the property of sharing with the aggregated process N_1 the same stationary distribution. What can be said is that if N_1 is Markov for some initial distribution of X, then it is *that* birth-and-death process. This birth-and-death process is called the *pseudo-aggregation* of X with respect to the considered partition; it always exists, even if X is not weakly lumpable (see Chapter 4 for the details).

To clarify the point, consider again the model given in Figure 1.10. We have seen that the aggregation of this chain with respect to the partition $\{1\}, \{2, 3\}$ is never Markovian, whatever the initial distribution of X. The steady-state distribution of X is $\pi = (0.4, 0.3, 0.3)$. We can *build* a new chain Z on the state space $\{a, b\}$ with transition probability matrix $\tilde{P}$ as follows:

$$\tilde{P} = \begin{pmatrix} \frac{1}{4} & \frac{3}{4} \\ x & 1-x \end{pmatrix},$$

and choose x such that if $\tilde{\pi}$ is the steady-state distribution of Z, then $\tilde{\pi}_a = \pi_1$ and $\tilde{\pi}_b = \pi_2 + \pi_3$. We have

$$x = \frac{\pi_3}{\pi_2 + \pi_3} = \frac{1}{2},$$

and we then see that Z actually has the dynamics depicted in Figure 1.9. We refer again to Chapter 4 for in-depth discussions of these points.

Let us look now at another typical situation. Consider the queuing model $M/PH/c/N$, where the service time distribution has $K \geq 1$ phases and the total storage capacity is $N \geq c$. When several servers are available to a customer to be served, any of them is chosen (say, the one with the smallest index). When dealing with this system to evaluate, for instance, the mean number of customers in equilibrium, we can define a Markov chain X over states of the form $(m; h_1, h_2, \dots)$ where m is the number of customers in the system (including those in service, if any), h_1 is the phase of the service of the first customer in service, for instance in FIFO order (or in any other order), h_2 is the phase of the service of the second customer being served, etc. Formally,

$$S = \{0\} \cup A \cup B,$$

where state 0 represents the empty system,

$$A = \bigcup_{j=1}^{c-1} \{ (j; h_1, \dots, h_j) \mid h_1, \dots, h_j \in \{1, 2, \dots, K\} \}$$

is the set of states where the system is not empty and there are strictly less than c customers in it, and

$$B = \bigcup_{j=c}^{N} \{ (j; h_1, \dots, h_c) \mid h_1, \dots, h_c \in \{1, 2, \dots, K\} \}$$

is the set of states where there are at least c customers in the system. We have

$$|S| = f(K, c, N)$$
$$= 1 + K + K^2 + \cdots + K^{c-1} + (N - c + 1)K^c$$
$$= \frac{K^{c+1} - 1}{K - 1} + (N - c)K^c, \quad \text{if } K \geq 2.$$

If $K = 1$, then $|S| = N + 1$. Other people will solve the same problem by defining a Markov chain Y over the state space

$$S' = \{ (n; m_1, \dots, m_K) \mid 0 \leq n \leq N, \, 0 \leq m_k \leq c, \, 1 \leq k \leq K,$$
$$\text{and } m_1 + \cdots + m_K = \min(c, n) \},$$

where n is again the total number of customers in the system and m_k is the number of customers in service whose phase is k. We have

$$|S'| = g(K,c,N) - \sum_{n=0}^{c-1} \binom{K+n-1}{n} + (N\ c+1)\binom{K+c-1}{c}.$$

Both ways are correct. This is because Y is the aggregation of X with respect to the appropriate partition and that strong lumpability holds here. The partition associates with state $(n;m_1,\ldots,m_K)$ of Y the following subset of S:

$$\left\{ (n;h_1,\ldots,h_{n\wedge c}) \in S \mid \forall k \in \{1,\ldots,K\}, \sum_{i=1}^{n\wedge c} 1_{\{h_i=k\}} = m_k \right\},$$

where $n \wedge c = \min(n,c)$. The mean number of customers in the model can be computed from the stationary distribution of X or from the stationary distribution of Y. Using chain Y is in a sense better, because we can easily check algebraically that $|S'| \leq |S|$. Actually, if $K \geq 2$ or if $c \geq 2$, $|S'| < |S|$. For instance,

$$f(4,10,20) = 11\,883\,861 \quad \text{and} \quad g(4,10,20) = 3\,861.$$

1.3.3 Lumping and dependability models

The same observations from the previous subsection also hold in a dependability context. Observe that there is always a first natural partition of the state space of the model into the subsets of up and down states. But the previous remarks can be made about many dependability models where the partition has more than two classes; just think of Machine Repairman Models with many classes of components, complex routing procedures, etc., and where we are interested in the behavior of some specific repairing system, or in the relationship between dependability properties of the system and on a specific class of components.

Let us now come back to the concept of up or operational periods (or the corresponding down periods). Considering again the example in Figure 1.7, a quick way of obtaining the distributions of the operational periods in this example is using Laplace transforms. Let us denote by $\tilde{U}_k$ the Laplace transform of U_k, when this random variable exists; and denote by $\tilde{f}$ the Laplace transform of the holding time of X in state 2, and by $\tilde{g}$ the corresponding transform associated with state 1, that is, $\tilde{f}(s) = 2\lambda/(s+2\lambda)$ and $\tilde{g}(s) = (\lambda+\mu)/(s+\lambda+\mu)$. To derive expressions of these $\tilde{U}_k$s, we condition with respect to the number N of repairs of a failed unit until the system's failure, that is, until the failure of both units. Using the notation $p = \lambda/(\lambda+\mu) = 1 - q$, we have $\mathbb{P}\{N = j\} = pq^j$, $j \geq 0$, and thus,

$$\tilde{U}_1(s) = \sum_{j=0}^{\infty} pq^j [\tilde{f}(s)\tilde{g}(s)]^{j+1} = \frac{p\tilde{f}(s)\tilde{g}(s)}{1 - q\tilde{f}(s)\tilde{g}(s)} = \frac{2\lambda^2}{2\lambda^2 + (3\lambda+\mu)s + s^2}.$$

A similar argument for any $\tilde{U}_k$, $k \geq 2$, gives

$$\tilde{U}_k(s) = \sum_{j=0}^{\infty} pq^j \tilde{g}(s) [\tilde{f}(s)\tilde{g}(s)]^j = \frac{p\tilde{g}(s)}{1 - q\tilde{f}(s)\tilde{g}(s)} = \frac{\lambda(2\lambda + s)}{2\lambda^2 + (3\lambda + \mu)s + s^2}.$$

In this case, these sojourn times are independent of each other. This is a consequence of the fact that we are in a Markov model and that we start the first sojourn in U with state 2, and any other operational period with state 1. Inverting the Laplace transform we obtain the density of the corresponding operational period. For instance, let us look at $\mathbb{P}\{U_k > t\}$ for $k \geq 2$, that we call here $R_2(t)$. The Laplace transform of R_2 is

$$\tilde{R}_2(s) = \frac{1 - \tilde{U}_k(s)}{s} = \frac{2\lambda + \mu + s}{2\lambda^2 + (3\lambda + \mu)s + s^2}.$$

Inverting, we have

$$R_2(t) = \frac{G - \lambda - \mu}{2G}e^{-a_1 t} + \frac{G + \lambda + \mu}{2G}e^{-a_2 t},$$

where G, a_1, and a_2 were defined on page 6, Subsection 1.2.2.

In the previous example, the lengths of the successive operational periods starting from the second one are i.i.d. Consider now a system composed of two components having the same failure rate λ, but with different repair times. Assume there is only one repair facility, and that the repair time of component i has rate μ_i, $i = 1,2$. A Markov model X representing the evolution of this system is given in Figure 1.12, where the caption explains the meaning of the states.

In this example, we just look at the expectations $\mathbb{E}\{U_k\}$ for $k \geq 1$. In Chapter 5 a more complete analysis of these types of objects is performed. Let us denote by h_x the mean holding time of X in state x. We have $h_{1,1} = 1/(2\lambda)$, $h_{0,1} = 1/(\lambda + \mu_1)$, and

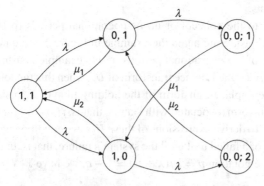

Figure 1.12   Two components with the same failure rate λ and different repair rates μ_1, μ_2; one repair facility; in state $(1,1)$, both components are up; in state $(0,1)$, component 1 is down and component 2 is up; in state $(1,0)$, 1 is up, 2 is down; state $(0,0;i)$ means that both components are down and that the one being repaired is component i; the set U of up states is $\{(1,1),(1,0),(0,1)\}$

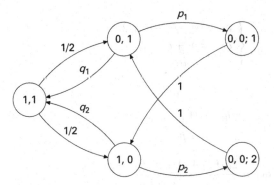

Figure 1.13 The canonical discrete-time Markov chain embedded in the Markov model of Figure 1.12

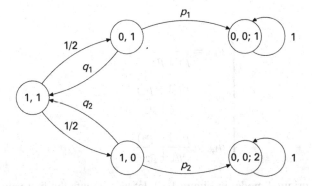

Figure 1.14 Auxiliary discrete-time Markov chain Z built from the model in Figure 1.13 in which states $(0,0;1)$ and $(0,0;2)$ are now absorbing

$h_{1,0} = 1/(\lambda + \mu_2)$. Denote also $p_i = \lambda/(\lambda + \mu_i)$ and $q_i = \mu_i/(\lambda + \mu_i)$, $i = 1,2$. The canonical discrete-time Markov chain embedded in X is shown in Figure 1.13.

Last, consider the auxiliary discrete-time Markov chain built from the previous one by making states $(0,0;1)$ and $(0,0;2)$ absorbing, as shown in Figure 1.14, which is denoted by Z.

Let us denote by T the time to absorption of X (Figure 1.12), that is, $T = \inf\{t > 0 \mid X_t \in \{(0,0;1),(0,0;2)\}\}$. Introduce also $\tau_x = \mathbb{E}\{T \mid X_0 = x\}$ and $a_x = \mathbb{P}\{X_T = (0,0;1) \mid X_0 = x\}$, $x \in \{(1,1),(0,1),(1,0)\}$. The a_xs are called absorption probabilities (or more appropriately here, conditional absorption probabilities). The linear systems that allow us to compute the τ_xs and the a_xs are given below:

$$\begin{cases} \tau_{1,1} = h_{1,1} + \tau_{0,1}/2 + \tau_{1,0}/2, \\ \tau_{0,1} = h_{0,1} + q_1\tau_{1,1}, \\ \tau_{1,0} = h_{1,0} + q_2\tau_{1,1} \end{cases}$$

and

$$\begin{cases} a_{1,1} = a_{0,1}/2 + a_{1,0}/2, \\ a_{0,1} = p_1 + q_1 a_{1,1}, \\ a_{1,0} = q_2 a_{1,1}. \end{cases}$$

Solving them, we get

$$\begin{cases} \tau_{1,1} = \dfrac{2h_{1,1} + h_{0,1} + h_{1,0}}{p_1 + p_2}, \\[2mm] \tau_{0,1} = \dfrac{2q_1 h_{1,1} + (1+p_2)h_{0,1} + q_1 h_{1,0}}{p_1 + p_2}, \\[2mm] \tau_{1,0} = \dfrac{2q_2 h_{1,1} + q_2 h_{0,1} + (1+p_1)h_{1,0}}{p_1 + p_2} \end{cases}$$

and

$$\begin{cases} a_{1,1} = \dfrac{p_1}{p_1 + p_2}, \\[2mm] a_{0,1} = \dfrac{p_1(1+p_2)}{p_1 + p_2}, \\[2mm] a_{1,0} = \dfrac{p_1(1-p_2)}{p_1 + p_2}. \end{cases}$$

Going back to our original model in Figure 1.12, let us compute the first values in the sequence $(\mathbb{E}\{U_k\})_{k \geq 1}$. For the first one, we have $\mathbb{E}\{U_1\} = \tau_{1,1}$. Now, for the mean of the second operational period, we see that it can start in state $(0,1)$ or state $(1,0)$. It will start in state $(1,0)$ if the first operational period ends in state $(0,1)$, that is, if the auxiliary absorbing model Z is absorbed by state $(0,0;1)$ starting in state $(1,1)$; finally, the second up period will start in state $(0,1)$ if the auxiliary chain is absorbed by $(0,0;2)$ (always starting in $(1,1)$). This leads to

$$\mathbb{E}\{U_2\} = a_{1,1}\tau_{1,0} + (1-a_{1,1})\tau_{0,1}.$$

Going one more step in the sequence, we need again to condition on the state in which the second operational period ends:

$$\mathbb{E}\{U_3\} = a_{11}\big[a_{1,0}\tau_{1,0} + (1-a_{1,0})\tau_{0,1}\big] + (1-a_{1,1})\big[a_{0,1}\tau_{1,0} + (1-a_{0,1})\tau_{0,1}\big],$$

which can be written

$$\mathbb{E}\{U_3\} = A_{3;1,0}\tau_{1,0} + A_{3;0,1}\tau_{0,1},$$

where

$$A_{3;1,0} = a_{1,1}a_{1,0} + (1-a_{1,1})a_{0,1},$$

and

$$A_{3;0,1} = a_{1,1}(1 - a_{1,0}) + (1 - a_{1,1})(1 - a_{0,1}).$$

Continuing in this way, we can see that the expectations of the successive operational periods are all different. To be more specific, denoting $\psi_i = \mu_i/\lambda$ and $i = 1, 2$, we have, after some algebra, for the two first moments,

$$\mathbb{E}\{U_1\} = \frac{3 + 2\psi_1 + 2\psi_2 + \psi_1\psi_2}{\lambda(2 + \psi_1 + \psi_2)}$$

and

$$\mathbb{E}\{U_2\} = \frac{4 + 5\psi_1 + 5\psi_2 + 4\psi_1\psi_2 + 2\psi_1^2 + 2\psi_2^2 + \psi_1\psi_2^2 + \psi_1^2\psi_2}{\lambda(2 + \psi_1 + \psi_2)^2}.$$

This kind of analysis, not only for the first moments but for the distributions as well, is performed in detail in Chapter 5, together with obvious questions such as the possible convergence of these moments or of the corresponding distributions. To show an example of a model where the nth operational period may not exist, just consider the model in Figure 1.5. We see that there is an nth operational period if and only if the first $n - 1$ repairs are successful, which happens with probability c^{n-1}. Observe that if the first n operational periods exist, their lengths are i.i.d. with common exponential distribution, with rate λ. Similar remarks hold for the periods where the system is being repaired (that is, all the down periods except the last one).

1.3.4 Using lumping to evaluate numerical procedures

Suppose you have designed a new procedure to evaluate a complex metric, for instance, the distribution of the Interval Availability $IA(t)$ on any interval $[0, t]$, for any finite Markov chain where the state space is partitioned into two classes U and D. You want to check your procedure, or to analyze its behavior as the size of the model increases, or as its *stiffness* grows. The stiffness of the model is the maximal ratio between pairs of transition rates. A model with a high stiffness may lead to numerical problems when solving it. This is developed in detail in Chapter 6.

One simple procedure to do this is to build a family of Markovian models where some parameters control the size, or the stiffness, or both, and to make it strongly lumpable with respect to some partition $\{U, D\}$. If such a model X is built, then the distribution of the Interval Availability is the same as the distribution of the same metric in the example of Figure 1.2. That is, for any X in the considered family of models, the distribution of the Interval Availability is given by any of the expressions (1.1), (1.2), or (1.3). As an example, consider the family of models represented in Figure 1.15.

If Q is the transition rate matrix of this chain, then $Q_{1,2}, Q_{2,3}, \ldots, Q_{M-1,M}$ and $Q_{M+1,M+2}, Q_{M+2,M+3}, \ldots, Q_{M+N-1,M+N}$ can be any positive real numbers, the following statement will always be valid: computing the distribution of $IA(t)$ in the model of Figures 1.15 and 1.2 gives exactly the same result (that is, the distribution of $IA(t)$ in the model of Figure 1.15 is given by (1.1) or (1.2) or (1.3)). We can use the rates inside U

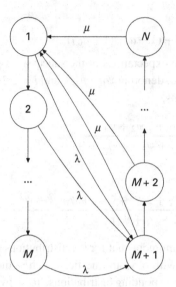

Figure 1.15   Model X: a single component Coxian lifetime (M phases) and Coxian repair time (N phases); $U = \{1, 2, \ldots, M\}$ and $D = \{M+1, M+2, \ldots, M+N\}$; this chain is strongly lumpable with respect to the partition $\{U, D\}$ and the aggregated process Y is the model in Figure 1.2

and/or D to control the stiffness of the model. For instance, assume $M \geq 3$. We choose a real $a > 1$ and we set $Q_{i,i+1} = 1 + (i-1)(a-1)/(M-2)$ for $i = 1, 2, \ldots, M-1$. The largest ratio between rates inside set U is thus equal to a. We can thus control the stiffness of the model through this parameter a. The same operation can be done on the rates inside D. We used this procedure and similar ones to check several algorithms of the type described in this book. Observe that we are free to build any transition graph topology inside sets U and D: provided X is irreducible, strong lumpability will hold. Playing with M and N allows us to control the size of the model, and choosing the value of a allows us to control the stiffness. For studying the accuracy of the method being examined, expressions (1.1), (1.2), or (1.3) provide the exact value of the distribution.

As a last comment, observe also that we can follow the same procedure with more complex models than the two-state one used as the aggregation of the model in Figure 1.15. In other words, we can start from another simple model, for instance the one in Figure 1.5, and define a Markov chain having arbitrary size and stiffness, such that the strong lumpability property holds and that the aggregated process is the one in Figure 1.5.

1.4   The contents of this book

The content of the remaining chapters of the book is described now.

In Chapter 2 we study the behavior of finite discrete-time homogeneous Markov chains. We analyze the transience, the recurrence, the irreducibility, and the periodicity properties that are needed in the analysis of the stationary distribution of these chains.

We consider the total number of visits to a fixed state and we show how the distribution of this number of visits is related to the transient or recurrent characteristic of this state. The convergence to steady-state is analyzed and we give a proof of the ergodic theorem. Absorbing Markov chains, that are of great importance in the dependability evaluation of systems, are also analyzed in detail and we describe an accurate algorithm for the computation of the solution to linear systems involved in various dependability measures.

Chapter 3 is devoted to the study of finite continuous-time Markov chains. After a classic review of the main properties of the states successively visited and of the transition function matrix, we show how to obtain the backward and forward Kolmogorov equations in both differential and integral versions. Usually, in almost all the books on Markov chains, the analysis of the continuous-time case is related to the discrete-time case through the use of the embedded discrete-time Markov chain at the instants of transition of the continuous-time Markov chain. In this book, we choose to analyze the continuous-time case using the uniformization technique, which consists of the construction of a stochastically equivalent chain whose transitions occur at the jump instants of a Poisson process. Using this construction, we obtain, as in the discrete-time case, the stationary behavior of these processes, the convergence to equilibrium, and the ergodic theorem. We also analyze, as in Chapter 2, the behavior of absorbing continuous-time Markov chains.

Chapter 4 studies the conditions under which the aggregated process of a finite homogeneous Markov chain with respect to a given partition of its state space is also Markov homogeneous. This is done first in discrete time, and then in continuous time. The analysis is different in the irreducible and the absorbing cases. In the former, the stationary distribution plays a central role; this role is played by the quasi-stationary distribution in the latter. One of the important results of the chapter is a finite characterization of the weak lumpability situation, given in a constructive way. Together with the study and characterization of the weak and the strong lumpability properties, we define what we call pseudo-aggregation of Markov chains (and that other authors call *exact aggregation*, see Chapter 10), an important concept in applications. It is actively used in that chapter, and it also appears in Chapter 5.

Chapter 5 studies the time a given Markov chain spends in a subset of its state space. This is done first for discrete-time chains, then in continuous time. We analyze the distribution of the nth sojourn of the chain in this subset of states, when it exists, and its moments. We also look at the asymptotic behavior of these quantities as n tends to infinity. Last, we make a connection between these sojourn times and the holding times of the pseudo-aggregation of the chain, defined and studied in Chapter 4, when one of the members of the associated partition is the considered subset of states. The results given here appear in other parts of the book, and are also at the heart of many important dependability metrics.

In Chapter 6, we consider again the occupation time, that is, the time spent by the process in a given subset of the state space, but now during a finite interval of time. This random variable is called interval availability in dependability. We first obtain its distribution in the discrete-time case and then we generalize this result to the continuous-time case using order statistics from the uniform distribution and the conditional distribution of the jumps in a Poisson process. By means of the uniformization technique, we explicitly show how the distribution of the occupation time in the continuous-time case is related to that in the discrete-time case.

In Chapter 7, we consider the distribution of weighted sums of occupation times. This random variable is called performability in the dependability analysis of fault-tolerant systems. A real number, called reward, is associated with each state of the Markov chain. The performability is then defined by the accumulated reward during a finite interval of time. In this chapter, we obtain the forward and backward partial differential equations describing the behavior of the distribution of this random variable. We then show how these equations lead, using the uniformization technique, to very simple expressions of that distribution. Based on these expressions, we describe an algorithm to compute this distribution, and a formal example illustrates these results.

Point availability and expected interval availability are dependability measures defined by the probability that a system is in operation at a given instant and by the mean percentage of time during which a system is in operation over a finite observation period, respectively. We consider a repairable system and we assume that the system is modeled by a finite Markov chain. We present in Chapter 8 an algorithm to compute these two availability measures. Again, this algorithm is based on the uniformization technique in which a test to detect the stationary behavior of the Markov chain is used to stop the computation if stationarity is approximately reached. In that case, the algorithm gives not only the transient availability measures, but also the steady-state availability measures, with significant computational savings, especially when the time at which measures are needed is large. In the case where the stationarity is not reached, the algorithm provides the transient availability measures and bounds for the steady-state availability. We also show how the algorithm can be extended to the computation of performability measures.

In Chapter 9, we deal with the problem of estimating dependability metrics on large systems' models. The usual situation is that these models are quite complex, precluding the use of analytic techniques (in general, models lack enough structure to allow this) and the use of numerical methods if the model's size is too large. The universal solution in these cases is simulation, which is conceptually a straightforward approach: build a representation of the system on the computer, focusing on the aspect you are interested in (a specific model of the system) and then build trajectories of that representation, from which statistics will be evaluated. However, there is a case where this often fails: if the underlying event of interest (typically the system's failure, in the context of this book) is rare, which means that it occurs with a very small probability, then simulation can be too expensive, or just impossible to apply, since we can not evaluate metrics concerning the event if we can not observe it frequently enough in the trajectories. In these cases we must use more sophisticated techniques to perform the estimations. This chapter illustrates two of them, addressing the two most important metrics in the area, the Mean Time To Failure of the system (the MTTF), and the reliability at time t, that is the probability $R(t)$ that the system works during the whole interval $[0, t]$.

The last chapter, Chapter 10, describes how to obtain bounds of typical dependability or performability metrics in two different situations having in common the fact that the chain is large (it has a large state space) and stiff (the ratio between different transition rates reaches very high values), one in the irreducible case, the other using an absorbing model. In the first case, we analyze the asymptotic availability (or, more generally, the mean asymptotic performability), and the Mean Time To Failure in the second. The techniques developed for obtaining the bounds are quite different in the two cases, and they make use of results coming from Chapters 4 and 5. In the case of a very stiff model, which is a frequent situation in dependability analysis when the underlying system is highly reliable, and when the characteristics of the model allow the use of the procedures described in the chapter, the obtained bounds can be so tight that the techniques can be seen as belonging to the numerical analysis area. The chapter illustrates the computation of the bounds with several examples coming from dependability analysis.

Some effort has been put into making the different chapters understandable without continuously moving backwards and forwards in the book. For instance, we almost always restate the basic notation for the chains, their state spaces, etc.

2   Discrete-time Markov chains

In this chapter we consider a sequence of random variables $X = \{X_n, \ n \in \mathbb{N}\}$ over a discrete state space S which forms a Markov chain, that is a stochastic process with the so-called Markovian property presented in our first definition below. In the rest of this book we use DTMC for Discrete-Time Markov Chain(s).

2.1   Definitions and properties

We start with the basic definitions and characterizations of discrete-time Markov chains.

DEFINITION 2.1.1. *A stochastic process $X = \{X_n, \ n \in \mathbb{N}\}$ over the state space S is a discrete-time Markov chain if*

- *for every $n \geq 0$, $X_n \in S$,*
- *for every $n \in \mathbb{N}$ and for all $i_n, i_{n-1}, \ldots, i_0 \in S$, we have*

$$\mathbb{P}\{X_n = i_n \mid X_{n-1} = i_{n-1}, \ldots, X_0 = i_0\} = \mathbb{P}\{X_n = i_n \mid X_{n-1} = i_{n-1}\},$$

when both conditional probabilities are defined.

DEFINITION 2.1.2. *The Markov chain $X = \{X_n, \ n \in \mathbb{N}\}$ is homogeneous if for all $n, k \in \mathbb{N}$ and for all $i, j \in S$, we have*

$$\mathbb{P}\{X_{n+k} = j \mid X_k = i\} = \mathbb{P}\{X_n = j \mid X_0 = i\}.$$

All the Markov chains considered here are homogeneous Markov chains with a finite state space S. We denote by $P = (P_{i,j})$ the transition probability matrix of X defined, for all $i, j \in S$, by

$$P_{i,j} = \mathbb{P}\{X_n = j \mid X_{n-1} = i\}, \quad \text{for all } n \geq 1.$$

We have

$$P_{i,j} \geq 0, \quad \text{for all } i, j \in S,$$

$$\sum_{j \in S} P_{i,j} = 1 \quad \text{for all } i \in S.$$

A matrix having these two properties is called a stochastic matrix. We denote by $\alpha = (\alpha_i,\ i \in S)$ the initial probability distribution of X, i.e. $\alpha_i = \mathbb{P}\{X_0 = i\}$.

THEOREM 2.1.3. *The process $X = \{X_n,\ n \in \mathbb{N}\}$ over the finite state space S is a Markov chain with initial distribution α and transition probability matrix P if and only if for every $n \geq 0$ and for every $i_n, \ldots, i_0 \in S$, we have*

$$\mathbb{P}\{X_n = i_n, \ldots, X_0 = i_0\} = \alpha_{i_0} P_{i_0,i_1} \ldots P_{i_{n-1},i_n}. \tag{2.1}$$

Proof. If $X = \{X_n,\ n \in \mathbb{N}\}$ is a Markov chain, we have, by definition,

$$\mathbb{P}\{X_n = i_n, \ldots, X_0 = i_0\} = P_{i_{n-1},i_n} \mathbb{P}\{X_{n-1} = i_{n-1}, \ldots, X_0 = i_0\}.$$

Repeating this argument $n - 1$ times on $\mathbb{P}\{X_{n-1} = i_{n-1}, \ldots, X_0 = i_0\}$, we obtain

$$\mathbb{P}\{X_n = i_n, \ldots, X_0 = i_0\} = \mathbb{P}\{X_0 = i_0\} P_{i_0,i_1} \ldots P_{i_{n-1},i_n} = \alpha_{i_0} P_{i_0,i_1} \ldots P_{i_{n-1},i_n}.$$

Conversely, if Relation (2.1) holds, we have, for $n = 0$, $\mathbb{P}\{X_0 = i_0\} = \alpha_{i_0}$ and for $n \geq 1$,

$$\begin{aligned}
\mathbb{P}\{X_n = i_n \mid X_{n-1} = i_{n-1}, \ldots, X_0 = i_0\} &= \frac{\mathbb{P}\{X_n = i_n, \ldots, X_0 = i_0\}}{\mathbb{P}\{X_{n-1} = i_{n-1}, \ldots, X_0 = i_0\}} \\
&= \frac{\alpha_{i_0} P_{i_0,i_1} \ldots P_{i_{n-1},i_n}}{\alpha_{i_0} P_{i_0,i_1} \ldots P_{i_{n-2},i_{n-1}}} \\
&= P_{i_{n-1},i_n} \\
&= \mathbb{P}\{X_n = i_n \mid X_{n-1} = i_{n-1}\},
\end{aligned}$$

which means that X is a Markov chain with initial distribution α and transition probability matrix P. ∎

This result shows that a Markov chain is completely characterized by its initial probability distribution and by its transition probability matrix.

For every $n \geq 0$, we denote by $(P^n)_{i,j}$ the (i, j) entry of matrix P^n, where $P^0 = I$ and I is the identity matrix whose dimension is given by the context.

THEOREM 2.1.4. *If X is a Markov chain with initial distribution α and transition probability matrix P, then*

(i) $\mathbb{P}\{X_n = j \mid X_0 = i\} = (P^n)_{i,j}$
(ii) $\mathbb{P}\{X_n = j\} = (\alpha P^n)_j$
(iii) *For every $n \geq 0$, matrix P^n is stochastic.*

Proof.

(i) We have, from Theorem 2.1.3,

$$\mathbb{P}\{X_n = j \mid X_0 = i\} = \sum_{i_1, \ldots, i_{n-1} \in S} \mathbb{P}\{X_n = j, X_{n-1} = i_{n-1}, \ldots, X_1 = i_1 \mid X_0 = i\}$$

$$= \sum_{i_1,\ldots,i_{n-1}\in S} \frac{\mathbb{P}\{X_n =j, X_{n-1} = i_{n-1},\ldots,X_1 = i_1, X_0 = i\}}{\mathbb{P}\{X_0 = i\}}$$

$$= \sum_{i_1,\ldots,i_{n-1}\in S} \frac{\alpha_i P_{i,i_1}\ldots P_{i_{n-1},j}}{\alpha_i}$$

$$= (P^n)_{i,j}.$$

(*ii*) Using (*i*), we can write

$$\mathbb{P}\{X_n = j\} = \sum_{i\in S} \mathbb{P}\{X_n = j \mid X_0 = i\}\mathbb{P}\{X_0 = i\}$$

$$= \sum_{i\in S} \alpha_i (P^n)_{i,j}$$

$$= (\alpha P^n)_j.$$

(*iii*) The result follows by summing over $j \in S$ in (*i*). ∎

THEOREM 2.1.5.   *If X is a Markov chain, then, for every $n \geq 0$, $0 \leq k \leq n$, $m \geq 1$, $i_k,\ldots,i_n \in S$ and $j_1,\ldots,j_m \in S$, we have*

$$\mathbb{P}\{X_{n+m} = j_m,\ldots,X_{n+1} = j_1 \mid X_n = i_n,\ldots,X_k = i_k\}$$
$$= \mathbb{P}\{X_m = j_m,\ldots,X_1 = j_1 \mid X_0 = i_n\}.$$

Proof.   From Theorem 2.1.3, we have

$$\mathbb{P}\{X_{n+m} = j_m,\ldots,X_{n+1} = j_1 \mid X_n = i_n,\ldots,X_k = i_k\}$$

$$= \frac{\mathbb{P}\{X_{n+m} = j_m,\ldots,X_{n+1} = j_1, X_n = i_n,\ldots,X_k = i_k\}}{\mathbb{P}\{X_n = i_n,\ldots,X_k = i_k\}}$$

$$= \frac{\displaystyle\sum_{i_0,\ldots,i_{k-1}\in S} \mathbb{P}\{X_{n+m} = j_m,\ldots,X_{n+1} = j_1, X_n = i_n,\ldots,X_0 = i_0\}}{\displaystyle\sum_{i_0,\ldots,i_{k-1}\in S} \mathbb{P}\{X_n = i_n,\ldots,X_0 = i_0\}}$$

$$= \frac{\left(\displaystyle\sum_{i_0,\ldots,i_{k-1}\in S} \alpha_{i_0} P_{i_0,i_1}\ldots P_{i_{k-1},i_k}\right) P_{i_k,i_{k+1}}\ldots P_{i_{n-1},i_n} P_{i_n,j_1}\ldots P_{j_{m-1},j_m}}{\left(\displaystyle\sum_{i_0,\ldots,i_{k-1}\in S} \alpha_{i_0} P_{i_0,i_1}\ldots P_{i_{k-1},i_k}\right) P_{i_k,i_{k+1}}\ldots P_{i_{n-1},i_n}}$$

$$= P_{i_n,j_1}\ldots P_{j_{m-1},j_m}$$

$$= \mathbb{P}\{X_m = j_m,\ldots,X_1 = j_1 \mid X_0 = i_n\},$$

which completes the proof. ∎

2.2 Strong Markov property

Let $X = \{X_n, n \in \mathbb{N}\}$ be a Markov chain over the finite state space S, defined on the probability space $(\Omega, \mathcal{F}, \mathbb{P})$. We first show that if we choose any nonnegative integer n and if we eliminate from X the variables $X_0, X_1, \ldots, X_{n-1}$, the remaining process is a Markov chain with the same dynamics as X.

THEOREM 2.2.1. *If $X = \{X_n, n \in \mathbb{N}\}$ is a Markov chain with transition probability matrix P, then, for every $n \geq 0$, $\{X_{n+p}, p \in \mathbb{N}\}$ is a Markov chain whose initial distribution is the distribution of X_n and whose transition probability matrix is P.*

Proof. The proof is straightforward, defining process $Y = \{Y_p, p \in \mathbb{N}\}$ by $Y_p = X_{n+p}$ and writing first $Y_0 = X_n$, we observe that the initial distribution of process Y is that of X_n. Then,

$$\mathbb{P}\{Y_{k+1} = j \mid Y_k = i, Y_{k-1} = i_{k-1}, \ldots, Y_0 = i_0\}$$
$$= \mathbb{P}\{X_{n+k+1} = j \mid X_{n+k} = i, X_{n+k-1} = i_{k-1}, \ldots, X_n = i_0\}$$
$$= \mathbb{P}\{X_{n+k+1} = j \mid X_{n+k} = i\}$$
$$= P_{i,j},$$

where the second equality follows because X is a Markov chain, proving that Y is also a Markov chain and the last one because X is homogeneous, proving that so is Y. At the same time, this proves that P is also the transition probability matrix of Y. ∎

Another useful property of Markov chains states that "given the present, the future and the past are independent". We formalize here the following version of this result. First, for every $n \geq 0$, we denote by $\mathcal{F}_n$ the σ-algebra of the events that can be expressed in terms of $X_0, \ldots, X_n$, i.e.

$$\mathcal{F}_n = \{\{\omega \in \Omega \mid (X_0(\omega), \ldots, X_n(\omega)) \in B_n\}, B_n \in \mathcal{P}(S^{n+1})\},$$

where $\mathcal{P}(U)$ is the set of all subsets of U.

THEOREM 2.2.2. *If $X = \{X_n, n \in \mathbb{N}\}$ is a Markov chain, then, for every $n \geq 0$, $A \in \mathcal{F}_n$, $m \geq 1$ and $i, j_1, \ldots, j_m \in S$, we have*

$$\mathbb{P}\{X_{n+m} = j_m, \ldots, X_{n+1} = j_1, A \mid X_n = i\}$$
$$= \mathbb{P}\{X_m = j_m, \ldots, X_1 = j_1 \mid X_0 = i\} \mathbb{P}\{A \mid X_n = i\}. \tag{2.2}$$

Moreover, if (2.2) holds, then X is a Markov chain.

Proof. In the informal terminology above, the "future" is the event $\{X_{n+m} = j_m, \ldots, X_{n+1} = j_1\}$, the "past" is represented by A and event $\{X_n = i\}$ is the "present".
It is sufficient to prove the result in the case where $A = \{X_n = i_n, X_{n-1} = i_{n-1}, \ldots, X_0 = i_0\}$. Indeed, A is a denumerable union of disjoint events of this type, so the general

case is easily deduced by the σ-additivity property. Moreover it is sufficient to consider the case where $i_n = i$, since in the opposite case, both quantities are zero.

So, let $A = \{X_n = i, X_{n-1} = i_{n-1}, \ldots, X_0 = i_0\}$. From the Markov property and applying Theorem 2.1.5, we have

$$\mathbb{P}\{X_{n+m} = j_m, \ldots, X_{n+1} = j_1, A \mid X_n = i\}$$

$$= \mathbb{P}\{X_{n+m} = j_m, \ldots, X_{n+1} = j_1 \mid X_n = i, A\}\mathbb{P}\{A \mid X_n = i\}$$

$$= \mathbb{P}\{X_m = j_m, \ldots, X_1 = j_1 \mid X_0 = i\}\mathbb{P}\{A \mid X_n = i\},$$

which proves (2.2).

Assume now that (2.2) holds. Then, taking $A = \{X_n = i, X_{n-1} = i_{n-1}, \ldots, X_0 = i_0\}$,

$$\mathbb{P}\{X_{n+1} = j \mid X_n = i, A\} = \frac{\mathbb{P}\{X_{n+1} = j, X_n = i, A\}}{\mathbb{P}\{X_n = i, A\}}$$

$$= \frac{\mathbb{P}\{X_{n+1} = j, A \mid X_n = i\}\mathbb{P}\{X_n = i\}}{\mathbb{P}\{A \mid X_n = i\}\mathbb{P}\{X_n = i\}}$$

$$= \frac{\mathbb{P}\{X_{n+1} = j \mid X_n = i\}\mathbb{P}\{A \mid X_n = i\}}{\mathbb{P}\{A \mid X_n = i\}}$$

$$= \mathbb{P}\{X_{n+1} = j \mid X_n = i\},$$

where the third equality follows from (2.2). This proves that X is a Markov chain. ∎

We introduce now the important concept of "stopping time".

DEFINITION 2.2.3.   *A random variable T taking values in $\mathbb{N} \cup \{\infty\}$ is called a stopping time for process X if for every $n \geq 0$, $\{T = n\} \in \mathcal{F}_n$.*

In the next section, we often make use of the random variable $\tau(j)$ counting the number of transitions needed to reach state j, i.e. defined by

$$\tau(j) = \inf\{n \geq 1 \mid X_n = j\},$$

where $\tau(j) = \infty$ if this set is empty. Clearly, for every $j \in S$, $\tau(j)$ is a stopping time since

$$\{\tau(j) = n\} = \{X_n = j, X_k \neq j, 1 \leq k \leq n - 1\} \in \mathcal{F}_n.$$

Let T be a stopping time and $\mathcal{F}_T$ be the σ-algebra of the events that can be expressed in terms of $X_0, \ldots, X_T$, i.e.

$$\mathcal{F}_T = \{B \in \mathcal{F} \mid \forall n \in \mathbb{N}, B \cap \{T = n\} \in \mathcal{F}_n\}.$$

For every $i \in S$, we denote by $\delta^i = (\delta^i_j, j \in S)$ the unit mass distribution at state i, defined by $\delta^i_j = 1_{\{i=j\}}$.

THEOREM 2.2.4.   (Strong Markov property) *If $X = \{X_n, n \in \mathbb{N}\}$ is a Markov chain and T is a stopping time for X then, for every $i \in S$, conditionally to $\{T < \infty\} \cap \{X_T = i\}$,*

$\{X_{T+n}, n \in \mathbb{N}\}$ *is a Markov chain with initial distribution* δ^i *and transition probability matrix P, and is independent of* $(X_0, \ldots, X_T)$, *i.e. for every* $A \in \mathcal{F}_T$, $m \geq 1$ *and* $j_1, \ldots, j_m \in S$, *we have*

$$\mathbb{P}\{X_{T+m} = j_m, \ldots, X_{T+1} = j_1, A \mid T < \infty, X_T = i\}$$
$$= \mathbb{P}\{X_m = j_m, \ldots, X_1 = j_1 \mid X_0 = i\}\mathbb{P}\{A \mid T < \infty, X_T = i\}.$$

Proof. Observe that the first part of the Theorem is related to the result given in Theorem 2.2.1. Here, a stopping time replaces a fixed integer value.

We have

$$\mathbb{P}\{X_{T+m} = j_m, \ldots, X_{T+1} = j_1, A \mid T < \infty, X_T = i\}$$

$$= \frac{\mathbb{P}\{X_{T+m} = j_m, \ldots, X_{T+1} = j_1, A, T < \infty, X_T = i\}}{\mathbb{P}\{T < \infty, X_T = i\}}$$

$$= \frac{\sum_{n=0}^{\infty} \mathbb{P}\{X_{T+m} = j_m, \ldots, X_{T+1} = j_1, A, T = n, X_T = i\}}{\mathbb{P}\{T < \infty, X_T = i\}}$$

$$= \frac{\sum_{n=0}^{\infty} \mathbb{P}\{X_{n+m} = j_m, \ldots, X_{n+1} = j_1, A, T = n, X_n = i\}}{\mathbb{P}\{T < \infty, X_T = i\}}$$

$$= \frac{\sum_{n=0}^{\infty} \mathbb{P}\{X_{n+m} = j_m, \ldots, X_{n+1} = j_1, A, T = n \mid X_n = i\}\mathbb{P}\{X_n = i\}}{\mathbb{P}\{T < \infty, X_T = i\}}$$

$$= \frac{\sum_{n=0}^{\infty} \mathbb{P}\{X_m = j_m, \ldots, X_1 = j_1 \mid X_0 = i\}\mathbb{P}\{A, T = n, X_n = i\}}{\mathbb{P}\{T < \infty, X_T = i\}}$$

$$= \mathbb{P}\{X_m = j_m, \ldots, X_1 = j_1 \mid X_0 = i\}\frac{\sum_{n=0}^{\infty} \mathbb{P}\{A, T = n, X_T = i\}}{\mathbb{P}\{T < \infty, X_T = i\}}$$

$$= \mathbb{P}\{X_m = j_m, \ldots, X_1 = j_1 \mid X_0 = i\}\frac{\mathbb{P}\{A, T < \infty, X_T = i\}}{\mathbb{P}\{T < \infty, X_T = i\}}$$

$$= \mathbb{P}\{X_m = j_m, \ldots, X_1 = j_1 \mid X_0 = i\}\mathbb{P}\{A \mid T < \infty, X_T = i\},$$

where the fifth equality results from Theorem 2.2.2 since $\{A, T = n\} \in \mathcal{F}_n$. ∎

2.3   Recurrent and transient states

For every $i, j \in S$ and for every $n \geq 1$, we define

$$f_{i,j}^{(n)} = \mathbb{P}\{\tau(j) = n \mid X_0 = i\} = \mathbb{P}\{X_n = j, X_k \neq j, 1 \leq k \leq n-1 \mid X_0 = i\}.$$

The quantity $f_{i,i}^{(n)}$ is the probability, starting from state i, that the first return to state i occurs at instant n. For $i \neq j$, $f_{i,j}^{(n)}$ is the probability, starting from state i, that the first visit to state j occurs at instant n.

THEOREM 2.3.1.   *For every $i, j \in S$ and $n \geq 1$, we have*

$$(P^n)_{i,j} = \sum_{k=1}^{n} f_{i,j}^{(k)} (P^{n-k})_{j,j}. \tag{2.3}$$

Recall that $(P^0)_{i,j} = 1_{\{i=j\}}$.

Proof.   Fix $i, j \in S$ and $n \geq 1$. Since $X_n = j \Longrightarrow \tau(j) \leq n$, we have

$$(P^n)_{i,j} = \mathbb{P}\{X_n = j \mid X_0 = i\}$$

$$= \mathbb{P}\{X_n = j, \tau(j) \leq n \mid X_0 = i\}$$

$$= \sum_{k=1}^{n} \mathbb{P}\{X_n = j, \tau(j) = k \mid X_0 = i\}$$

$$= \sum_{k=1}^{n} \mathbb{P}\{X_n = j \mid \tau(j) = k, X_0 = i\} \mathbb{P}\{\tau(j) = k \mid X_0 = i\}$$

$$= \sum_{k=1}^{n} f_{i,j}^{(k)} \mathbb{P}\{X_n = j \mid X_k = j, \tau(j) = k, X_0 = i\}$$

$$= \sum_{k=1}^{n} f_{i,j}^{(k)} \mathbb{P}\{X_n = j \mid X_k = j\}$$

$$= \sum_{k=1}^{n} f_{i,j}^{(k)} (P^{n-k})_{j,j},$$

where the fifth equality is due to the fact that $\{\tau(j) = k\} = \{X_k = j, \tau(j) = k\}$ and the sixth one uses the Markov property, since $\tau(j)$ is a stopping time.   ∎

For every $i, j \in S$, we define $f_{i,j}$ as

$$f_{i,j} = \mathbb{P}\{\tau(j) < \infty \mid X_0 = i\} = \sum_{n=1}^{\infty} f_{i,j}^{(n)}.$$

The quantity $f_{i,i}$ is the probability, starting from state i, that the return time to state i is finite and, for $i \neq j$, $f_{i,j}$ is the probability, starting from state i, that the first visit to state j occurs in a finite time.

The computation of the $f_{i,j}^{(n)}$ and $f_{i,j}$ can be done using the following result.

THEOREM 2.3.2. *For every $i,j \in S$ and $n \geq 1$, we have*

$$
f_{i,j}^{(n)} =
\begin{cases}
P_{i,j} & \text{if } n = 1 \\[2mm]
\displaystyle\sum_{\ell \in S \setminus \{j\}} P_{i,\ell} f_{\ell,j}^{(n-1)} & \text{if } n \geq 2
\end{cases}
\tag{2.4}
$$

and

$$
f_{i,j} = P_{i,j} + \sum_{\ell \in S \setminus \{j\}} P_{i,\ell} f_{\ell,j}.
\tag{2.5}
$$

Proof. By definition of $f_{i,j}^{(n)}$ we have, for $n = 1$, $f_{i,j}^{(1)} = P_{i,j}$. For $n \geq 2$, we have

$$
\begin{aligned}
f_{i,j}^{(n)} &= \mathbb{P}\{X_n = j, X_k \neq j, 1 \leq k \leq n-1 \mid X_0 = i\} \\
&= \sum_{\ell \in S \setminus \{j\}} \mathbb{P}\{X_n = j, X_k \neq j, 2 \leq k \leq n-1, X_1 = \ell \mid X_0 = i\} \\
&= \sum_{\ell \in S \setminus \{j\}} \mathbb{P}\{X_1 = \ell \mid X_0 = i\} \\
&\quad \times \mathbb{P}\{X_n = j, X_k \neq j, 2 \leq k \leq n-1 \mid X_1 = \ell, X_0 = i\} \\
&= \sum_{\ell \in S \setminus \{j\}} P_{i,\ell} \mathbb{P}\{X_n = j, X_k \neq j, 2 \leq k \leq n-1 \mid X_1 = \ell, X_0 = i\} \\
&= \sum_{\ell \in S \setminus \{j\}} P_{i,\ell} \mathbb{P}\{X_n = j, X_k \neq j, 2 \leq k \leq n-1 \mid X_1 = \ell\} \\
&= \sum_{\ell \in S \setminus \{j\}} P_{i,\ell} \mathbb{P}\{X_{n-1} = j, X_k \neq j, 1 \leq k \leq n-2 \mid X_0 = \ell\} \\
&= \sum_{\ell \in S \setminus \{j\}} P_{i,\ell} f_{\ell,j}^{(n-1)},
\end{aligned}
$$

where the antepenultimate and the penultimate equalities come from the Markov property and the homogeneity of the Markov chain X, respectively. The sum over index n leads to the second relation. ∎

DEFINITION 2.3.3. *A state $i \in S$ is called recurrent if $f_{i,i} = 1$ and transient if $f_{i,i} < 1$. A Markov chain is called recurrent (resp. transient) if all its states are recurrent (resp. transient).*

DEFINITION 2.3.4. *A state $i \in S$ is called absorbing if $P_{i,i} = 1$.*

If i is an absorbing state, we have by definition $f_{i,i}^{(n)} = 1_{\{n=1\}}$ and so $f_{i,i} = 1$, which means that every absorbing state is recurrent.

THEOREM 2.3.5.   *State $j \in S$ is recurrent if and only if*

$$\sum_{n=1}^{\infty} (P^n)_{j,j} = \infty.$$

Proof.   Consider Relation (2.3) for $i = j$, that is,

$$(P^n)_{j,j} = \sum_{k=1}^{n} f_{j,j}^{(k)} (P^{n-k})_{j,j}.$$

Since $(P^0)_{j,j} = 1$, the sum over index n leads, using Fubini's Theorem, to

$$\sum_{n=1}^{\infty} (P^n)_{j,j} = \sum_{n=1}^{\infty} \sum_{k=1}^{n} f_{j,j}^{(k)} (P^{n-k})_{j,j}$$

$$= \sum_{k=1}^{\infty} f_{j,j}^{(k)} \sum_{n=k}^{\infty} (P^{n-k})_{j,j}$$

$$= f_{j,j} \sum_{n=0}^{\infty} (P^n)_{j,j}$$

$$= f_{j,j} \left(1 + \sum_{n=1}^{\infty} (P^n)_{j,j} \right).$$

Thus, if $\sum_{n=1}^{\infty} (P^n)_{j,j} < \infty$ then

$$f_{j,j} = \frac{\displaystyle\sum_{n=1}^{\infty} (P^n)_{j,j}}{1 + \displaystyle\sum_{n=1}^{\infty} (P^n)_{j,j}} < 1,$$

which means, by definition, that state j is transient.

Conversely, suppose that $\sum_{n=1}^{\infty} (P^n)_{j,j} = \infty$. We then have, for every $N \geq 1$,

$$\sum_{n=1}^{N} (P^n)_{j,j} = \sum_{n=1}^{N} \sum_{k=1}^{n} f_{j,j}^{(k)} (P^{n-k})_{j,j}$$

$$= \sum_{k=1}^{N} f_{j,j}^{(k)} \sum_{n=k}^{N} (P^{n-k})_{j,j}$$

$$\leq \sum_{k=1}^{N} f_{j,j}^{(k)} \sum_{n=0}^{N} (P^n)_{j,j}$$

$$\leq f_{j,j} \left(1 + \sum_{n=1}^{N} (P^n)_{j,j} \right),$$

and thus we get

$$f_{j,j} \geq \frac{\displaystyle\sum_{n=1}^{N} (P^n)_{j,j}}{1 + \displaystyle\sum_{n=1}^{N} (P^n)_{j,j}} \longrightarrow 1 \text{ when } N \longrightarrow \infty,$$

which means that $f_{j,j} = 1$, that is, that state j is recurrent. ∎

DEFINITION 2.3.6. *State $j \in S$ is said to be accessible from state $i \in S$ if there exists $n \geq 0$ such that $(P^n)_{i,j} > 0$. This means, in particular, that each state is accessible from itself. We say that states i and j communicate if i is accessible from j and j is accessible from i.*

LEMMA 2.1. *State j is accessible from state i, for $i \neq j$, if and only if $f_{i,j} > 0$.*

Proof. Assume first that state j is accessible from state i. That is, for some $n \geq 0$ we have $(P^n)_{i,j} > 0$. Using Relation (2.3), this means that for some k, $1 \leq k \leq n$, $f_{i,j}^{(k)} > 0$, and since $f_{i,j} = \sum_{\ell \geq 1} f_{i,j}^{(\ell)}$, we have $f_{i,j} > 0$.

Conversely, assume $f_{i,j} > 0$. This means that for some $m \geq 1$, we have $f_{i,j}^{(m)} > 0$. Using again Relation (2.3), now for this value m, we have

$$(P^m)_{i,j} = \sum_{k=1}^{m} f_{i,j}^{(k)} (P^{m-k})_{j,j}$$

$$= f_{i,j}^{(m)} + \sum_{k=1}^{m-1} f_{i,j}^{(k)} (P^{m-k})_{j,j}$$

$$\geq f_{i,j}^{(m)} > 0,$$

which completes the proof. ∎

This communication relation is obviously an equivalence relation (that is, it has the reflexivity, symmetry, and transitivity properties) which leads to a partition of the state space S into equivalence classes in such a way that all the states of a given class communicate. These equivalence classes are also called communication classes.

DEFINITION 2.3.7. *The Markov chain X is said to be irreducible if there exists only one communication class.*

COROLLARY 2.3.8.   *(i) If state j is recurrent and if states i and j communicate, then state i is recurrent. (ii) If state j is transient and if states i and j communicate, then state i is transient.*

Proof. Since i and j communicate, there exist integers $\ell \geq 0$ and $m \geq 0$ such that $(P^\ell)_{i,j} > 0$ and $(P^m)_{j,i} > 0$. For every $n \geq 0$, we have

$$(P^{\ell+n+m})_{i,i} = \sum_{k \in S} (P^\ell)_{i,k}(P^{n+m})_{k,i} \geq (P^\ell)_{i,j}(P^{n+m})_{j,i},$$

and

$$(P^{n+m})_{j,i} = \sum_{k \in S} (P^n)_{j,k}(P^m)_{k,i} \geq (P^n)_{j,j}(P^m)_{j,i},$$

thus,

$$(P^{\ell+m+n})_{i,i} \geq (P^\ell)_{i,j}(P^n)_{j,j}(P^m)_{j,i}.$$

If j is recurrent, the sum over index n leads, by Theorem 2.3.5, to

$$\sum_{n=1}^{\infty}(P^n)_{i,i} \geq \sum_{n=1}^{\infty}(P^{\ell+n+m})_{i,i} \geq (P^\ell)_{i,j}(P^m)_{j,i}\sum_{n=1}^{\infty}(P^n)_{j,j} = \infty,$$

which means that state i is recurrent.

If j is transient and states i and j communicate, then i can not be recurrent since, by (i), this would mean that j is recurrent. ■

This result shows that recurrence and transience are class properties.

COROLLARY 2.3.9.   *If state j is transient then, for every state i,*

$$\sum_{n=1}^{\infty}(P^n)_{i,j} < \infty,$$

and, thus,

$$\lim_{n \to \infty}(P^n)_{i,j} = 0.$$

Proof. Let us consider again Relation (2.3), that is,

$$(P^n)_{i,j} = \sum_{k=1}^{n} f_{i,j}^{(k)}(P^{n-k})_{j,j}.$$

Since $(P^0)_{j,j} = 1$, the sum over index n leads, using Fubini's Theorem, to

$$\sum_{n=1}^{\infty} (P^n)_{i,j} = \sum_{n=1}^{\infty} \sum_{k=1}^{n} f_{i,j}^{(k)} (P^{n-k})_{j,j}$$

$$= \sum_{k=1}^{\infty} f_{i,j}^{(k)} \sum_{n=k}^{\infty} (P^{n-k})_{j,j}$$

$$= f_{i,j} \sum_{n=0}^{\infty} (P^n)_{j,j}$$

$$= f_{i,j} \left(1 + \sum_{n=1}^{\infty} (P^n)_{j,j} \right).$$

If state j is transient then, from Theorem 2.3.5, we obtain $\sum_{n=1}^{\infty} (P^n)_{j,j} < \infty$ and so $\sum_{n=1}^{\infty} (P^n)_{i,j} < \infty$ which implies that $\lim_{n \to \infty} (P^n)_{i,j} = 0$. ∎

COROLLARY 2.3.10. *The state space S being finite,*

- *a Markov chain cannot be transient,*
- *an irreducible Markov chain is recurrent.*

Proof. If the Markov chain is transient then we have from Corollary 2.3.9, for every $i, j \in S$,

$$\sum_{n=1}^{\infty} (P^n)_{i,j} < \infty,$$

and thus, since S is finite,

$$\sum_{j \in S} \sum_{n=1}^{\infty} (P^n)_{i,j} < \infty.$$

But we have

$$\sum_{j \in S} \sum_{n=1}^{\infty} (P^n)_{i,j} = \sum_{n=1}^{\infty} \sum_{j \in S} (P^n)_{i,j} = \sum_{n=1}^{\infty} 1 = \infty,$$

which leads to a contradiction. The second item follows immediately from the first one, since recurrence is a class property. ∎

2.4 Visits to a fixed state

For every $j \in S$, we denote by N_j the total number of visits the chain makes to state j, not counting the initial state, i.e.

$$N_j = \sum_{n=1}^{\infty} 1_{\{X_n = j\}}.$$

It is easily checked that for every $j \in S$, we have $\{\tau(j) < \infty\} = \{N_j > 0\}$.

THEOREM 2.4.1. *For every $i, j \in S$ and $\ell \geq 0$, we have*

$$\mathbb{P}\{N_j > \ell \mid X_0 = i\} = f_{i,j}(f_{j,j})^\ell.$$

Proof. Let us consider the random variable $N_{j,m}$ which counts the number of visits to state j from instant m, i.e.

$$N_{j,m} = \sum_{n=m}^{\infty} 1_{\{X_n = j\}}.$$

Remark that $N_j = N_{j,1}$. Since, for every $\ell \geq 1$, $\{N_j > \ell\} \subseteq \{\tau(j) < \infty\}$, we obtain

$$\mathbb{P}\{N_j > \ell \mid X_0 = i\} = \sum_{k=1}^{\infty} \mathbb{P}\{N_j > \ell, \tau(j) = k \mid X_0 = i\}$$

$$= \sum_{k=1}^{\infty} \mathbb{P}\{N_{j,k+1} > \ell - 1, \tau(j) = k \mid X_0 = i\}$$

$$= \sum_{k=1}^{\infty} \mathbb{P}\{N_{j,k+1} > \ell - 1 \mid \tau(j) = k, X_0 = i\}\mathbb{P}\{\tau(j) = k \mid X_0 = i\}$$

$$= \sum_{k=1}^{\infty} f_{i,j}^{(k)}\mathbb{P}\{N_{j,k+1} > \ell - 1 \mid X_{\tau(j)} = j, \tau(j) = k, X_0 = i\}$$

$$= \sum_{k=1}^{\infty} f_{i,j}^{(k)}\mathbb{P}\{N_{j,k+1} > \ell - 1 \mid X_k = j, \tau(j) = k, X_0 = i\}$$

$$= \sum_{k=1}^{\infty} f_{i,j}^{(k)}\mathbb{P}\{N_{j,k+1} > \ell - 1 \mid X_k = j\}$$

$$= \sum_{k=1}^{\infty} f_{i,j}^{(k)}\mathbb{P}\{N_j > \ell - 1 \mid X_0 = j\}$$

$$= f_{i,j}\mathbb{P}\{N_j > \ell - 1 \mid X_0 = j\},$$

where we use the definition of $\tau(j)$ and the fact that $X_{\tau(j)} = j$ when $\tau(j) < \infty$. In the sixth equality we use the Markov property, since $\tau(j)$ is a stopping time, and in the seventh we use the homogeneity of X.

Taking $i = j$, we obtain, for $\ell \geq 1$,

$$\mathbb{P}\{N_j > \ell \mid X_0 = j\} = f_{j,j}\mathbb{P}\{N_j > \ell - 1 \mid X_0 = j\},$$

that is, for $\ell \geq 0$,

$$\mathbb{P}\{N_j > \ell \mid X_0 = j\} = (f_{j,j})^\ell \mathbb{P}\{N_j > 0 \mid X_0 = j\}$$
$$= (f_{j,j})^\ell \mathbb{P}\{\tau(j) < \infty \mid X_0 = j\}$$
$$= (f_{j,j})^{\ell+1},$$

which leads, for every $i, j \in S$ and $\ell \geq 0$, to

$$\mathbb{P}\{N_j > \ell \mid X_0 = i\} = f_{i,j}(f_{j,j})^\ell,$$

which completes the proof.   ∎

COROLLARY 2.4.2.

$$\mathbb{E}\{N_j \mid X_0 = i\} = \begin{cases} 0 & \text{if } f_{i,j} = 0, \\ \dfrac{f_{i,j}}{1 - f_{j,j}} & \text{if } f_{i,j} \neq 0 \text{ and } f_{j,j} < 1, \\ \infty & \text{if } f_{i,j} \neq 0 \text{ and } f_{j,j} = 1. \end{cases}$$

Proof. From Theorem 2.4.1, if $f_{i,j} = 0$, we have $N_j = 0$ and thus, $\mathbb{E}\{N_j \mid X_0 = i\} = 0$. If $f_{i,j} \neq 0$ and $f_{j,j} = 1$, then

$$\mathbb{P}\{N_j > \ell \mid X_0 = i\} = f_{i,j} > 0,$$

meaning that $\mathbb{E}\{N_j \mid X_0 = i\} = \infty$.
    If $f_{i,j} \neq 0$ and $f_{j,j} < 1$, then

$$\mathbb{E}\{N_j \mid X_0 = i\} = \sum_{\ell=0}^{\infty} \mathbb{P}\{N_j > \ell \mid X_0 = i\}$$

$$= f_{i,j} \sum_{\ell=0}^{\infty} (f_{j,j})^\ell$$

$$= \frac{f_{i,j}}{1 - f_{j,j}},$$

which completes the proof.   ∎

COROLLARY 2.4.3.   *For every state $j \in S$, we have*

- *State j is recurrent if and only if $\mathbb{P}\{N_j = \infty \mid X_0 = j\} = 1$;*
- *State j is transient if and only if $\mathbb{P}\{N_j < \infty \mid X_0 = j\} = 1$.*

Proof. From Theorem 2.4.1. we have

$$\mathbb{P}\{N_j > \ell \mid X_0 = j\} = (f_{j,j})^{\ell+1}.$$

If state j is recurrent then $f_{j,j} = 1$ and we have $\mathbb{P}\{N_j > \ell \mid X_0 = j\} = 1$ for every $\ell \geq 0$, which means that $\mathbb{P}\{N_j = \infty \mid X_0 = j\} = 1$. Conversely, if $\mathbb{P}\{N_j = \infty \mid X_0 = j\} = 1$, then $\mathbb{P}\{N_j > \ell \mid X_0 = i\} = 1$ for every $\ell \geq 0$, so, $f_{j,j} = 1$.

If j is transient then $f_{j,j} < 1$ and thus, $\mathbb{P}\{N_j > \ell \mid X_0 = j\} \longrightarrow 0$ when $\ell \longrightarrow \infty$. Conversely, if $\mathbb{P}\{N_j < \infty \mid X_0 = j\} = 1$, then since

$$1 = \mathbb{P}\{N_j < \infty \mid X_0 = j\} = \lim_{\ell \to \infty} \left(1 - (f_{j,j})^{\ell+1}\right),$$

we get $f_{j,j} < 1$.   ∎

LEMMA 2.4.4.   *For every $i, j \in S$, we have*

$$\mathbb{E}\{N_j \mid X_0 = i\} = \sum_{n=1}^{\infty} (P^n)_{i,j}.$$

Proof.   By the monotone convergence theorem, we have

$$\mathbb{E}\{N_j \mid X_0 = i\} = \mathbb{E}\left\{\sum_{n=1}^{\infty} 1_{\{X_n = j\}} \,\Big|\, X_0 = i\right\}$$

$$= \sum_{n=1}^{\infty} \mathbb{E}\{1_{\{X_n = j\}} \mid X_0 = i\}$$

$$= \sum_{n=1}^{\infty} \mathbb{P}\{X_n = j \mid X_0 = i\}$$

$$= \sum_{n=1}^{\infty} (P^n)_{i,j},$$

which completes the proof.   ∎

THEOREM 2.4.5.   *If X is an irreducible Markov chain then, for every $i, j \in S$ we have $f_{i,j} = 1$ and $\mathbb{P}\{\tau(j) < \infty\} = 1$.*

Proof.   Recall that $f_{i,j} = \mathbb{P}\{\tau(j) < \infty \mid X_0 = i\}$. From Lemma 2.1, X being irreducible, we have $f_{i,j} > 0$ for every $i \neq j$. From Corollary 2.3.10, X is recurrent, so we have $f_{j,j} = 1$ for every $j \in S$. From Theorem 2.4.1, we obtain, for every $\ell \geq 0$, $\mathbb{P}\{N_j > \ell \mid X_0 = i\} = f_{i,j} > 0$. Taking the limit as $\ell \longrightarrow \infty$, we obtain

$$\mathbb{P}\{N_j = \infty \mid X_0 = i\} = f_{i,j} > 0.$$

Since state i is recurrent, from Corollary 2.4.3 we have $\mathbb{P}\{N_i = \infty \mid X_0 = i\} = 1$ and thus, for every $m \geq 1$, we also have $\mathbb{P}\{N_{i,m} = \infty \mid X_0 = i\} = 1$, where $N_{i,m}$ is defined in the proof of Theorem 2.4.1. The same reasoning as used in the proof of Theorem 2.4.1

leads to

$$1 = \sum_{k=1}^{\infty} \mathbb{P}\{\tau(j) = k \mid X_0 = i\} + \mathbb{P}\{\tau(j) = \infty \mid X_0 = i\}$$

$$= \sum_{k=1}^{\infty} \mathbb{P}\{\tau(j) = k, N_{i,k+1} = \infty \mid X_0 = i\} + 1 - f_{i,j}$$

$$= \sum_{k=1}^{\infty} \mathbb{P}\{N_{i,k+1} = \infty \mid \tau(j) = k, X_0 = i\} \mathbb{P}\{\tau(j) = k \mid X_0 = i\} + 1 - f_{i,j}$$

$$= \sum_{k=1}^{\infty} f_{i,j}^{(k)} \mathbb{P}\{N_{i,k+1} = \infty \mid X_{\tau(j)} = j, \tau(j) = k, X_0 = i\} + 1 - f_{i,j}$$

$$= \sum_{k=1}^{\infty} f_{i,j}^{(k)} \mathbb{P}\{N_{i,k+1} = \infty \mid X_k = j\} + 1 - f_{i,j}$$

$$= \sum_{k=1}^{\infty} f_{i,j}^{(k)} \mathbb{P}\{N_i = \infty \mid X_0 = j\} + 1 - f_{i,j}$$

$$= f_{i,j} \mathbb{P}\{N_i = \infty \mid X_0 = j\} + 1 - f_{i,j}$$

$$= f_{i,j} f_{j,i} + 1 - f_{i,j}.$$

We thus have

$$1 = f_{i,j} f_{j,i} + 1 - f_{i,j},$$

that is,

$$f_{i,j} = f_{i,j} f_{j,i}.$$

Since $f_{i,j} > 0$, we obtain that $f_{j,i} = 1$ for every $i, j \in S$, which means that $f_{i,j} = \mathbb{P}\{\tau(j) < \infty \mid X_0 = i\} = 1$. Finally, for every $j \in S$, we have

$$\mathbb{P}\{\tau(j) < \infty\} = \sum_{i \in S} \mathbb{P}\{X_0 = i\} \mathbb{P}\{\tau(j) < \infty \mid X_0 = i\} = \sum_{i \in S} \mathbb{P}\{X_0 = i\} = 1,$$

which completes the proof. ∎

2.5 Invariant measures and irreducible Markov chains

DEFINITION 2.5.1.   *A row vector $v = (v_j, j \in S)$ is called a measure on S if $0 \le v_j < \infty$. The Markov chain X is said to have an invariant measure v if v is a measure on S and $v = vP$. The measure v is said to be positive if $v_j > 0$ for every $j \in S$.*

THEOREM 2.5.2.   *If the Markov chain X is irreducible then it has a unique positive invariant measure up to a multiplicative constant.*

Proof.   **Existence.** Let γ_j^i be the expected number of visits to state j, starting from state i, until the first return to i, that is

$$\gamma_j^i = \mathbb{E}\left\{\sum_{n=1}^{\tau(i)} 1_{\{X_n=j\}} \,\Big|\, X_0 = i\right\}. \tag{2.6}$$

Note that $\gamma_i^i = 1$.

From Theorem 2.4.5, we have $\tau(i) < \infty$ with probability 1, so, using Fubini's Theorem,

$$\gamma_j^i = \sum_{k=1}^{\infty} \mathbb{E}\left\{1_{\{\tau(i)=k\}} \sum_{n=1}^{\tau(i)} 1_{\{X_n=j\}} \,\Big|\, X_0 = i\right\}$$

$$= \sum_{k=1}^{\infty}\sum_{n=1}^{k} \mathbb{P}\{X_n = j, \tau(i) = k \mid X_0 = i\}$$

$$= \sum_{n=1}^{\infty} \mathbb{P}\{X_n = j, \tau(i) \geq n \mid X_0 = i\} \tag{2.7}$$

$$= \sum_{\ell \in S}\sum_{n=1}^{\infty} \mathbb{P}\{X_{n-1} = \ell, \tau(i) \geq n, X_n = j \mid X_0 = i\}$$

$$= \sum_{\ell \in S}\sum_{n=1}^{\infty} \mathbb{P}\{X_n = j \mid X_{n-1} = \ell, \tau(i) > n-1, X_0 = i\}$$

$$\times \mathbb{P}\{X_{n-1} = \ell, \tau(i) \geq n \mid X_0 = i\}$$

$$= \sum_{\ell \in S}\sum_{n=1}^{\infty} \mathbb{P}\{X_{n-1} = \ell, \tau(i) \geq n \mid X_0 = i\}P_{\ell,j}, \tag{2.8}$$

where the sixth equality uses the Markov property, since $\tau(i)$ is a stopping time, and the homogeneity of X.

Since $X_{\tau(i)} = i$, we obtain, when $X_0 = i$,

$$\sum_{n=1}^{\tau(i)} 1_{\{X_n=j\}} = \sum_{n=0}^{\tau(i)-1} 1_{\{X_n=j\}}.$$

We then have, again by Fubini's Theorem,

$$\gamma_j^i = \mathbb{E}\left\{\sum_{n=0}^{\tau(i)-1} 1_{\{X_n=j\}} \,\Big|\, X_0 = i\right\}$$

$$= \sum_{k=1}^{\infty} \mathbb{E}\left\{1_{\{\tau(i)=k\}} \sum_{n=0}^{\tau(i)-1} 1_{\{X_n=j\}} \,\Big|\, X_0 = i\right\}$$

$$= \sum_{k=1}^{\infty} \sum_{n=0}^{k-1} \mathbb{P}\{X_n = j, \tau(i) = k \mid X_0 = i\}$$

$$= \sum_{k=1}^{\infty} \sum_{n=1}^{k} \mathbb{P}\{X_{n-1} = j, \tau(i) = k \mid X_0 = i\}$$

$$= \sum_{n=1}^{\infty} \mathbb{P}\{X_{n-1} = j, \tau(i) \geq n \mid X_0 = i\}. \tag{2.9}$$

Let us introduce the row vector $\gamma^i = (\gamma_j^i, j \in S)$. Combining Relations (2.8) and (2.9), we obtain

$$\gamma_j^i = \sum_{\ell \in S} \gamma_\ell^i P_{\ell,j},$$

that is

$$\gamma^i = \gamma^i P.$$

From the irreducibility of X, there exist n and m such that $(P^n)_{i,j} > 0$ and $(P^m)_{j,i} > 0$. Observing that if μ is a measure satisfying $\mu = \mu P$ then for any $k \geq 0$, $\mu = \mu P^k$, and that, by definition of γ_j^i, $\gamma_i^i = 1$, we get

$$0 < (P^n)_{i,j} = \gamma_i^i (P^n)_{i,j} \leq \sum_{\ell \in S} \gamma_\ell^i (P^n)_{\ell,j} = (\gamma^i P^n)_j = \gamma_j^i$$

and

$$\gamma_j^i (P^m)_{j,i} \leq \sum_{\ell \in S} \gamma_\ell^i (P^m)_{\ell,i} = (\gamma^i P^m)_i = \gamma_i^i = 1,$$

so $\gamma_j^i \leq 1/(P^m)_{j,i} < \infty$. This means that γ^i is a positive invariant measure.

Uniqueness. Let us fix $i \in S$ and let λ be an invariant measure such that $\lambda_i = 1$. For every $N \geq 1$ and since $\lambda = \lambda P$, we have by induction, for every $j \in S$,

$$\lambda_j = P_{i,j} + \sum_{i_1 \neq i} \lambda_{i_1} P_{i_1,j}$$

$$= P_{i,j} + \sum_{i_1 \neq i} P_{i,i_1} P_{i_1,j} + \sum_{i_1 \neq i} \sum_{i_2 \neq i} \lambda_{i_2} P_{i_2,i_1} P_{i_1,j}$$

$$= \cdots$$

$$= P_{i,j} + \sum_{n=1}^{N} \sum_{i_1 \neq i,\dots,i_n \neq i} P_{i,i_n} P_{i_n,i_{n-1}} \dots P_{i_1,j}$$

$$+ \sum_{i_1 \neq i,\dots,i_{N+1} \neq i} \lambda_{i_{N+1}} P_{i_{N+1},i_N} P_{i_N,i_{N-1}} \dots P_{i_1,j}$$

$$\geq P_{i,j} + \sum_{n=1}^{N} \sum_{i_1 \neq i,\dots,i_n \neq i} P_{i,i_n} P_{i_n,i_{n-1}} \dots P_{i_1,j}$$

$$= P_{i,j} + \sum_{n=1}^{N} \mathbb{P}\{X_{n+1} = j, X_\ell \neq i, 1 \leq \ell \leq n \mid X_0 = i\}$$

$$= P_{i,j} + \sum_{n=1}^{N} \mathbb{P}\{X_{n+1} = j, \tau(i) \geq n+1 \mid X_0 = i\}$$

$$= \sum_{n=0}^{N} \mathbb{P}\{X_{n+1} = j, \tau(i) \geq n+1 \mid X_0 = i\}$$

$$= \sum_{n=1}^{N+1} \mathbb{P}\{X_n = j, \tau(i) \geq n \mid X_0 = i\}.$$

When $N \longrightarrow \infty$, using Relation (2.7), we obtain

$$\lambda_j \geq \sum_{n=1}^{\infty} \mathbb{P}\{X_n = j, \tau(i) \geq n \mid X_0 = i\} = \gamma_j^i.$$

So, we have shown that the measure $\mu = \lambda - \gamma^i$ is an invariant measure with $\mu_i = 0$. The Markov chain being irreducible, for every $j \in S$, there exists m such that $(P^m)_{j,i} > 0$. We then have, for every $j \in S$,

$$0 \leq \mu_j (P^m)_{j,i} \leq \sum_{\ell \in S} \mu_\ell (P^m)_{\ell,i} = (\mu P^m)_i = \mu_i = 0.$$

Thus $\mu_j = 0$ for every $j \in S$, i.e. $\lambda = \gamma^i$.   ■

Recall that a state i is recurrent if $f_{i,i} = \mathbb{P}\{\tau(i) < \infty \mid X_0 = i\} = 1$. Let m_i be the expected return time to state i, that is,

$$m_i = \mathbb{E}\{\tau(i) \mid X_0 = i\}.$$

By definition of $\tau(i)$, if i is recurrent, we have

$$m_i = \sum_{n=1}^{\infty} n f_{i,i}^{(n)},$$

and if i is transient, then $m_i = \infty$. If in Relation (2.6), we sum over $j \in S$, we also have

$$m_i = \sum_{j \in S} \gamma_j^i. \tag{2.10}$$

We denote by $\mathbb{1}$ the column vector indexed by S with all its entries equal to 1.

DEFINITION 2.5.3. *An invariant measure $v = (v_j, j \in S)$ is called an invariant prob-ability on S if $v\mathbb{1} = 1$. The invariant probability v is positive if $v_j > 0$ for every $j \in S$.*

COROLLARY 2.5.4. *If X is an irreducible Markov chain then it has a unique positive invariant probability $\pi = (\pi_j, j \in S)$ given by $\pi_j = 1/m_j$.*

Proof. If X is an irreducible Markov chain then, by Theorem 2.5.2 and since the state space is finite, from (2.10) all the m_i are finite and X has a unique positive invariant probability π on S given by

$$\pi_j = \frac{\gamma_j^i}{m_i},$$

which is independent of i since all the measures γ^i are proportional. Taking $i = j$, we obtain

$$\pi_j = \frac{\gamma_j^j}{m_j} = \frac{1}{m_j},$$

which completes the proof. ∎

COROLLARY 2.5.5. *If $\pi = (\pi_j, j \in S)$ is a positive invariant probability on S of the Markov chain X, then X is recurrent.*

Proof. Let $\pi = (\pi_j, j \in S)$ be a positive invariant probability on S of X. For every $j \in S$ and $n \geq 0$, we have

$$\pi_j = \sum_{i \in S} \pi_i (P^n)_{i,j}.$$

If state j is transient then, when n tends to ∞, we obtain $\pi_j = 0$ by Corollary 2.3.9. This contradicts the fact that π is a positive invariant probability, which means that every state j is recurrent. ∎

2.6 Aperiodic Markov chains

Let U be a finite set of positive integers. We denote by $\gcd(U)$ the greatest common divisor of U, i.e. the largest positive integer dividing all the elements of U. Let $\{u_n, n \geq 1\}$ be a sequence of positive integers. For $k \geq 1$, the sequence $d_k = \gcd\{u_1, \ldots, u_k\}$ is non-increasing and bounded below by 1, so it converges to a limit $d \geq 1$ called the gcd of the sequence $\{u_n, n \geq 1\}$. Since the d_k are integers, the limit d is reached in a finite number of steps and, thus there exists a positive integer k_0 such that $d = \gcd\{u_1, \ldots, u_k\}$ for all $k \geq k_0$.

DEFINITION 2.6.1.   *The period, $d(i)$, of a state i is defined by*

$$d(i) = \gcd\{n \geq 1 \mid (P^n)_{i,i} > 0\},$$

with the convention $d(i) = 0$ if $(P^n)_{i,i} = 0$ for every $n \geq 1$. If $d(i) = 1$ then state i is said to be aperiodic.

Observe that a state i such that $P_{i,i} > 0$ is aperiodic.

THEOREM 2.6.2.   *If states i and j communicate then $d(i) = d(j)$.*

Proof.   If $i = j$ the claim is obviously true. So we suppose that $i \neq j$. If states i and j communicate and since they are different, there exist integers $\ell \geq 1$ and $m \geq 1$ such that $(P^\ell)_{i,j} > 0$ and $(P^m)_{j,i} > 0$. We thus have

$$(P^{\ell+m})_{i,i} = \sum_{k \in S} (P^\ell)_{i,k}(P^m)_{k,i} \geq (P^\ell)_{i,j}(P^m)_{j,i} > 0,$$

which shows that $\ell + m$ is a multiple of $d(i)$. We have, in the same way

$$(P^{\ell+m})_{j,j} = \sum_{k \in S} (P^m)_{j,k}(P^\ell)_{k,j} \geq (P^m)_{j,i}(P^\ell)_{i,j} > 0,$$

which shows that $\ell + m$ is a multiple of $d(j)$.

Let $E_i = \{n \geq 1 \mid (P^n)_{i,i} > 0\}$. By definition $d(i) = \gcd(E_i)$ and if $n \in E_i$ then n is a multiple of $d(i)$. Let $\mathbb{N}^*$ be the set of positive integers. We introduce the set $d(i)\mathbb{N}^*$ made up of all the multiples of $d(i)$. Clearly, $E_i \subseteq d(i)\mathbb{N}^*$ and $d(i) = \gcd(d(i)\mathbb{N}^*)$. We have shown that $E_i \neq \emptyset$, since $\ell + m \in E_i$. So let $n \in E_i$. Since $(P^n)_{i,i} > 0$, we have

$$(P^{n+\ell+m})_{j,j} \geq (P^m)_{j,i}(P^n)_{i,i}(P^\ell)_{i,j} > 0,$$

which implies that $n + \ell + m \in E_j$ and thus $n + \ell + m \in d(j)\mathbb{N}^*$. Moreover, we have shown that $\ell + m \in d(j)\mathbb{N}^*$, so $n \in d(j)\mathbb{N}^*$, which proves that $E_i \subseteq d(j)\mathbb{N}^*$. Taking the gcd of both sets, we obtain $d(j) \leq d(i)$.

The roles played by states i and j being symmetric, we also have $d(i) \leq d(j)$, that is $d(i) = d(j)$.   ∎

We deduce that periodicity is a class property, i.e. all the states of the same class have the same period.

DEFINITION 2.6.3.   *A Markov chain is said to be aperiodic if all its states have a period of 1.*

Before going further, we recall some results about the greatest common divisor of integers. We denote by $\mathbb{Z}$ the set of all integers.

DEFINITION 2.6.4.   *A non-empty subset I of $\mathbb{Z}$ is an ideal of $\mathbb{Z}$ if*

- $x \in I$ and $y \in I$ implies $x + y \in I$,
- $x \in I$ and $\lambda \in \mathbb{Z}$ implies $\lambda x \in I$.

Observe that if I is an ideal of $\mathbb{Z}$ then $0 \in I$ and if $x \in I$ then $-x \in I$.

For every $a \in \mathbb{Z}$, we denote by $a\mathbb{Z}$ the set made up of all the multiples of a, i.e. $a\mathbb{Z} = \{\lambda a, \lambda \in \mathbb{Z}\}$. It is easily checked that $a\mathbb{Z}$ is an ideal of $\mathbb{Z}$.

DEFINITION 2.6.5. *An ideal I of $\mathbb{Z}$ is said to be principal if there exists a unique integer $a \geq 0$ such that $I = a\mathbb{Z}$.*

LEMMA 2.6.6. *Every ideal of $\mathbb{Z}$ is principal.*

Proof. Let I be an ideal of $\mathbb{Z}$. We have to show that there exists a unique integer $a \geq 0$ such that $I = a\mathbb{Z}$.

If $I = \{0\}$ we have $I = 0\mathbb{Z}$, which means that I is principal. We suppose now that $I \neq \{0\}$.

Existence. $I \neq \{0\}$ means that I contains positive integers since if $x \in I$ then $-x \in I$. We denote by a the smallest positive integer of I. By definition of an ideal, the multiples of one of its elements also belong to it. Thus we have $a\mathbb{Z} \subseteq I$.

Conversely, if $x \geq 0$ and $x \in I$ then the Euclidian division of x by a gives

$$x = aq + r \quad \text{with} \quad 0 \leq r \leq a - 1.$$

Since $x \in I$ and $-aq \in I$, we have $r = x - aq \in I$, which means that $r = 0$ because a is the smallest positive integer of I. We thus have $x = aq$, i.e. $x \in a\mathbb{Z}$. Finally, if $x \leq 0$ and $x \in I$ then $-x \in I$. We have shown that in this case, we have $-x \in a\mathbb{Z}$ and thus $x \in a\mathbb{Z}$.

So, $I \subseteq a\mathbb{Z}$. We have then shown that $I = a\mathbb{Z}$.

Uniqueness. If a and b are two positive integers such that $a\mathbb{Z} = b\mathbb{Z}$ then a and b are multiples of each other, so $a = b$. ∎

LEMMA 2.6.7. *Let $n_1, \ldots, n_k$ be positive integers. There exists a common divisor, d, to these k integers, of the form $d = \lambda_1 n_1 + \cdots + \lambda_k n_k$, where $\lambda_1, \ldots, \lambda_k$ are integers. Such a common divisor is a multiple of every other common divisor to these k integers. That is, $d = gcd\{n_1, \ldots, n_k\}$.*

Proof. Let $I = \{\lambda_1 n_1 + \cdots + \lambda_k n_k, \lambda_1, \ldots, \lambda_k \in \mathbb{Z}\}$ be the set of all linear combinations, with coefficients in $\mathbb{Z}$, of the integers $n_1, \ldots, n_k$. It is easily checked that I is an ideal of $\mathbb{Z}$. By Lemma 2.6.6, it is a principal ideal, so there exists a unique integer $d \geq 0$ such that $I = d\mathbb{Z}$. In particular, $d \in I$, so it can be written as $d = \lambda_1 n_1 + \cdots + \lambda_k n_k$.

But, for every $j = 1, \ldots, k$, taking $\lambda_j = 1$ and all the others equal to 0, we obtain $n_j \in I$, which proves that, for every $j = 1, \ldots, k$, n_j is a multiple of d. We conclude that d is a common divisor to integers $n_1, \ldots, n_k$.

Let x be another common divisor to integers $n_1, \ldots, n_k$. Thus, $n_1, \ldots, n_k$ and $\lambda_1 n_1 + \cdots + \lambda_k n_k$ are multiples of x, for every $\lambda_1, \ldots, \lambda_k$. So d is also a multiple of x, which proves that $d = gcd\{n_1, \ldots, n_k\}$. ∎

LEMMA 2.6.8. *Let $n_1,\ldots,n_k$ be positive integers and let $d = \gcd\{n_1,\ldots,n_k\}$. There exists an integer $N > 0$ such that $n \geq N$ implies that there exist nonnegative integers $c_1,\ldots,c_k$ such that $nd = c_1 n_1 + \cdots + c_k n_k$.*

Proof. By Lemma 2.6.7, there exist integers $\lambda_1,\ldots,\lambda_k$ such that $d = \lambda_1 n_1 + \cdots + \lambda_k n_k$. Defining

$$N_1 = \sum_{j|\lambda_j>0} \lambda_j n_j \quad \text{and} \quad N_2 = \sum_{j|\lambda_j<0} (-\lambda_j) n_j, \tag{2.11}$$

we have $d = N_1 - N_2$, with $N_1 > 0$ and $N_2 \geq 0$ since $d \geq 1$. In the proof of Lemma 2.6.7, we have shown that $n_1,\ldots,n_k$ are all multiples of d. So N_1 and N_2 are also multiples of d. Let $N = N_2^2/d$. An integer $n \geq N$ can be written as $n = N + \ell$ and the Euclidian division of ℓ by integer N_2/d allows us to write $\ell = \delta N_2/d + b$ with $\delta \geq 0$ and $0 \leq b < N_2/d$. So, we have

$$nd = Nd + \ell d = N_2^2 + \delta N_2 + bd.$$

Replacing d with $N_1 - N_2$, we obtain

$$nd = (N_2 - b + \delta)N_2 + bN_1.$$

Since $b < N_2/d$, we have $N_2 > bd$ and since $d \geq 1$, we obtain $N_2 - b > 0$, which shows, using the expressions of N_1 and N_2 given in (2.11), that nd is written as a linear combination, with nonnegative coefficients, of the integers n_j, for all $n \geq N$. ∎

THEOREM 2.6.9. *If state i has period $d(i)$ then there exists an integer, N, such that for every $n \geq N$, we have*

$$(P^{nd(i)})_{i,i} > 0.$$

Proof. If $d(i) = 0$ the result is immediate, so we suppose that $d(i) \geq 1$. By definition of period $d(i)$, there exists a finite number of positive integers n_ℓ, $\ell = 1,\ldots,k$ such that $(P^{n_\ell})_{i,i} > 0$ and $d(i) = \gcd\{n_1,\ldots,n_k\}$. By Lemma 2.6.8, there exists an integer $N > 0$ such that $n \geq N$ implies that there exist nonnegative integers $c_1,\ldots,c_k$ such that $nd(i) = c_1 n_1 + \cdots + c_k n_k$.

We thus have, for every $n \geq N$,

$$(P^{nd(i)})_{i,i} = (P^{c_1 n_1 + \cdots + c_k n_k})_{i,i} \geq \prod_{\ell=1}^{k} (P^{c_\ell n_\ell})_{i,i} \geq \prod_{\ell=1}^{k} \left((P^{n_\ell})_{i,i}\right)^{c_\ell} > 0,$$

which completes the proof. ∎

2.7 Convergence to steady-state

THEOREM 2.7.1. *Let X be an irreducible and aperiodic Markov chain and let π be its unique positive invariant probability. For every $j \in S$, we have*

$$\lim_{n\to\infty} \mathbb{P}\{X_n = j\} = \pi_j,$$

for every initial distribution. In particular, for every $i, j \in S$,

$$\lim_{n\to\infty} (P^n)_{i,j} = \pi_j.$$

Proof. We denote by α the initial distribution of X. This proof is based on a coupling argument. Let $Y = \{Y_n, n \in \mathbb{N}\}$ be a Markov chain on the same state space S as X, with initial distribution π, with the same transition probability matrix P as X and independent of X. Let ℓ be an arbitrary state of S. We introduce the time T defined by

$$T = \inf\{n \geq 1 \mid X_n = Y_n = \ell\}.$$

Step 1. We prove that $\mathbb{P}\{T < \infty\} = 1$.
The process $W = \{W_n, n \in \mathbb{N}\}$ defined by $W_n = (X_n, Y_n)$ is a Markov chain on the state space $S \times S$, with initial distribution β given by

$$\beta_{(i,k)} = \alpha_i \pi_k,$$

and transition probability matrix $\tilde{P}$ given by

$$\tilde{P}_{(i,k),(j,l)} = P_{i,j} P_{k,l}.$$

It is easily checked that W has a positive invariant probability $\tilde{\pi}$ given by

$$\tilde{\pi}_{(i,k)} = \pi_i \pi_k$$

and, by recurrence, that for every $n \geq 0$, we have

$$(\tilde{P}^n)_{(i,k),(j,l)} = (P^n)_{i,j}(P^n)_{k,l}.$$

Time T can be written as $T = \inf\{n \geq 1 \mid W_n = (\ell, \ell)\}$. Let i, j, k, and l be four states of S. X being irreducible, there exist integers m and h such that $(P^m)_{i,j} > 0$ and $(P^h)_{k,l} > 0$. X being aperiodic, from Theorem 2.6.9, there exist integers N_j and N_l such that, for every $n \geq N = \max\{N_j, N_l\}$, $(P^n)_{j,j} > 0$ and $(P^n)_{l,l} > 0$. We thus have, for every $n \geq N$,

$$(\tilde{P}^{n+m+h})_{(i,k),(j,l)} = (P^{n+m+h})_{i,j}(P^{n+m+h})_{k,l}$$

$$\geq (P^m)_{i,j}(P^{n+h})_{j,j}(P^h)_{k,l}(P^{n+m})_{l,l}$$

$$> 0,$$

which proves that W is irreducible. From Theorem 2.4.5, we have $\mathbb{P}\{T < \infty\} = 1$.

Step 2. Concatenation of X and Y.

Time T being finite with probability 1, we define the process $Z = \{Z_n, n \in \mathbb{N}\}$, by

$$Z_n = \begin{cases} X_n & \text{if } n < T, \\ Y_n & \text{if } n \geq T. \end{cases}$$

We show in this step that Z is a Markov chain. We clearly have, by definition of T, that $Z_0 = X_0$. Let $n \geq 1$, $1 \leq k \leq n$ and $i_k, \ldots, i_n \in S$. We have

$$\mathbb{P}\{Y_n = i_n, \ldots, Y_k = i_k \mid W_k = (\ell, \ell)\}$$

$$= \sum_{h_n, \ldots, h_{k+1} \in S} \mathbb{P}\{W_n = (h_n, i_n), \ldots, W_k = (\ell, i_k) \mid W_k = (\ell, \ell)\}$$

$$= 1_{\{i_k = \ell\}} \sum_{h_n, \ldots, h_{k+1} \in S} \tilde{P}_{(\ell, \ell),(h_{k+1}, i_{k+1})} \cdots \tilde{P}_{(h_{n-1}, i_{n-1}),(h_n, i_n)}$$

$$= 1_{\{i_k = \ell\}} \sum_{h_n, \ldots, h_{k+1} \in S} P_{\ell, h_{k+1}} P_{\ell, i_{k+1}} \cdots P_{h_{n-1}, h_n} P_{i_{n-1}, i_n}$$

$$= 1_{\{i_k = \ell\}} P_{\ell, i_{k+1}} \cdots P_{i_{n-1}, i_n}.$$

Proceeding in the same way for the Markov chain X, we obtain

$$\mathbb{P}\{Y_n = i_n, \ldots, Y_k = i_k \mid W_k = (\ell, \ell)\} = \mathbb{P}\{X_n = i_n, \ldots, X_k = i_k \mid W_k = (\ell, \ell)\}. \quad (2.12)$$

Using the facts that $\{T = k\} = \{W_k = (\ell, \ell), W_{k-1} \neq (\ell, \ell), \ldots, W_1 \neq (\ell, \ell)\}$ and $\{T = k\} \subseteq \{W_k = (\ell, \ell)\}$, we obtain

$$\mathbb{P}\{Z_n = i_n, \ldots, Z_0 = i_0, T = k\}$$

$$= \mathbb{P}\{Y_n = i_n, \ldots, Y_k = i_k, X_{k-1} = i_{k-1}, \ldots, X_0 = i_0, T = k\}$$

$$= \mathbb{P}\{Y_n = i_n, \ldots, Y_k = i_k, W_k = (\ell, \ell), T = k, X_{k-1} = i_{k-1}, \ldots, X_0 = i_0\}$$

$$= \mathbb{P}\{Y_n = i_n, \ldots, Y_k = i_k \mid W_k = (\ell, \ell)\} \mathbb{P}\{T = k, X_{k-1} = i_{k-1}, \ldots, X_0 = i_0\}$$

$$= \mathbb{P}\{X_n = i_n, \ldots, X_k = i_k \mid W_k = (\ell, \ell)\} \mathbb{P}\{T = k, X_{k-1} = i_{k-1}, \ldots, X_0 = i_0\}$$

$$= \mathbb{P}\{X_n = i_n, \ldots, X_k = i_k, W_k = (\ell, \ell), T = k, X_{k-1} = i_{k-1}, \ldots, X_0 = i_0\}$$

$$= \mathbb{P}\{X_n = i_n, \ldots, X_0 = i_0, T = k\},$$

where the third and the fifth equalities use the Markov property, since T is a stopping time for W and where the fourth one uses Relation (2.12). This relation being true for every $1 \leq k \leq n$, we have

$$\mathbb{P}\{Z_n = i_n, \ldots, Z_0 = i_0, T \leq n\} = \mathbb{P}\{X_n = i_n, \ldots, X_0 = i_0, T \leq n\}.$$

In addition, by definition of Z, we have

$$\mathbb{P}\{Z_n = i_n, \ldots, Z_0 = i_0, T > n\} = \mathbb{P}\{X_n = i_n, \ldots, X_0 = i_0, T > n\}.$$

Adding these two equalities, we obtain

$$\mathbb{P}\{Z_n = i_n, \ldots, Z_0 = i_0\} = \mathbb{P}\{X_n = i_n, \ldots, X_0 = i_0\}.$$

This proves, from Theorem 2.1.3, that Z is, like X, a Markov chain with initial distribution α and transition probability matrix P.

Step 3. Convergence.

From step 2, we have $\mathbb{P}\{X_n = j\} = \mathbb{P}\{Z_n = j\}$ and, by definition of Z,

$$\mathbb{P}\{Z_n = j\} = \mathbb{P}\{X_n = j, T > n\} + \mathbb{P}\{Y_n = j, T \leq n\}.$$

Thus, since $\mathbb{P}\{Y_n = j\} = \pi_j$,

$$
\begin{aligned}
|\mathbb{P}\{X_n = j\} - \pi_j| &= |\mathbb{P}\{Z_n = j\} - \mathbb{P}\{Y_n = j\}| \\
&= |\mathbb{P}\{X_n = j, T > n\} + \mathbb{P}\{Y_n = j, T \leq n\} - \mathbb{P}\{Y_n = j\}| \\
&= |\mathbb{P}\{X_n = j, T > n\} - \mathbb{P}\{Y_n = j, T > n\}| \\
&\leq \max\left(\mathbb{P}\{X_n = j, T > n\}, \mathbb{P}\{Y_n = j, T > n\}\right) \\
&\leq \mathbb{P}\{T > n\},
\end{aligned}
$$

and $\mathbb{P}\{T > n\} \longrightarrow 0$ when $n \longrightarrow \infty$, since T is finite with probability 1. ■

2.8 Ergodic theorem

Let j be a fixed state of S. We start by analyzing the time elapsed between two successive visits of X to j.

Recall that $\tau(j)$ is the time needed to reach state j, i.e.

$$\tau(j) = \inf\{n \geq 1 \mid X_n = j\}.$$

In the same way, we define the sequence $\tau_\ell(j)$ of the successive passage times to state j by

$$\tau_0(j) = 0 \quad \text{and} \quad \tau_\ell(j) = \inf\{n > \tau_{\ell-1}(j) \mid X_n = j\}, \quad \text{for } \ell \geq 1,$$

with the convention $\inf \emptyset = \infty$. With this notation, we have $\tau(j) = \tau_1(j)$. The length $S_\ell(j)$ of the ℓth excursion to state j is defined, for $\ell \geq 1$, by

$$S_\ell(j) = \begin{cases} \tau_\ell(j) - \tau_{\ell-1}(j) & \text{if } \tau_{\ell-1}(j) < \infty, \\ 0 & \text{otherwise.} \end{cases}$$

LEMMA 2.8.1. *For $\ell \geq 2$, conditional on $\tau_{\ell-1}(j) < \infty$, the random variable $S_\ell(j)$ is independent of $\{X_0, \ldots, X_{\tau_{\ell-1}(j)}\}$ and*

$$\mathbb{P}\{S_\ell(j) = n \mid \tau_{\ell-1}(j) < \infty\} = \mathbb{P}\{\tau(j) = n \mid X_0 = j\}.$$

Proof. We apply the strong Markov property to the stopping time $T = \tau_{\ell-1}(j)$. We have $X_T = j$ when $T < \infty$. So, conditional on $T < \infty$, $\{X_{T+n}, n \in \mathbb{N}\}$ is a Markov chain with initial distribution δ^j and transition probability matrix P, which is independent of $\{X_0, \ldots, X_T\}$, by Theorem 2.2.4. But we have

$$S_\ell(j) = \inf\{n \geq 1 \mid X_{T+n} = j\},$$

so $S_\ell(j)$ is the time needed by the Markov chain $\{X_{T+n}, n \in \mathbb{N}\}$ to return to state j. This means that

$$\mathbb{P}\{S_\ell(j) = n \mid \tau_{\ell-1}(j) < \infty\} = \mathbb{P}\{\tau(j) = n \mid X_0 = j\},$$

which concludes the proof. ∎

COROLLARY 2.8.2. *If X is irreducible then, for every $j \in S$ and $\ell \geq 2$, the random variables $S_1(j), \ldots, S_\ell(j)$ are independent. Moreover, conditional on $X_0 = j$, they are identically distributed with mean $\mathbb{E}\{S_1(j) \mid X_0 = j\} = 1/\pi_j$, where π is the unique positive invariant probability of X.*

Proof. The Markov chain being irreducible, from Theorem 2.4.5, the stopping times, $\tau_n(j)$, are all finite with probability 1, for every $j \in S$ and $n \geq 1$. We thus have $X_{\tau_n(j)} = j$. For every $k_1, \ldots, k_\ell \geq 1$, we have, from Lemma 2.8.1, for every $\ell \geq 2$,

$$\mathbb{P}\{S_\ell(j) = k_\ell \mid S_{\ell-1}(j) = k_{\ell-1}, \ldots, S_1(j) = k_1\} = \mathbb{P}\{S_\ell(j) = k_\ell\}$$
$$= \mathbb{P}\{\tau(j) = k_\ell \mid X_0 = j\}$$
$$= f_{j,j}^{(k_\ell)}.$$

Repeating the same argument and using $S_1(j) = \tau(j)$, we obtain

$$\mathbb{P}\{S_\ell(j) = k_\ell, \ldots, S_1(j) = k_1\} = \mathbb{P}\{S_\ell(j) = k_\ell\} \ldots \mathbb{P}\{S_1(j) = k_1\}$$
$$= f_{j,j}^{(k_\ell)} f_{j,j}^{(k_{\ell-1})} \ldots f_{j,j}^{(k_2)} \mathbb{P}\{\tau(j) = k_1\}.$$

We thus have, from Corollary 2.5.4,

$$\mathbb{E}\{S_1(j) \mid X_0 = j\} = \mathbb{E}\{\tau(j) \mid X_0 = j\} = m_j = 1/\pi_j,$$

which concludes the proof. ∎

THEOREM 2.8.3. (Strong law of large numbers) *Let $Y_1, Y_2, \ldots$ be a sequence of random variables independent and identically distributed with mean $\mathbb{E}\{Y_1\} = \mu < \infty$. We have*

$$\lim_{n \to \infty} \frac{1}{n} \sum_{k=1}^{n} Y_k = \mu \text{ with probability 1.}$$

Proof. See for instance [111]. ∎

For every $n \geq 1$ and $j \in S$, we denote by $V_j(n)$ the number of visits of X to state j before the nth transition, i.e.

$$V_j(n) = \sum_{k=0}^{n-1} 1_{\{X_k = j\}}.$$

THEOREM 2.8.4. (Ergodic theorem) *Let X be an irreducible Markov chain. For every $j \in S$, we have*

$$\lim_{n \to \infty} \frac{1}{n} \sum_{k=0}^{n-1} 1_{\{X_k = j\}} = \pi_j \text{ with probability 1,}$$

where π is the unique positive invariant probability of X.
Moreover, for every function $r : S \longrightarrow \mathbb{R}$, we have

$$\lim_{n \to \infty} \frac{1}{n} \sum_{k=0}^{n-1} r_{X_k} = \sum_{j \in S} r_j \pi_j \text{ with probability 1.}$$

Proof. Since X is irreducible, there exists, from Corollary 2.5.4, a unique positive invariant probability denoted by π. Let α be the initial distribution of X and j be a fixed state of S. The Markov chain being irreducible, by Theorem 2.4.5 the time $\tau(j)$ is finite with probability 1.

If $X_0 = i \neq j$, we have, for every $n \geq 1$, by definition of $S_\ell(j)$,

$$S_1(j) + \cdots + S_{V_j(n)}(j) \leq n - 1 < S_1(j) + \cdots + S_{V_j(n)+1}(j),$$

where the left-hand side is equal to 0 when $V_j(n) = 0$. Indeed, the left-hand side is the instant of the last visit to state j before instant n and the right-hand side is the instant of the first visit to state j after instant $n - 1$.

We thus obtain

$$\frac{S_1(j)}{V_j(n)} + \frac{S_2(j) + \cdots + S_{V_j(n)}(j)}{V_j(n)} < \frac{n}{V_j(n)} \leq \frac{S_1(j)}{V_j(n)} + \frac{S_2(j) + \cdots + S_{V_j(n)+1}(j)}{V_j(n)}. \quad (2.13)$$

From Corollary 2.8.2, the random variables $S_2(j), \ldots$ are independent and identically distributed, with mean $1/\pi_j$. We have seen that $S_1(j) = \tau(j)$ is finite with probability 1. From Corollaries 2.3.10 and 2.4.3, we have

$$\lim_{n \to \infty} V_j(n) = \infty \text{ with probability 1.}$$

Thus, we have $\lim_{n \to \infty} S_1(j)/V_j(n) = 0$ with probability 1 and, from Theorem 2.8.3, we have

$$\lim_{n \to \infty} \frac{1}{n} \sum_{k=2}^{n} S_k(j) = \mathbb{E}\{S_1(j) \mid X_0 = j\} = \frac{1}{\pi_j} \text{ with probability 1.}$$

Taking the limit in (2.13), leads to

$$\lim_{n \to \infty} \frac{n}{V_j(n)} = \frac{1}{\pi_j} \text{ with probability 1,}$$

that is

$$\lim_{n \to \infty} \frac{V_j(n)}{n} = \pi_j \text{ with probability 1.}$$

If $X_0 = j$, we have, for every $n \geq 1$, by definition of the $S_\ell(j)$,

$$S_1(j) + \cdots + S_{V_j(n)-1}(j) \leq n - 1 < S_1(j) + \cdots + S_{V_j(n)}(j),$$

where the left-hand side is equal to 0 when $V_j(n) = 1$.

We thus obtain

$$\frac{S_1(j) + \cdots + S_{V_j(n)-1}(j)}{V_j(n)} < \frac{n}{V_j(n)} \leq \frac{S_1(j) + \cdots + S_{V_j(n)}(j)}{V_j(n)}. \tag{2.14}$$

Again, from Corollary 2.8.2, the random variables $S_1(j), S_2(j), \ldots$ are independent and identically distributed, with mean $1/\pi_j$ and the same argument as used above leads to

$$\lim_{n \to \infty} \frac{V_j(n)}{n} = \pi_j \text{ with probability 1.}$$

So, we have shown that, for every $i, j \in S$,

$$\mathbb{P}\{\lim_{n \to \infty} V_j(n)/n = \pi_j \mid X_0 = i\} = 1.$$

Unconditioning on the initial state completes the first part of the proof.

Let r be a function from S to $\mathbb{R}$. We have

$$\left| \frac{1}{n} \sum_{k=0}^{n-1} r_{X_k} - \sum_{j \in S} r_j \pi_j \right| = \left| \frac{1}{n} \sum_{j \in S} r_j \sum_{k=0}^{n-1} 1_{\{X_k = j\}} - \sum_{j \in S} r_j \pi_j \right|$$

$$= \left| \sum_{j \in S} r_j \left(\frac{V_j(n)}{n} - \pi_j \right) \right|$$

$$\leq \sum_{j \in S} |r_j| \left| \frac{V_j(n)}{n} - \pi_j \right|,$$

which tends to 0 when n tends to ∞, using the first part of the proof and the fact that the state space is finite. ∎

COROLLARY 2.8.5.   *Let X be an irreducible Markov chain. For every $j \in S$ and for every initial distribution α, we have*

$$\lim_{n \to \infty} \frac{1}{n} \sum_{k=0}^{n-1} (\alpha P^k)_j = \pi_j.$$

In particular, for every $i, j \in S$, we have

$$\lim_{n \to \infty} \frac{1}{n} \sum_{k=0}^{n-1} (P^k)_{i,j} = \pi_j.$$

Proof.   From Theorem 2.8.4, we have, with probability 1,

$$\lim_{n \to \infty} \frac{1}{n} \sum_{k=0}^{n-1} 1_{\{X_k = j\}} = \pi_j,$$

and since

$$0 \le \frac{1}{n} \sum_{k=0}^{n-1} 1_{\{X_k = j\}} \le 1,$$

by the dominated convergence theorem, we obtain

$$\lim_{n \to \infty} \mathbb{E} \left\{ \frac{1}{n} \sum_{k=0}^{n-1} 1_{\{X_k = j\}} \right\} = \pi_j,$$

that is,

$$\lim_{n \to \infty} \frac{1}{n} \sum_{k=0}^{n-1} \mathbb{P}\{X_k = j\} = \pi_j.$$

The second part of the proof is obtained by taking $\alpha = \delta^i$.   ∎

2.9   Absorbing Markov chains

In this section we consider Markov chains with absorbing states, i.e. states that can never be left.

DEFINITION 2.9.1.   *A discrete-time Markov chain is called absorbing if it has at least one absorbing state, at least one transient state and no other recurrent state than the absorbing ones.*

Let X be an absorbing Markov chain with transition probability matrix P, over the finite state space, S. The state space can be decomposed as $S = B \cup H$, where B is the set of transient states and H is the set of absorbing states. By definition B and H are both non-empty. Following this decomposition, matrix P can be written as

$$P = \begin{pmatrix} P_B & P_{BH} \\ 0 & I \end{pmatrix}.$$

where submatrix P_B contains the transition probabilities between transient states and P_{BH} contains the transition probabilities from states of B to states of H. Matrix 0 here is the null matrix with dimension $(|H|, |B|)$ and I is the identity matrix with dimension $|H|$. Without any loss of generality, we suppose that for every absorbing state, j, there exists a transient state, i, such that $P_{i,j} \neq 0$. This means that matrix P_{BH} has no null column. If this is not the case, it would mean that there exists an absorbing state which remains alone, that is not accessible from any other state.

LEMMA 2.9.2. *The matrix $I - P_B$ is invertible and we have*

$$(I - P_B)^{-1} = \sum_{n=0}^{\infty} (P_B)^n.$$

Proof. The states of B are transient. From Theorem 2.3.9, we have, for every $i, j \in B$,

$$\sum_{n=1}^{\infty} (P^n)_{i,j} < \infty.$$

This means that

$$\sum_{n=0}^{\infty} (P_B)^n < \infty \quad \text{and thus that} \quad \lim_{n \to \infty} (P_B)^n = 0.$$

Now, for every $N \geq 0$, we have

$$(I - P_B) \sum_{n=0}^{N} (P_B)^n = I - (P_B)^{N+1}.$$

When $N \longrightarrow \infty$, we have

$$(I - P_B) \sum_{n=0}^{\infty} (P_B)^n = I,$$

which completes the proof. ∎

The initial probability distribution, α, of X can be decomposed through subsets B and H as

$$\alpha = (\alpha_B \; \alpha_H),$$

where row subvector α_B (resp. α_H) contains the initial probabilities of states of B (resp. H).

For every $j \in H$, the hitting time of state j has been defined by

$$\tau(j) = \inf\{n \geq 1 \mid X_n = j\},$$

where $\tau(j) = \infty$ if this set is empty. With this definition, we have $\tau(j) > 0$ and, for $n \geq 1$,

$$\tau(j) \leq n \iff X_n = j.$$

Since matrix P is block triangular, we have, for $n \geq 0$,

$$P^n = \begin{pmatrix} (P_B)^n & \sum_{\ell=0}^{n-1} (P_B)^\ell P_{BH} \\ 0 & I \end{pmatrix}, \qquad (2.15)$$

where as usual, an empty sum (when $n = 0$ here) is equal to 0. We thus obtain, using Theorem 2.1.4,

$$\mathbb{P}\{\tau(j) = 0\} = 0,$$

$$\mathbb{P}\{\tau(j) \leq n\} = \mathbb{P}\{X_n = j\} = (\alpha P^n)_j, \quad \text{for } n \geq 1.$$

We denote by v_j the column vector containing the jth column of matrix P_{BH}. Using the decomposition in (2.15), we obtain, for $n \geq 0$,

$$\mathbb{P}\{\tau(j) \leq n\} = \left(\alpha_B \sum_{\ell=0}^{n-1} (P_B)^\ell P_{BH} + \alpha_H \right)_j - \alpha_j 1_{\{n=0\}}$$

$$= \alpha_B \sum_{\ell=0}^{n-1} (P_B)^\ell v_j + \alpha_j 1_{\{n \geq 1\}}$$

$$= \alpha_B (I - (P_B)^n)(I - P_B)^{-1} v_j + \alpha_j 1_{\{n \geq 1\}}.$$

In particular, we have, since $\lim_{n \to \infty} (P_B)^n = 0$,

$$\mathbb{P}\{\tau(j) < \infty\} = \alpha_B (I - P_B)^{-1} v_j + \alpha_j. \qquad (2.16)$$

The expected time, $\mathbb{E}\{\tau(j)\}$, to hit state $j \in H$ is thus equal to ∞ when $\alpha_B(I - P_B)^{-1} v_j + \alpha_j < 1$. When $\alpha_B(I - P_B)^{-1} v_j + \alpha_j = 1$, we have, for every $n \geq 0$,

$$\mathbb{P}\{\tau(j) > n\} = 1 - \mathbb{P}\{\tau(j) \leq n\}$$

$$= 1 - \alpha_B(I - P_B)^{-1} v_j - \alpha_j 1_{\{n \geq 1\}} + \alpha_B(P_B)^n(I - P_B)^{-1} v_j$$

$$= \alpha_j 1_{\{n=0\}} + \alpha_B(P_B)^n(I - P_B)^{-1} v_j,$$

and so

$$\mathbb{E}\{\tau(j)\} = \sum_{n=0}^{\infty} \mathbb{P}\{\tau(j) > n\} = \alpha_j + \alpha_B(I - P_B)^{-2} v_j. \qquad (2.17)$$

Note that when $|H| = 1$, i.e. $H = \{j\}$, we have

$$(I - P_B)^{-1}v_j = (I - P_B)^{-1}P_{BH}\mathbb{1} = (I - P_B)^{-1}(I - P_B)\mathbb{1} = \mathbb{1},$$

where we recall that $\mathbb{1}$ is a column vector with all its entries equal to 1, with the right dimension. This means that $\alpha_B(I - P_B)^{-1}v_j + \alpha_j = 1$. We thus obtain, in this case,

$$\mathbb{E}\{\tau(j)\} = \alpha_j + \alpha_B(I - P_B)^{-1}\mathbb{1}.$$

We consider now the time, T, needed to hit the subset, H, of absorbing states, also called the time to absorption, defined by

$$T = \inf\{n \geq 1 \mid X_n \in H\}.$$

The distribution of T is obtained in the same way as the one for $\tau(j)$. Indeed, we have $T > 0$ and, for $n \geq 1$,

$$T \leq n \iff X_n \in H.$$

This leads, for $n \geq 0$, to

$$\mathbb{P}\{T \leq n\} = \mathbb{P}\{X_n \in H\}1_{\{n \geq 1\}}$$

$$= \sum_{j \in H}\left(\alpha_B \sum_{\ell=0}^{n-1}(P_B)^\ell v_j + \alpha_j 1_{\{n \geq 1\}}\right)$$

$$= \alpha_B \sum_{\ell=0}^{n-1}(P_B)^\ell P_{BH}\mathbb{1} + \alpha_H \mathbb{1}1_{\{n \geq 1\}}$$

$$= \alpha_B(I - (P_B)^n)(I - P_B)^{-1}P_{BH}\mathbb{1} + \alpha_H \mathbb{1}1_{\{n \geq 1\}}$$

$$= \alpha_B(I - (P_B)^n)\mathbb{1} + \alpha_H \mathbb{1}1_{\{n \geq 1\}}$$

$$= \alpha_B\mathbb{1} + \alpha_H \mathbb{1}1_{\{n \geq 1\}} - \alpha_B(P_B)^n\mathbb{1},$$

or equivalently,

$$\mathbb{P}\{T > n\} = \alpha_B(P_B)^n\mathbb{1} + \alpha_H \mathbb{1}1_{\{n=0\}}. \tag{2.18}$$

This means, in particular, that T is finite with probability 1, since

$$\lim_{n \to \infty}\mathbb{P}\{T > n\} = 0.$$

The mean time to absorption, $\mathbb{E}\{T\}$, is obtained from Relation (2.18) by

$$\mathbb{E}\{T\} = \sum_{n=0}^{\infty}\mathbb{P}\{T > n\} = \alpha_H\mathbb{1} + \alpha_B(I - P_B)^{-1}\mathbb{1}.$$

The probability mass function of T is then given, for $n \geq 0$, by

$$\mathbb{P}\{T = n\} = \begin{cases} 0 & \text{if } n = 0, \\ 1 - \alpha_B P_B \mathbb{1} & \text{if } n = 1, \\ \alpha_B(P_B)^{n-1}(I - P_B)\mathbb{1} & \text{if } n \geq 2. \end{cases}$$

We obtain, for every $j \in H$,

$$\mathbb{P}\{T = 0, X_T = j\} = 0.$$

$$\begin{aligned}
\mathbb{P}\{T = 1, X_T = j\} &= \mathbb{P}\{T = 1, X_1 = j\} \\
&= \mathbb{P}\{X_0 \in B, X_1 = j\} + \mathbb{P}\{X_0 = j, X_1 = j\} \\
&= \sum_{\ell \in B} \alpha_\ell P_{\ell,j} + \alpha_j P_{j,j} \\
&= \alpha_B v_j + \alpha_j.
\end{aligned}$$

In the same way, we obtain for $n \geq 2$,

$$\begin{aligned}
\mathbb{P}\{T = n, X_T = j\} &= \mathbb{P}\{T = n, X_n = j\} \\
&= \mathbb{P}\{X_{n-1} \in B, X_n = j\} \\
&= \sum_{\ell \in B} \mathbb{P}\{X_{n-1} = \ell, X_n = j\} \\
&= \sum_{\ell \in B} \mathbb{P}\{X_{n-1} = \ell\} \mathbb{P}\{X_n = j \mid X_{n-1} = \ell\} \\
&= \sum_{\ell \in B} (\alpha P^{n-1})_\ell P_{\ell,j} \\
&= \sum_{\ell \in B} (\alpha_B (P_B)^{n-1})_\ell P_{\ell,j} \\
&= \alpha_B (P_B)^{n-1} v_j,
\end{aligned}$$

where the fifth equality is due to Theorem 2.1.4 and the sixth comes from the decomposition in (2.15).

For every $j \in H$, we denote by p_j the probability of the Markov chain X being absorbed in state j. This probability is given by

$$\begin{aligned}
p_j &= \mathbb{P}\{X_T = j\} \\
&= \mathbb{P}\{T > 0, X_T = j\} \\
&= \sum_{n=1}^{\infty} \mathbb{P}\{T = n, X_T = j\} \\
&= \alpha_B v_j + \alpha_j + \sum_{n=2}^{\infty} \alpha_B (P_B)^{n-1} v_j \\
&= \alpha_j + \sum_{n=1}^{\infty} \alpha_B (P_B)^{n-1} v_j \\
&= \alpha_j + \alpha_B (I - P_B)^{-1} v_j.
\end{aligned}$$

Note that we could have observed that $X_T = j \iff \tau(j) < \infty$, its probability being given by Relation (2.16).

For every $j \in H$, the expected time to absorption in state j is given by

$$\mathbb{E}\{T1_{\{X_T=j\}}\} = \sum_{n=0}^{\infty} \mathbb{P}\{T > n, X_T = j\}$$

$$= \sum_{n=0}^{\infty} \sum_{k=n+1}^{\infty} \mathbb{P}\{T = k, X_T = j\}$$

$$= \sum_{n=0}^{\infty} \sum_{k=n+1}^{\infty} \left(\alpha_B (P_B)^{k-1} v_j + \alpha_j 1_{\{k=1\}}\right)$$

$$= \sum_{n=0}^{\infty} \left(\alpha_B (P_B)^n (I - P_B)^{-1} v_j + \alpha_j 1_{\{n=0\}}\right)$$

$$= \alpha_B (I - P_B)^{-2} v_j + \alpha_j.$$

Note that this last relation is also equal to $\mathbb{E}\{\tau(j)1_{\{\tau(j)<\infty\}}\}$ which was obtained in Relation (2.17). The expected time to absorption given that absorption occurs in state j is given by

$$\mathbb{E}\{T \mid X_T = j\} = \frac{\mathbb{E}\{T1_{\{X_T=j\}}\}}{p_j} = \frac{\alpha_B (I - P_B)^{-2} v_j + \alpha_j}{\alpha_B (I - P_B)^{-1} v_j + \alpha_j},$$

which is also equal to $\mathbb{E}\{\tau(j) \mid \tau(j) < \infty\}$.

2.9.1   Application to irreducible Markov chains

Let X be an irreducible Markov chain on the finite state space S with transition probability matrix P and initial probability distribution α. We are interested in the computation of the distribution of the time, $\tau(j)$, needed to reach a given state, j. This time has already been defined as

$$\tau(j) = \inf\{n \geq 1 \mid X_n = j\},$$

and Theorem 2.3.2 gives its distribution, i.e.

$$f_{i,j}^{(n)} = \begin{cases} P_{i,j} & \text{if } n = 1, \\ \displaystyle\sum_{\ell \in S - \{j\}} P_{i,\ell} f_{\ell,j}^{(n-1)} & \text{if } n \geq 2, \end{cases} \tag{2.19}$$

where $f_{i,j}^{(n)} = \mathbb{P}\{\tau(j) = n \mid X_0 = i\}$.

From Theorem 2.4.5, we have $f_{i,j} = 1$ for every $i, j \in S$. We introduce matrix $Q^{(j)}$ obtained from matrix P by replacing the jth column by zeros, that is

$$Q_{i,\ell}^{(j)} = \begin{cases} P_{i,\ell} & \text{if } \ell \neq j, \\ 0 & \text{if } \ell = j, \end{cases}$$

and column vector P_j containing the jth column of matrix P, i.e. $P_j(i) = P_{i,j}$, for every $i \in S$. Equation (2.19) can then be written in matrix notation as

$$f_j^{(n)} = \begin{cases} P_j & \text{if } n = 1, \\ Q^{(j)} f_j^{(n-1)} & \text{if } n \geq 2, \end{cases} \tag{2.20}$$

where $f_j^{(n)}$ is the column vector whose ith entry $f_j^{(n)}(i)$ is given, for every $i \in S$, by

$$f_j^{(n)}(i) = f_{i,j}^{(n)}.$$

We thus have, for every $j \in S$ and $n \geq 1$,

$$f_j^{(n)} = (Q^{(j)})^{n-1} f_j^{(1)} = (Q^{(j)})^{n-1} P_j,$$

which leads, since $Q^{(j)} \mathbb{1} + P_j = \mathbb{1}$, to

$$\mathbb{P}\{\tau(j) = n\} = \alpha f_j^{(n)} = \alpha (Q^{(j)})^{n-1} P_j = \alpha (Q^{(j)})^{n-1} (I - Q^{(j)}) \mathbb{1}$$

or equivalently to

$$\mathbb{P}\{\tau(j) > n\} = \alpha (Q^{(j)})^n \mathbb{1}. \tag{2.21}$$

Let us consider the Markov chain Y over the state space $S \cup \{a\}$, where a is an absorbing state, with transition probability matrix M defined by

$$M = \begin{pmatrix} Q^{(j)} & P_j \\ 0 & 1 \end{pmatrix}$$

and initial probability distribution β defined by $\beta = (\alpha \; 0)$. If X is irreducible then it is easy to check that the states of S are transient for the Markov chain Y, so, from Lemma 2.9.2, matrix $I - Q^{(j)}$ is invertible for every $j \in S$ and we have

$$(I - Q^{(j)})^{-1} = \sum_{n=0}^{\infty} (Q^{(j)})^n.$$

The expected time $\mathbb{E}\{\tau(j)\}$ to reach state j is then obtained by

$$\mathbb{E}\{\tau(j)\} = \sum_{n=0}^{\infty} \mathbb{P}\{\tau(j) > n\} = \sum_{n=0}^{\infty} \alpha (Q^{(j)})^n \mathbb{1} = \alpha (I - Q^{(j)})^{-1} \mathbb{1}.$$

2.9.2 Computational aspects

The inverse of matrices $I - P_B$ or $I - Q^{(j)}$ is needed in several formulas. This inverse doesn't arise alone, but generally it is multiplied on both left and right sides by another rectangular matrix, possibly a row or column vector. From a computational point of view, the evaluation of such quantities is not a simple task. It consists of finding the solution of a linear system. There is a huge amount of literature on this subject, mainly dealing with the methods for solving linear systems. Here we would like to point out that the result we need is always a positive quantity, which can be an expectation or a value between 0 and 1, in this case corresponding to the probabilly of an event.

We present here an algorithm to compute such matrices. This algorithm has the property of being stable, since essentially only additions and multiplications of positive numbers are required. The method is based on two pioneering papers that appeared in 1985, Sheskin [107] and Grassmann, Taksar, and Heyman [44], that independently proposed practically the same technique to calculate the invariant distribution.

Let Q be a sub-stochastic square matrix over m transient states $\{1,2,\dots,m\}$. The matrix $I - Q$ is thus invertible. For $h \geq 1$, let H be an (m,h) matrix with non-negative entries. We propose an algorithm to compute the (m,h) matrix G defined by $G = (I - Q)^{-1}H$.

Relation $G = (I - Q)^{-1}H$ can also be written as $G = H + QG$, i.e.

$$G_{i,j} = H_{i,j} + \sum_{k=1}^{m} Q_{i,k} G_{k,j}.$$

Note that, since state m is transient, we cannot have $Q_{m,m} = 1$. For $i = m$, we have

$$G_{m,j} = H_{m,j} + \sum_{k=1}^{m-1} Q_{m,k} G_{k,j} + Q_{m,m} G_{m,j},$$

which leads to

$$G_{m,j} = \frac{H_{m,j}}{1 - Q_{m,m}} + \sum_{k=1}^{m-1} \frac{Q_{m,k}}{1 - Q_{m,m}} G_{k,j}.$$

Putting this relation in the previous expression of $G_{i,j}$, we obtain, for $i = 1,\dots,m-1$ and $j = 1,\dots,h$,

$$G_{i,j} = H_{i,j} + \sum_{k=1}^{m-1} Q_{i,k} G_{k,j} + Q_{i,m} G_{m,j}$$

$$= H_{i,j} + \sum_{k=1}^{m-1} Q_{i,k} G_{k,j} + Q_{i,m} \left(\frac{H_{m,j}}{1 - Q_{m,m}} + \sum_{k=1}^{m-1} \frac{Q_{m,k}}{1 - Q_{m,m}} G_{k,j} \right)$$

$$= H_{i,j} + \frac{Q_{i,m} H_{m,j}}{1 - Q_{m,m}} + \sum_{k=1}^{m-1} \left(Q_{i,k} + \frac{Q_{i,m} Q_{m,k}}{1 - Q_{m,m}} \right) G_{k,j}.$$

Introduce the matrices $Q^{(m)}$ and $H^{(m)}$ with dimensions $(m-1, m-1)$ and $(m-1, h)$, respectively, defined by

$$Q_{i,k}^{(m)} = Q_{i,k} + \frac{Q_{i,m} Q_{m,k}}{1 - Q_{m,m}}, \quad \text{for } i,k = 1,\ldots,m-1,$$

$$H_{i,j}^{(m)} = H_{i,j} + \frac{Q_{i,m} H_{m,j}}{1 - Q_{m,m}}, \quad \text{for } i = 1,\ldots,m-1 \text{ and } j = 1,\ldots,h.$$

Then, for $i = 1,\ldots,m-1$ and $j = 1,\ldots,h$, we have

$$G_{i,j} = H_{i,j}^{(m)} + \sum_{k=1}^{m-1} Q_{i,k}^{(m)} G_{k,j} \tag{2.22}$$

$$G_{m,j} = \frac{H_{m,j}}{1 - Q_{m,m}} + \sum_{k=1}^{m-1} \frac{Q_{m,k}}{1 - Q_{m,m}} G_{k,j}. \tag{2.23}$$

So, the numerical process consists of solving system (2.22) and then, using its solution, in computing $G_{m,j}$ using (2.23). In order to compute the solution of (2.22), we reuse the same process where the dimension is now $m-1$ instead of m. At step n, we compute the matrices $Q^{(n)}$ and $H^{(n)}$ with dimensions $(n-1, n-1)$ and $(n-1, h)$, respectively, defined by

$$Q_{i,k}^{(n)} = Q_{i,k}^{(n+1)} + \frac{Q_{i,n}^{(n+1)} Q_{n,k}^{(n+1)}}{1 - Q_{n,n}^{(n+1)}}, \quad \text{for } i,k = 1,\ldots,n-1,$$

$$H_{i,j}^{(n)} = H_{i,j}^{(n+1)} + \frac{Q_{i,n}^{(n+1)} H_{n,j}^{(n+1)}}{1 - Q_{n,n}^{(n+1)}}, \quad \text{for } i = 1,\ldots,n-1 \text{ and } j = 1,\ldots,h,$$

where $Q^{(m+1)}$ and $H^{(m+1)}$ are the initial matrices Q and H, respectively. Then for $i = 1,\ldots,n-1$ and $j = 1,\ldots,h$, we have, for $i = 1,\ldots,n-1$ and $j = 1,\ldots,h$,

$$G_{i,j} = H_{i,j}^{(n)} + \sum_{k=1}^{n-1} Q_{i,k}^{(n)} G_{k,j}, \tag{2.24}$$

and for $i = n,\ldots,m$ and $j = 1,\ldots,h$,

$$G_{i,j} = \frac{H_{i,j}^{(i+1)}}{1 - Q_{i,i}^{(i+1)}} + \sum_{k=1}^{i-1} \frac{Q_{i,k}^{(i+1)}}{1 - Q_{i,i}^{(i+1)}} G_{k,j}. \tag{2.25}$$

The algorithm for computing the matrix $G = (I - Q)^{-1} H$ is thus:

input : (m,m) matrix Q and (m,h) matrix H
output : $G = (I - Q)^{-1} H$
for $n = m$ **downto** 2 **do**
    $s = 1 - Q_{n,n}$

```
for i = 1 to n − 1 do
    Q_{i,n} = Q_{i,n}/s
    for k = 1 to n − 1 do
        Q_{i,k} = Q_{i,k} + Q_{i,n}Q_{n,k}
    endfor
    for j = 1 to h do
        H_{i,j} = H_{i,j} + Q_{i,n}H_{n,j}
    endfor
endfor
endfor
s = 1 − Q_{1,1}
for j = 1 to h do
    G_{1,j} = H_{1,j}/s
endfor
for i = 2 to m do
    s = 1 − Q_{i,i}
    for j = 1 to h do
        G_{i,j} = H_{i,j}
        for k = 1 to i − 1 do
            G_{i,j} = G_{i,j} + Q_{i,k}G_{k,j}
        endfor
        G_{i,j} = G_{i,j}/s
    endfor
endfor
```

3 Continuous-time Markov chains

Continuous-Time Markov Chain(s) are abbreviated as CTMC in this book.

3.1 Definitions and properties

A continuous-time stochastic process, $X = \{X_t, \ t \in \mathbb{R}^+\}$, on a finite state space, S, is a collection of random variables, X_t, indexed by $\mathbb{R}^+$, the set of nonnegative real numbers, taking their values in S.

A trajectory, ω, of the process is a function $t \longmapsto X_t(\omega)$ from $\mathbb{R}^+$ to S. The state space being discrete, the trajectories are piecewise constant functions. We suppose that the process is right continuous, i.e. for every trajectory w, and $t \geq 0$, there exists $\varepsilon > 0$ such that

$$X_s(\omega) = X_t(\omega) \text{ for every } t \leq s \leq t + \varepsilon.$$

We denote by $T_1, T_2, \ldots$ the jump instants of process X and by $S_1, S_2, \ldots$ the sequence of the times spent in the successive states visited by X. Setting $T_0 = 0$, we have

$$T_{n+1} = \inf\{t \geq T_n \mid X_t \neq X_{T_n}\},$$

for every $n \in \mathbb{N}$, with the convention $\inf \emptyset = \infty$, and, for every $n \geq 1$,

$$S_n = \begin{cases} T_n - T_{n-1} & \text{if } T_{n-1} < \infty, \\ \infty & \text{otherwise.} \end{cases}$$

The trajectories being right continuous, we have $S_n > 0$ for every $n \geq 1$, which means that there are no instantaneous states.

DEFINITION 3.1.1. *A stochastic process, $X = \{X_t, \ t \in \mathbb{R}^+\}$, with values in a finite state space, S, is a continuous-time Markov chain if for every $n \geq 0$, for all instants $0 \leq s_0 < \cdots < s_n < s < t$ and for all states $i_0, \ldots, i_n, i, j \in S$, we have*

$$\mathbb{P}\{X_t = j \mid X_s = i, X_{s_n} = i_n, \ldots X_{s_0} = i_0\} = \mathbb{P}\{X_t = j \mid X_s = i\}.$$

DEFINITION 3.1.2. *A continuous-time Markov chain, $X = \{X_t, t \in \mathbb{R}^+\}$, is homogeneous if for all $t,s \geq 0$ and for all $i,j \in S$, we have*

$$\mathbb{P}\{X_{t+s} = j \mid X_s = i\} = \mathbb{P}\{X_t = j \mid X_0 = i\}.$$

All the Markov chains considered in the following are homogeneous.

DEFINITION 3.1.3. *A random variable T, with values in $[0,\infty]$ is called a stopping time for X if the event $\{T \leq t\}$ is completely determined by $\{X_u, 0 \leq u \leq t\}$.*

For instance, the jump times T_n are stopping times.

THEOREM 3.1.4. (Strong Markov property). *If $X = \{X_t, t \in \mathbb{R}^+\}$ is a Markov chain and T is a stopping time for X then, for every $i \in S$, conditional on $\{T < \infty\} \cap \{X_T = i\}$, $\{X_{T+t}, t \in \mathbb{R}^+\}$ is a Markov chain with initial probability distribution δ^i, and is independent of $\{X_s, s \leq T\}$.*

Proof. The proof of this theorem is outside the scope of this book. We refer, for instance, to [2] for a detailed proof.   ■

THEOREM 3.1.5. *For every $i \in S$ and $t \geq 0$, we have*

$$\mathbb{P}\{T_1 > t \mid X_0 = i\} = e^{-\lambda_i t},$$

for some nonnegative real number λ_i.

Proof. Let $i \in S$ such that $X_0 = i$. If $T_1 = \infty$, then the theorem holds by taking $\lambda_i = 0$. Such a state i will later be called *absorbing*. Assume now that $T_1 < \infty$. By definition of T_1, we have, for $s,t \geq 0$,

$$T_1 > t+s \Longrightarrow T_1 > t \Longrightarrow X_t = i.$$

We thus have

$$\mathbb{P}\{X_t = i \mid T_1 > t, X_0 = i\} = 1 \tag{3.1}$$

and thus

$$
\begin{aligned}
\mathbb{P}\{T_1 &> t+s \mid X_0 = i\} \\
&= \mathbb{P}\{T_1 > t+s, T_1 > t \mid X_0 = i\} \\
&= \mathbb{P}\{T_1 > t+s \mid T_1 > t, X_0 = i\}\mathbb{P}\{T_1 > t \mid X_0 = i\} \\
&= \mathbb{P}\{T_1 > t+s, X_t = i \mid T_1 > t, X_0 = i\}\mathbb{P}\{T_1 > t \mid X_0 = i\} \\
&= \mathbb{P}\{T_1 > t+s \mid X_t = i\}\mathbb{P}\{T_1 > t \mid X_0 = i\} \\
&= \mathbb{P}\{T_1 > s \mid X_0 = i\}\mathbb{P}\{T_1 > t \mid X_0 = i\},
\end{aligned}
$$

where the fourth equality follows from the Markov property and from (3.1), since T_1 is a stopping time, and the fifth one follows from the homogeneity of X. The function

$f(t) = \mathbb{P}\{T_1 > t \mid X_0 = i\}$ is decreasing, right continuous, and satisfies the equation $f(t+s) = f(t)f(s)$. This ensures that there exists a positive real number, λ_i, such that $f(t) = e^{-\lambda_i t}$. ∎

If $\lambda_i > 0$, then T_1 is an exponential random variable with rate λ_i.

Let us determine now the distribution of the pair (X_{T_1}, T_1). We denote by $P = (P_{i,j})_{i,j \in S}$ the matrix whose entries are given by

$$P_{i,j} = \mathbb{P}\{X_{T_1} = j \mid X_0 = i\}.$$

If state i is not absorbing, we have $P_{i,i} = 0$ and if state i is absorbing, we have $P_{i,j} = 1_{\{i=j\}}$. Matrix P is thus a stochastic matrix.

THEOREM 3.1.6. *For every $i, j \in S$ and $t \geq 0$, we have*

$$\mathbb{P}\{X_{T_1} = j, T_1 > t \mid X_0 = i\} = P_{i,j} e^{-\lambda_i t}.$$

Proof. See [105]. ∎

We define the discrete-time process, $Y = \{Y_n, n \in \mathbb{N}\}$, by $Y_n = X_{T_n}$ for every $n \geq 0$.

THEOREM 3.1.7. *The process Y is a discrete-time Markov chain on the state space S, with transition probability matrix P, and, for every $n \geq 0$, $i_0, \ldots, i_n \in S$ and $t_1, \ldots, t_n \in \mathbb{R}^+$, we have*

$$\mathbb{P}\{T_n - T_{n-1} > t_n, \ldots, T_1 - T_0 > t_1 \mid Y_n = i_n, \ldots, Y_0 = i_0\} = \prod_{\ell=1}^{n} e^{-\lambda_{i_{\ell-1}} t_\ell}.$$

Proof. For every $n \geq 0$, we have $Y_n \in S$. Using the strong Markov property at the stopping time T_{n-1}, we have

$$\mathbb{P}\{X_{T_n} = i_n, T_n - T_{n-1} > t_n, \ldots, X_{T_1} = i_1, T_1 - T_0 > t_1 \mid X_0 = i_0\}$$
$$= \mathbb{P}\{X_{T_{n-1}} = i_{n-1}, T_{n-1} - T_{n-2} > t_{n-1}, \ldots, X_{T_1} = i_1, T_1 - T_0 > t_1 \mid X_0 = i_0\}$$
$$\times \mathbb{P}\{X_{T_n} = i_n, T_n - T_{n-1} > t_n \mid X_{T_{n-1}} = i_{n-1}\}$$
$$= \mathbb{P}\{X_{T_{n-1}} = i_{n-1}, T_{n-1} - T_{n-2} > t_{n-1}, \ldots, X_{T_1} = i_1, T_1 - T_0 > t_1 \mid X_0 = i_0\}$$
$$\times \mathbb{P}\{X_{T_1} = i_n, T_1 > t_n \mid X_0 = i_{n-1}\}$$
$$= \mathbb{P}\{X_{T_{n-1}} = i_{n-1}, T_{n-1} - T_{n-2} > t_{n-1}, \ldots, X_{T_1} = i_1, T_1 - T_0 > t_1 \mid X_0 = i_0\}$$
$$\times P_{i_{n-1}, i_n} e^{-\lambda_{i_{n-1}} t_n},$$

where the first equality follows from the strong Markov property, the second one follows from the homogeneity of X, and the third one is due to Theorem 3.1.6.

By induction, we easily obtain

$$\mathbb{P}\{X_{T_n} = i_n, T_n - T_{n-1} > t_n, \ldots, X_{T_1} = i_1, T_1 - T_0 > t_1 \mid X_0 = i_0\}$$

$$= \prod_{\ell=1}^{n} P_{i_{\ell-1}, i_\ell} e^{-\lambda_{i_{\ell-1}} t_\ell},$$

or

$$\mathbb{P}\{Y_n = i_n, T_n - T_{n-1} > t_n, \ldots, Y_1 = i_1, T_1 - T_0 > t_1 \mid Y_0 = i_0\}$$

$$= \prod_{\ell=1}^{n} P_{i_{\ell-1}, i_\ell} e^{-\lambda_{i_{\ell-1}} t_\ell}. \tag{3.2}$$

Taking $t_1 = \cdots = t_n = 0$ in (3.2), we get

$$\mathbb{P}\{Y_n = i_n, \ldots, Y_1 = i_1 \mid Y_0 = i_0\} = \prod_{\ell=1}^{n} P_{i_{\ell-1}, i_\ell},$$

which means, from Theorem 2.1.3, that Y is a Markov chain with transition probability matrix P.

Conditioning on $\{Y_n = i_n, \ldots, Y_1 = i_1\}$ in (3.2) and using the fact that Y is a Markov chain with transition probability matrix P, we obtain

$$\mathbb{P}\{T_n - T_{n-1} > t_n, \ldots, T_1 - T_0 > t_1 \mid Y_n = i_n, \ldots, Y_0 = i_0\} = \prod_{\ell=1}^{n} e^{-\lambda_{i_{\ell-1}} t_\ell},$$

which completes the proof. ∎

This theorem shows in particular that the successive sojourn times in the states of X are independent and exponentially distributed. These sojourn times are called *holding times*. Chain Y is called the *embedded* chain at jump times (or at transition instants) of X, or simply the canonically embedded chain when X is implicit.

LEMMA 3.1.8.   *Let $n \geq 1$ and $U_1, \ldots, U_n$ be n independent random variables with common exponential distribution with rate c. The sum $V_n = U_1 + \cdots + U_n$ has an Erlang distribution with n phases and rate c, i.e., for every $t \geq 0$,*

$$\mathbb{P}\{V_n > t\} = \sum_{k=0}^{n-1} e^{-ct} \frac{(ct)^k}{k!}.$$

Proof.   The result is true for $n = 1$ since we obtain the exponential distribution with rate c. For $n \geq 2$, the density function g_n of V_n is obtained by convolution as

$$g_n(x) = \int_0^x g_{n-1}(y) g_1(x - y) \, dy.$$

By induction, we obtain

$$g_n(x) = e^{-cx} \frac{c^n}{(n-2)!} \int_0^x y^{n-2} dy = \alpha e^{-cx} \frac{(cx)^{n-1}}{(n-1)!},$$

and the result follows by integration of g_n from t to ∞. ∎

DEFINITION 3.1.9. *The Markov chain X is said to be regular if* $\sup\{T_n, n \geq 0\} = \infty$ *with probability 1.*

THEOREM 3.1.10. *Every finite Markov chain is regular.*

Proof. Let $\lambda = \max\{\lambda_i, i \in S\}$. The state space S being finite, λ is finite. For every $n \geq 1$, we introduce the random variable U_n, defined by

$$U_n = \lambda_{Y_{n-1}}(T_n - T_{n-1}).$$

From Theorem 3.1.7, we obtain

$$\mathbb{P}\{U_n > t_n, \ldots, U_1 > t_1 \mid Y_n = i_n, \ldots, Y_0 = i_0\} = \prod_{\ell=1}^n e^{-t_\ell}$$

and unconditioning, we obtain

$$\mathbb{P}\{U_n > t_n, \ldots, U_1 > t_1\} = \prod_{\ell=1}^n e^{-t_\ell},$$

which means that the random variables $U_1, U_2, \ldots$ are independent with common exponential distribution with rate 1. Now, by definition of λ, we have, for every $n \geq 1$,

$$\sum_{k=1}^n U_k = \sum_{k=1}^n \lambda_{Y_{k-1}}(T_k - T_{k-1}) \leq \lambda \sum_{k=1}^n (T_k - T_{k-1}) = \lambda T_n.$$

This leads for every $t \geq 0$, from Lemma 3.1.8, to

$$\mathbb{P}\{T_n > t\} \geq \mathbb{P}\{U_1 + \cdots + U_n > \lambda t\} = \sum_{k=0}^{n-1} e^{-\lambda t} \frac{(\lambda t)^k}{k!}.$$

Since $0 < T_1 < \cdots$, we have, for every $t \geq 0$, $\{T_1 > t\} \subset \{T_2 > t\} \subset \ldots$, and thus, from the monotone convergence theorem, we obtain

$$\mathbb{P}\{\sup\{T_n, n \geq 0\} > t\} = \lim_{n \to \infty} \mathbb{P}\{T_n > t\} \geq \sum_{k=0}^{\infty} e^{-\lambda t} \frac{(\lambda t)^k}{k!} = 1,$$

which means that $\sup\{T_n, n \geq 0\} = \infty$ with probability 1 and so the Markov chain X is regular. ∎

We denote by $N(t)$ the number of transitions of the Markov chain X up to time t, i.e.

$$N(t) = \sup\{n \geq 0 \mid T_n \leq t\}.$$

LEMMA 3.1.11.   *For every $t \geq 0$, $N(t) < \infty$ with probability 1.*

Proof.   By definition of $N(t)$, we have, for every $t \geq 0$ and $n \in \mathbb{N}$,

$$N(t) = n \iff T_n \leq t < T_{n+1},$$

which can also be written as

$$N(t) \leq n \iff T_{n+1} > t,$$

which leads to

$$\lim_{n \to \infty} \mathbb{P}\{N(t) \leq n\} = \lim_{n \to \infty} \mathbb{P}\{T_{n+1} > t\} = 1,$$

from Theorem 3.1.10, since the state space S of X is finite. This proves that, for every $t \geq 0$, $N(t) < \infty$ with probability 1.   ∎

3.2   Transition function matrix

For all $i, j \in S$ and for every $t \geq 0$, we define the functions $P_{i,j}(t) = \mathbb{P}\{X_t = j \mid X_0 = i\}$ and the matrix $P(t) = (P_{i,j}(t))_{i,j \in S}$, which is called the transition function matrix of the Markov chain X.

THEOREM 3.2.1.   *The matrix $P(t)$ of a finite Markov chain X satisfies*

a) *For every $i, j \in S$ and $t \geq 0$, $P_{i,j}(t) \geq 0$.*
b) *For every $i \in S$ and $t \geq 0$, $\displaystyle\sum_{j \in S} P_{i,j}(t) = 1$.*
c) *For every $s, t \geq 0$, $P(t + s) = P(t)P(s)$.*

Proof.   a) is obvious since $P_{i,j}(t)$ are probabilities. To prove b) we observe that

$$\{X_t = j, N(t) = n\} = \{X_{T_n} = j, N(t) = n\} = \{Y_n = j, N(t) = n\},$$

and thus we have

$$\sum_{j \in S} P_{i,j}(t) = \sum_{j \in S} \mathbb{P}\{X_t = j \mid X_0 = i\}$$

$$= \sum_{j \in S} \sum_{n=0}^{\infty} \mathbb{P}\{X_t = j, N(t) = n \mid X_0 = i\}$$

$$= \sum_{j \in S} \sum_{n=0}^{\infty} \mathbb{P}\{Y_n = j, N(t) = n \mid X_0 = i\}$$

$$= \sum_{n=0}^{\infty} \sum_{j \in S} \mathbb{P}\{Y_n = j, N(t) = n \mid X_0 = i\}$$

$$= \sum_{n=0}^{\infty} \mathbb{P}\{N(t) = n \mid X_0 = i\}$$

$$= 1,$$

where we have used the fact that $N(t)$ is finite and that Y is a discrete-time Markov chain.

To prove c) we write, since from b) $X_t \in S$, for every $t \geq 0$,

$$P_{i,j}(t+s) = \mathbb{P}\{X_{t+s} = j \mid X_0 = i\}$$

$$= \sum_{k \in S} \mathbb{P}\{X_{t+s} = j, X_t = k \mid X_0 = i\}$$

$$= \sum_{k \in S} \mathbb{P}\{X_{t+s} = j \mid X_t = k, X_0 = i\} \mathbb{P}\{X_t = k \mid X_0 = i\}$$

$$= \sum_{k \in S} P_{i,k}(t) \mathbb{P}\{X_{t+s} = j \mid X_t = k\}$$

$$= \sum_{k \in S} P_{i,k}(t) \mathbb{P}\{X_s = j \mid X_0 = k\}$$

$$= \sum_{k \in S} P_{i,k}(t) P_{k,j}(s),$$

where the third equality is due to the Markov property and the fourth one follows from the homogeneity of X. ∎

LEMMA 3.2.2. *For every $i, j \in S$, the functions $P_{i,j}(t)$ are continuous on $\mathbb{R}^+$.*

Proof. We first show that for every $i, j \in S$, the functions $P_{i,j}(t)$ are right continuous at $t = 0$. Indeed, we have $0 \leq 1_{\{X_t = j\}} \leq 1$ and, since the trajectories are right continuous, we have

$$\lim_{t \to 0} 1_{\{X_t = j\}} = 1_{\{X_0 = j\}}.$$

The dominated convergence theorem allows us to conclude that

$$\lim_{t \to 0} \mathbb{E}\{1_{\{X_t = j\}} \mid X_0 = i\} = \mathbb{E}\{1_{\{X_0 = j\}} \mid X_0 = i\},$$

that is,

$$\lim_{t \to 0} P_{i,j}(t) = P_{i,j}(0) = 1_{\{i = j\}}.$$

Let $h > 0$. We have, from Theorem 3.2.1,

$$P_{i,j}(t+h) - P_{i,j}(t) = \sum_{k \in S} P_{i,k}(h)P_{k,j}(t) - P_{i,j}(t)$$

$$= \sum_{k \in S, k \neq i} P_{i,k}(h)P_{k,j}(t) - P_{i,j}(t)(1 - P_{i,i}(h)).$$

This leads to

$$P_{i,j}(t+h) - P_{i,j}(t) \geq -P_{i,j}(t)(1 - P_{i,i}(h)) \geq -(1 - P_{i,i}(h))$$

and

$$P_{i,j}(t+h) - P_{i,j}(t) \leq \sum_{k \in S, k \neq i} P_{i,k}(h)P_{k,j}(t) \leq \sum_{k \in S, k \neq i} P_{i,k}(h) \leq 1 - P_{i,i}(h).$$

We then have

$$\left| P_{i,j}(t+h) - P_{i,j}(t) \right| \leq 1 - P_{i,i}(h),$$

which proves the right continuity of $P_{i,j}(t)$ since $P_{i,j}(t)$ is right continuous at 0.

For the left continuity, we consider $P_{i,j}(t-h) - P_{i,j}(t)$ with $0 < h < t$. We then have, using the same argument as above,

$$\left| P_{i,j}(t-h) - P_{i,j}(t) \right| = \left| P_{i,j}(t) - P_{i,j}(t-h) \right|$$

$$\leq 1 - P_{i,i}(t - (t-h))$$

$$= 1 - P_{i,i}(h),$$

which proves that $P_{i,j}(t)$ is left continuous. ∎

3.3   Backward and forward equations

In this section we study the Kolmogorov backward and forward equations. The next theorem gives the Kolmogorov backward integral equation.

THEOREM 3.3.1.   *For every $t \geq 0$ and $i, j \in S$, we have*

$$P_{i,j}(t) = e^{-\lambda_i t} 1_{\{i=j\}} + \int_0^t \sum_{k \in S} P_{i,k} P_{k,j}(t-u)\lambda_i e^{-\lambda_i u} du. \tag{3.3}$$

Proof. If i is absorbing, then $\lambda_i = 0$ and $P_{i,j}(t) = 1_{\{i=j\}}$ for every $t \geq 0$, so Equation (3.3) holds. Suppose now that state i is such that $\lambda_i > 0$. Conditioning on the stopping time T_1, we have, from Theorem 3.1.5,

$$P_{i,j}(t) = \int_0^\infty \mathbb{P}\{X_t = j \mid T_1 = u, X_0 = i\}\lambda_i e^{-\lambda_i u} du.$$

If $T_1 = u > t$ and $X_0 = i$ then $X_t = i$. So, for every $u > t$, we have

$$\mathbb{P}\{X_t = j \mid T_1 = u, X_0 = i\} = 1_{\{i=j\}},$$

which leads to

$$P_{i,j}(t) = \int_0^t \mathbb{P}\{X_t = j \mid T_1 = u, X_0 = i\}\lambda_i e^{-\lambda_i u} du + 1_{\{i=j\}} \int_t^\infty \lambda_i e^{-\lambda_i u} du,$$

that is

$$P_{i,j}(t) = e^{-\lambda_i t} 1_{\{i=j\}} + \int_0^t \mathbb{P}\{X_t = j \mid T_1 = u, X_0 = i\}\lambda_i e^{-\lambda_i u} du. \tag{3.4}$$

From Theorem 3.1.6, we have

$$\mathbb{P}\{X_u = k \mid T_1 = u, X_0 = i\} = P_{i,k}.$$

If $u \leq t$, then using this relation, the Markov property, and the homogeneity of X, we obtain

$$\mathbb{P}\{X_t = j \mid T_1 = u, X_0 = i\}$$
$$= \sum_{k \in S} \mathbb{P}\{X_t = j, X_u = k \mid T_1 = u, X_0 = i\}$$
$$= \sum_{k \in S} \mathbb{P}\{X_t = j \mid X_u = k, T_1 = u, X_0 = i\}\mathbb{P}\{X_u = k \mid T_1 = u, X_0 = i\}$$
$$= \sum_{k \in S} P_{i,k}\mathbb{P}\{X_t = j \mid X_u = k\}$$
$$= \sum_{k \in S} P_{i,k}P_{k,j}(t - u).$$

Replacing this relation in (3.4), we obtain

$$P_{i,j}(t) = e^{-\lambda_i t} 1_{\{i=j\}} + \int_0^t \sum_{k \in S} P_{i,k}P_{k,j}(t - u)\lambda_i e^{-\lambda_i u} du,$$

which completes the proof. ∎

The variable change $v = t - u$ in Equation (3.3) leads to

$$P_{i,j}(t) = e^{-\lambda_i t}\left[1_{\{i=j\}} + \int_0^t \sum_{k \in S} P_{i,k}P_{k,j}(v)\lambda_i e^{\lambda_i v} dv\right]. \tag{3.5}$$

From Lemma 3.2.2, the integrand in (3.5) is continuous since S is finite. This means that function $P_{i,j}(t)$ is differentiable. Differentiating (3.5), we obtain

$$P'_{i,j}(t) = -\lambda_i P_{i,j}(t) + \sum_{k \in S} \lambda_i P_{i,k}P_{k,j}(t). \tag{3.6}$$

We denote by Λ the diagonal matrix containing the rates λ_i, $i \in S$, and by A the matrix defined by

$$A = -\Lambda(I - P),$$

where I is the identity matrix. Equation (3.6) then becomes, in matrix notation,

$$P'(t) = AP(t). \tag{3.7}$$

This equation is called the Kolmogorov backward differential equation. Matrix A, which is equal to $P'(0)$, is called the infinitesimal generator of the Markov chain X and has, by definition, the following properties:

a) For every $i, j \in S$, $i \neq j$, $A_{i,j} \geq 0$.
b) For every $i \in S$, $\sum_{j \in S} A_{i,j} = 0$.

Since $P_{i,j}(t)$ is differentiable, we can differentiate relation c) of Theorem 3.2.1 with respect to s to obtain

$$P'(t + s) = P(t)P'(s).$$

At point $s = 0$ and since $P'(0) = A$, we obtain the so-called Kolmogorov forward differential equation,

$$P'(t) = P(t)A. \tag{3.8}$$

The solution to these equations is given, for every $t \geq 0$, by

$$P(t) = e^{At}, \tag{3.9}$$

and it is the unique solution. To see this, let $M(t)$ be another solution to (3.7) with $M(0) = I$; we have

$$[e^{-At}M(t)]' = -e^{-At}AM(t) + e^{-At}M'(t) = -e^{-At}AM(t) + e^{-At}AM(t) = 0,$$

which means that $e^{-At}M(t)$ is constant. We thus have $e^{-At}M(t) = I$, which is its value at point $t = 0$. We thus have $M(t) = e^{At} = P(t)$. The same reasoning holds for the forward equation by considering $[M(t)e^{-At}]'$ instead of $[e^{-At}M(t)]'$.

We have obtained the backward differential equation from the backward integral equation. Conversely we can also obtain the backward integral equation from the backward differential equation. A similar comment holds for the forward equation. We just show here how the forward integral equation can be obtained from the forward differential equation.

From Equation (3.8), we have for every $v \geq 0$,

$$P'_{i,j}(v) = \sum_{k \in S} P_{i,k}(v)A_{k,j}.$$

Using the relation $A = -\Lambda(I - P)$, we obtain

$$P'_{i,j}(v) = -P_{i,j}(v)\lambda_j + \sum_{k \in S} P_{i,k}(v)\lambda_k P_{k,j}.$$

Multiplying both sides by $e^{\lambda_j v}$, we obtain

$$P'_{i,j}(v)e^{\lambda_j v} = -P_{i,j}(v)\lambda_j e^{\lambda_j v} + \sum_{k \in S} P_{i,k}(v)\lambda_k P_{k,j} e^{\lambda_j v},$$

that is

$$P'_{i,j}(v)e^{\lambda_j v} + P_{i,j}(v)\lambda_j e^{\lambda_j v} = \sum_{k \in S} P_{i,k}(v)\lambda_k P_{k,j} e^{\lambda_j v},$$

which can be written as

$$\left(P_{i,j}(v)e^{\lambda_j v}\right)' = \sum_{k \in S} P_{i,k}(v)\lambda_k P_{k,j} e^{\lambda_j v}.$$

Integrating both sides from 0 to t, we obtain

$$P_{i,j}(t)e^{\lambda_j t} - P_{i,j}(0) = \int_0^t \sum_{k \in S} P_{i,k}(v)\lambda_k P_{k,j} e^{\lambda_j v} dv.$$

Multiplying both parts by $e^{-\lambda_j t}$, we obtain

$$P_{i,j}(t) = 1_{\{i=j\}}e^{-\lambda_j t} + \int_0^t \sum_{k \in S} P_{i,k}(v)\lambda_k P_{k,j} e^{-\lambda_j(t-v)} dv,$$

which is the Kolmogorov forward integral equation.

Note that this forward equation could also have been obtained as the backward one has been obtained but by conditioning on the last transition instant just before t instead of the first transition instant just after 0.

We denote by $\pi(t) = (\pi_j(t), j \in S)$ the row vector containing the distribution of the Markov chain X at time t, i.e.

$$\pi_j(t) = \mathbb{P}\{X_t = j\}.$$

We then have for every $j \in S$ and $t \geq 0$,

$$\pi_j(t) = \sum_{i \in S} \pi_i(0)P_{i,j}(t) = (\pi(0)P(t))_j = (\pi(0)e^{At})_j,$$

that is

$$\pi(t) = \pi(0)e^{At}, \tag{3.10}$$

which means that the Markov chain X is completely characterized by its initial probability distribution $\pi(0)$ and its infinitesimal generator A.

3.4   Uniformization

We denote by λ a positive real number greater than or equal to the maximal ouput rate of process X, i.e.

$$\lambda \geq \max\{\lambda_i, \ i \in S\},$$

and we consider a Poisson process $\{N_t, \ t \geq 0\}$ with rate λ. By definition of λ, matrix U, defined by

$$U = I + \frac{1}{\lambda}A,$$

is a stochastic matrix. Obviously, matrix U is dependent on parameter λ and we have

$$A = -\lambda(I - U).$$

We introduce the discrete-time Markov chain $Z = \{Z_n, \ n \geq 0\}$ on the state space S, with one-step transition probability matrix U, such that $Z_0 = X_0$ in distribution, i.e. for every $i \in S$, $\pi_i(0) = \mathbb{P}\{Z_0 = i\}$. We then have

$$\mathbb{P}\{Z_n = j \mid Z_0 = i\} = (U^n)_{i,j}.$$

The processes $\{N_t\}$ and $\{Z_n\}$ are moreover constructed independently of one another. The process $\{Z_{N_t}, \ t \geq 0\}$ is referred to as the uniformized process of X with respect to the uniformization rate λ. The next theorem shows that both processes have the same probabilistic transient behavior. First, we need the following lemma.

LEMMA 3.4.1.   *For every $t \geq 0$ and $i, j \in S$, we have*

$$\mathbb{P}\{X_t = j \mid X_0 = i\} = \sum_{n=0}^{\infty} e^{-\lambda t}\frac{(\lambda t)^n}{n!}(U^n)_{i,j}. \tag{3.11}$$

Proof.   From Relation (3.9) and since $A = -\lambda(I - U)$, we have

$$\mathbb{P}\{X_t = j \mid X_0 = i\} = (e^{At})_{i,j}$$

$$= (e^{-\lambda(I-U)t})_{i,j}$$

$$= e^{-\lambda t}(e^{\lambda t U})_{i,j}$$

$$= \sum_{n=0}^{\infty} e^{-\lambda t}\frac{(\lambda t)^n}{n!}(U^n)_{i,j},$$

which completes the proof.   ∎

THEOREM 3.4.2. *For every $k \geq 1$, $0 < t_1 < \cdots < t_{k-1} < t_k$, and $j_1,\ldots,j_k \in S$, we have*

$$\mathbb{P}\{X_{t_k} = j_k, X_{t_{k-1}} = j_{k-1},\ldots,X_{t_1} = j_1\}$$
$$= \mathbb{P}\{Z_{N_{t_k}} = j_k, Z_{N_{t_{k-1}}} = j_{k-1},\ldots,Z_{N_{t_1}} = j_1\}.$$

Proof. We introduce the function $p_n(t)$ defined, for every $t \geq 0$ and $n \geq 0$, by

$$p_n(t) = \mathbb{P}\{N_t = n\} = e^{-\lambda t}\frac{(\lambda t)^n}{n!}.$$

We then have

$$\mathbb{P}\{Z_{N_{t_k}} = j_k, Z_{N_{t_{k-1}}} = j_{k-1},\ldots,Z_{N_{t_1}} = j_1 \mid Z_0 = i\}$$

$$= \sum_{n_1=0}^{\infty}\sum_{n_2=n_1}^{\infty}\cdots\sum_{n_k=n_{k-1}}^{\infty} \mathbb{P}\{Z_{n_k} = j_k, N_{t_k} = n_k,\ldots,Z_{n_1} = j_1, N_{t_1} = n_1 \mid Z_0 = i\}$$

$$= \sum_{n_1=0}^{\infty}\sum_{n_2=n_1}^{\infty}\cdots\sum_{n_k=n_{k-1}}^{\infty} \mathbb{P}\{N_{t_k} = n_k,\ldots,N_{t_1} = n_1\}$$

$$\times \mathbb{P}\{Z_{n_k} = j_k,\ldots,Z_{n_1} = j_1 \mid Z_0 = i\}$$

$$= \sum_{n_1=0}^{\infty}\sum_{n_2=n_1}^{\infty}\cdots\sum_{n_k=n_{k-1}}^{\infty} p_{n_1}(t_1)p_{n_2-n_1}(t_2 - t_1)\ldots p_{n_k-n_{k-1}}(t_k - t_{k-1})$$

$$\times \mathbb{P}\{Z_{n_k} = j_k,\ldots,Z_{n_1} = j_1 \mid Z_0 = i\}$$

$$= \sum_{n_1=0}^{\infty}\sum_{n_2=n_1}^{\infty}\cdots\sum_{n_k=n_{k-1}}^{\infty} p_{n_1}(t_1)p_{n_2-n_1}(t_2 - t_1)\ldots p_{n_k-n_{k-1}}(t_k - t_{k-1})$$

$$\times (P^{n_1})_{i,j_1}(P^{n_2-n_1})_{j_1 j_2}\cdots(P^{n_k-n_{k-1}})_{j_{k-1} j_k}$$

$$= \sum_{n_1=0}^{\infty} p_{n_1}(t_1)(P^{n_1})_{i,j_1}\sum_{n_2=0}^{\infty} p_{n_2}(t_2 - t_1)(P^{n_2})_{j_1 j_2}$$

$$\times \cdots \times \sum_{n_k=0}^{\infty} p_{n_k}(t_k - t_{k-1})(P^{n_k})_{j_{k-1} j_k}$$

$$= \mathbb{P}\{X_{t_1} = j_1 \mid X_0 = i\}\mathbb{P}\{X_{t_2-t_1} = j_2 \mid X_0 = j_1\}$$

$$\times \cdots \times \mathbb{P}\{X_{t_k-t_{k-1}} = j_k \mid X_0 = j_{k-1}\}$$

$$= \mathbb{P}\{X_{t_1} = j_1 \mid X_0 = i\}\mathbb{P}\{X_{t_2} = j_2 \mid X_{t_1} = j_1\}$$

$$\times \cdots \times \mathbb{P}\{X_{t_k} = j_k \mid X_{t_{k-1}} = j_{k-1}\}$$

$$= \mathbb{P}\{X_{t_k} = j_k, X_{t_{k-1}} = j_{k-1},\ldots,X_{t_1} = j_1 \mid X_0 = i\},$$

where the second equality follows from the independence of $\{N_t\}$ and $\{Z_n\}$, the third one follows because $\{N_t\}$ is a Poisson process, the fourth one follows because $\{Z_n\}$ is a homogeneous Markov chain, the sixth uses Relation (3.11), the seventh one follows because X is homogeneous, and the last one is due to the Markov property of X. As the two processes have the same initial distribution, we obtain the desired result.   ∎

3.5   Limiting behavior

DEFINITION 3.5.1.   *State $j \in S$ is said to be accessible from state $i \in S$ if there exists $t \geq 0$ such that $P_{i,j}(t) > 0$. This means, in particular, that each state is accessible from itself, since $P_{i,i}(0) = 1$. We say that states i and j communicate if i is accessible from j and j is accessible from i.*

DEFINITION 3.5.2.   *The Markov chain X is said to be irreducible if all states communicate.*

THEOREM 3.5.3.   *X is irreducible if and only if Z is irreducible.*

Proof.   Relation (3.11) states that

$$P_{i,j}(t) = \sum_{n=0}^{\infty} e^{-\lambda t} \frac{(\lambda t)^n}{n!} (U^n)_{i,j}.$$

Thus, if there exists $t \geq 0$ such that $P_{i,j}(t) > 0$, then there exists at least one $n \geq 0$ such that $(U^n)_{i,j} > 0$. Conversely, since all the terms in the sum are nonnegative, if there exists $n \geq 0$ such that $(U^n)_{i,j} > 0$ then for all $t > 0$, we have $P_{i,j}(t) > 0$.

So we have shown that state j is accessible from state i in X if and only if state j is accessible from state i in Z. Thus, X is irreducible if and only if Z is irreducible.   ∎

LEMMA 3.5.4.   *Let u_n, $n \geq 0$, be a sequence of real numbers converging to the real number $u \in \mathbb{R}$ when n tends to infinity, i.e., such that*

$$\lim_{n \longrightarrow \infty} u_n = u.$$

The function $f : \mathbb{R}^+ \longrightarrow \mathbb{R}$ defined by

$$f(x) = \sum_{n=0}^{\infty} e^{-x} \frac{x^n}{n!} u_n$$

satisfies

$$\lim_{x \longrightarrow \infty} f(x) = u.$$

Proof.   The convergence of sequence u_n implies that it is bounded, that is, there exists a positive number, M, such that

$$|u_n| \leq M.$$

Since $u_n \longrightarrow u$ as $n \longrightarrow \infty$, for any $\varepsilon > 0$ there exists an integer, N, such that

$$|u_n - u| \leq \frac{\varepsilon}{2}, \quad \text{for any } n \geq N.$$

Now, ε and N being fixed, there exists a positive number, T, such that

$$\sum_{n=0}^{N} e^{-x} \frac{x^n}{n!} \leq \frac{\varepsilon}{4M}, \quad \text{for any } x > T.$$

We then have, for $x > T$,

$$|f(x) - u| = \left| \sum_{n=0}^{\infty} e^{-x} \frac{x^n}{n!} u_n - u \right| = \left| \sum_{n=0}^{\infty} e^{-x} \frac{x^n}{n!} (u_n - u) \right| \leq \sum_{n=0}^{\infty} e^{-x} \frac{x^n}{n!} |u_n - u|$$

$$= \sum_{n=0}^{N} e^{-x} \frac{x^n}{n!} |u_n - u| + \sum_{n=N+1}^{\infty} e^{-x} \frac{x^n}{n!} |u_n - u| \leq 2M \sum_{n=0}^{N} e^{-x} \frac{x^n}{n!} + \frac{\varepsilon}{2}$$

$$\leq \frac{\varepsilon}{2} + \frac{\varepsilon}{2} = \varepsilon,$$

which completes the proof. ∎

THEOREM 3.5.5. *If X is irreducible then we have, for every $j \in S$,*

$$\lim_{t \to \infty} \mathbb{P}\{X_t = j\} = \pi_j,$$

for every initial distribution, where $\pi = (\pi_j, j \in S)$ is the unique positive invariant probability of Z. Note that π is independent of λ. In particular, for every $i, j \in S$, we have

$$\lim_{t \to \infty} P_{i,j}(t) = \pi_j \quad and \quad \pi A = 0.$$

Proof. If the Markov chain X is irreducible then, from Theorem 3.5.3 the Markov chain Z is also irreducible. Taking the uniformization rate $\lambda > \max\{\lambda_i, i \in S\}$ ensures that Z is aperiodic, since there exists a state, i, with a loop, i.e. such that $U_{i,i} > 0$. Now, from Theorem 2.7.1, we have

$$\lim_{n \to \infty} \mathbb{P}\{Z_n = j\} = \pi_j,$$

where $\pi = (\pi_j, j \in S)$ is the unique positive invariant probability of Z. From Theorem 3.4.2, we have

$$\mathbb{P}\{X_t = j\} = \sum_{n=0}^{\infty} e^{-\lambda t} \frac{(\lambda t)^n}{n!} \mathbb{P}\{Z_n = j\},$$

so using Lemma 3.5.4 we obtain

$$\lim_{t \to \infty} \mathbb{P}\{X_t = j\} = \pi_j,$$

which also proves that π is independent of λ.

Since π satisfies $\pi U = \pi$ and since $A = -\lambda(I - U)$, we obtain $\pi A = 0$. ∎

3.6   Recurrent and transient states

Recall that T_1 is the instant of the first jump for the Markov chain X. Define for every $j \in S$,

$$\tau_X(j) = \inf\{t \geq T_1 \mid X_t = j\},$$

with the convention $\inf \emptyset = \infty$. The instant $\tau_X(j)$ is thus the first instant where the Markov chain X enters (resp. reenters) state j if X doesn't start in j (resp. if X starts in j). Now, define

$$f_{i,j}^X = \mathbb{P}\{\tau_X(j) < \infty \mid X_0 = i\}.$$

DEFINITION 3.6.1.   *An absorbing state is called recurrent. A non-absorbing state $i \in S$, i.e. such that $\lambda_i > 0$, is called recurrent for X if $f_{i,i}^X = 1$ and transient if $f_{i,i}^X < 1$. A Markov chain is called recurrent (resp. transient) if all its states are recurrent (resp. transient).*

3.6.1   General case

To obtain the distribution of $\tau_X(j)$, we introduce the continuous-time Markov chain, $W = \{W_t, t \in \mathbb{R}^+\}$, over the state space $S \cup \{a\}$, where a is an absorbing state, with infinitesimal generator L defined by

$$L = \begin{pmatrix} A^{(j)} & V_j \\ 0 & 0 \end{pmatrix},$$

where matrix $A^{(j)}$ is obtained from matrix A by replacing its jth column by zeros except for entry $A_{j,j}$ which remains unchanged, that is

$$A_{i,\ell}^{(j)} = \begin{cases} A_{i,\ell} & \text{if } \ell \neq j, \\ 0 & \text{if } \ell = j \text{ and } i \neq j, \\ A_{j,j} & \text{if } \ell = j \text{ and } i = j, \end{cases}$$

and where column vector V_j contains the jth column of matrix A with the jth entry equal to 0, i.e.

$$V_j(i) = \begin{cases} A_{i,j} & \text{if } i \neq j, \\ 0 & \text{if } i = j. \end{cases}$$

The initial probability distribution, β, of W is defined by $\beta = (\alpha\ 0)$, which means that W doesn't start in the absorbing state a.

With these definitions, we have, for every $t \geq 0$,

$$\tau_X(j) > t \iff W_t \in S.$$

This leads to

$$\mathbb{P}\{\tau_X(j) > t\} = \mathbb{P}\{W_t \in S\} = \sum_{k \in S}\mathbb{P}\{W_t = k\} = \sum_{k \in S}(\beta e^{Lt})_k = \alpha e^{A^{(j)}t}\mathbb{1}, \tag{3.12}$$

where the third equality comes from Equation (3.10). Let λ be the uniformization rate associated with X. The rate λ is such that $\lambda \geq \max\{\lambda_i, i \in S\}$. The transition probability matrix, U, of the Markov chain, Z, obtained after uniformization of X is given by

$$U = I + A/\lambda.$$

For every $j \in S$, we introduce the matrix $T^{(j)}$ defined by

$$T^{(j)} = I + A^{(j)}/\lambda. \tag{3.13}$$

From Relation (3.12), we have

$$\mathbb{P}\{\tau_X(j) > t\} = \alpha e^{A^{(j)}t}\mathbb{1} = \sum_{n=0}^{\infty}e^{-\lambda t}\frac{(\lambda t)^n}{n!}\alpha\left(T^{(j)}\right)^n\mathbb{1}. \tag{3.14}$$

The results concerning absorbing discrete-time Markov chains that we obtained in the previous chapter can be translated here as follows.

The time, $\tau_Z(j)$, needed to reach a given state, j, on the discrete-time Markov chain, Z, is defined as

$$\tau_Z(j) = \inf\{n \geq 1 \mid Z_n = j\}.$$

Using Relation (2.21), we obtain

$$\mathbb{P}\{\tau_Z(j) > n\} = \alpha\left(Q^{(j)}\right)^n\mathbb{1}, \tag{3.15}$$

where matrix $Q^{(j)}$ is obtained from matrix U by replacing its jth column by zeros, that is

$$Q_{i,\ell}^{(j)} = \begin{cases} U_{i,\ell} & \text{if } \ell \neq j, \\ 0 & \text{if } \ell = j. \end{cases}$$

Matrices $T^{(j)}$ and $Q^{(j)}$ are related as follows. For every $i, \ell \in S$, we have

$$T_{i,\ell}^{(j)} = \mathbb{1}_{\{i=\ell\}} + \frac{A_{i,\ell}^{(j)}}{\lambda}$$

$$= \begin{cases} 1_{\{i=\ell\}} + \dfrac{A_{i,\ell}}{\lambda} & \text{if } \ell \neq j, \\ 0 & \text{if } \ell = j \text{ and } i \neq j, \\ 1 + \dfrac{A_{jj}}{\lambda} & \text{if } (i,\ell) = (j,j) \end{cases}$$

$$= Q_{i,\ell}^{(j)} + U_{jj} 1_{\{(i,\ell)=(j,j)\}}.$$

Note that $\left((Q^{(j)})^n \right)_{i,j} = 0$ for every $i \in S$ and $n \geq 1$. This relation leads, for every $n \geq 0$ and for $i \neq j$, to

$$\left((T^{(j)})^n \right)_{i,\ell} = \left((Q^{(j)})^n \right)_{i,\ell}$$

and

$$\left((T^{(j)})^n \right)_{j,\ell} = \sum_{k=0}^{n} \left((Q^{(j)})^k \right)_{j,\ell} (U_{jj})^{n-k}.$$

We then have, for $i \neq j$,

$$\left((T^{(j)})^n 1 \right)_i = \sum_{\ell \in S} \left((T^{(j)})^n \right)_{i,\ell} = \sum_{\ell \in S} \left((Q^{(j)})^n \right)_{i,\ell} = \left((Q^{(j)})^n 1 \right)_i$$

and

$$\left((T^{(j)})^n 1 \right)_j = \sum_{\ell \in S} \left((T^{(j)})^n \right)_{j,\ell} = \sum_{k=0}^{n} \left((Q^{(j)})^k 1 \right)_j (U_{jj})^{n-k}. \qquad (3.16)$$

Combining this first relation with (3.14), we obtain, for $i \neq j$,

$$\mathbb{P}\{\tau_X(j) > t \mid X_0 = i\} = \left(e^{A^{(j)}t} 1 \right)_i$$

$$= \sum_{n=0}^{\infty} e^{-\lambda t} \frac{(\lambda t)^n}{n!} \left((T^{(j)})^n 1 \right)_i$$

$$= \sum_{n=0}^{\infty} e^{-\lambda t} \frac{(\lambda t)^n}{n!} \left((Q^{(j)})^n 1 \right)_i. \qquad (3.17)$$

Defining, for every $i, j \in S$, $f_{i,j}^X(t) = \mathbb{P}\{\tau_X(j) \leq t \mid X_0 = i\}$ and $f_{i,j}^Z(n) = \mathbb{P}\{\tau_Z(j) \leq n \mid X_0 = i\}$, we obtain from (3.15),

$$f_{i,j}^Z(n) = 1 - \left((Q^{(j)})^n 1 \right)_i.$$

This leads, using (3.17), for $i \neq j$, to

$$f_{i,j}^X(t) = \sum_{n=0}^{\infty} e^{-\lambda t} \frac{(\lambda t)^n}{n!} f_{i,j}^Z(n).$$

Taking the limit when $t \longrightarrow \infty$, and using Lemma 3.5.4 we obtain

$$f_{i,j}^X = f_{i,j}^Z,$$

where $f_{i,j}^Z = \mathbb{P}\{\tau_Z(j) < \infty \mid X_0 = i\}$.

Combining (3.16) and (3.14), we obtain, for $i = j$,

$$\mathbb{P}\{\tau_X(j) > t \mid X_0 = j\} = \left(e^{A^{(j)}t}\mathbb{1}\right)_j$$

$$= \sum_{n=0}^{\infty} e^{-\lambda t} \frac{(\lambda t)^n}{n!} \left(\left(T^{(j)}\right)^n \mathbb{1}\right)_j$$

$$= \sum_{n=0}^{\infty} e^{-\lambda t} \frac{(\lambda t)^n}{n!} \sum_{k=0}^{n} \left(\left(Q^{(j)}\right)^k \mathbb{1}\right)_j (U_{j,j})^{n-k}. \qquad (3.18)$$

Again using (3.15),

$$f_{j,j}^X(t) = 1 - \sum_{n=0}^{\infty} e^{-\lambda t} \frac{(\lambda t)^n}{n!} \sum_{k=0}^{n} \left(1 - f_{j,j}^Z(k)\right) (U_{j,j})^{n-k}$$

$$= 1 - \sum_{n=0}^{\infty} e^{-\lambda t} \frac{(\lambda t)^n}{n!} \sum_{k=0}^{n} (U_{j,j})^k \left(1 - f_{j,j}^Z(n-k)\right). \qquad (3.19)$$

The next technical lemma is used to determine the limit of $f_{j,j}^X(t)$ when t tends to ∞.

LEMMA 3.6.2. *Let u_k and v_k be two sequences of real numbers such that*

$$u_k \geq 0, \quad \sum_{k=0}^{\infty} u_k = u > 0 \quad and \quad \lim_{k \to \infty} v_k = v.$$

Then

$$\lim_{n \to \infty} \sum_{k=0}^{n} u_k v_{n-k} = uv.$$

Proof. Let us introduce the sum, s_n, defined, for every $n \geq 0$, by

$$s_n = \sum_{k=0}^{n} u_k.$$

Since s_n converges to u when n tends to ∞, it is sufficient to show that

$$\lim_{n \to \infty} \frac{1}{s_n} \sum_{k=0}^{n} u_k v_{n-k} = v.$$

The convergence of sequence v_k implies that it is bounded, that is, there exists an integer, M, such that, for every $k \geq 0$,

$$|v_k| \leq M.$$

Moreover its convergence implies that, for any $\varepsilon > 0$, there exists an integer, K, such that for any $k \geq K$, we have

$$|v_k - v| \leq \frac{\varepsilon}{2}.$$

For $n \geq K$, we consider the sequence, w_n, defined by

$$w_n = \frac{1}{s_n} \sum_{k=n-K+1}^{n} u_k = 1 - \frac{s_{n-K}}{s_n}.$$

Since the sequence, s_n, converges when n tends to ∞, we have

$$\lim_{n \longrightarrow \infty} w_n = 0,$$

which means that there exists an integer, N, such that for any $n \geq \max\{K,N\}$, we have

$$w_n \leq \frac{\varepsilon}{4M}.$$

As a consequence, we recover, for $n \geq \max\{K,N\}$,

$$\left| \frac{1}{s_n} \sum_{k=0}^{n} u_k v_{n-k} - v \right| = \frac{1}{s_n} \left| \sum_{k=0}^{n} u_k(v_{n-k} - v) \right|$$

$$\leq \frac{1}{s_n} \left| \sum_{k=0}^{n-K} u_k(v_{n-k} - v) \right| + \frac{1}{s_n} \left| \sum_{k=n-K+1}^{n} u_k(v_{n-k} - v) \right|.$$

In the first sum, since $k \leq n - K$, we have $n - k \geq K$, so we can write

$$\frac{1}{s_n} \left| \sum_{k=0}^{n-K} u_k(v_{n-k} - v) \right| \leq \frac{1}{s_n} \sum_{k=0}^{n-K} u_k |v_{n-k} - v| \leq \frac{\varepsilon}{2s_n} \sum_{k=0}^{n-K} u_k \leq \frac{\varepsilon}{2s_n} \sum_{k=0}^{n} u_k = \frac{\varepsilon}{2}.$$

Concerning the second sum, since $|v_k - v| \leq 2M$, we have

$$\frac{1}{s_n} \left| \sum_{k=n-K+1}^{n} u_k(v_{n-k} - v) \right| \leq \frac{1}{s_n} \sum_{k=n-K+1}^{n} u_k |v_{n-k} - v|$$

$$\leq 2M \frac{1}{s_n} \sum_{k=n-K+1}^{n} u_k$$

$$= 2M w_n \leq \frac{\varepsilon}{2},$$

which means that for every $n \geq \max\{K, N\}$,

$$\left| \frac{1}{s_n} \sum_{k=0}^{n} u_k v_{n-k} - v \right| \leq \varepsilon,$$

which completes the proof. ∎

THEOREM 3.6.3. *A state j is recurrent (resp. transient) for X if and only if it is recurrent (resp. transient) for Z.*

Proof. If state j is absorbing for X then it is also absorbing for Z and conversely, so the result is true in that case. Suppose now that j is a non-absorbing state. This means that $U_{j,j} < 1$.

From Lemma 3.6.2, we have

$$\lim_{n \to \infty} \sum_{k=0}^{n} (U_{j,j})^k \left(1 - f_{j,j}^Z(n-k)\right) = \frac{1 - f_{j,j}^Z}{1 - U_{j,j}}.$$

By construction, matrix $T^{(j)}$ is a sub-stochastic matrix, i.e. $T^{(j)}\mathbb{1} \leq \mathbb{1}$ where the inequality between vectors is meant entrywise, so the previous limit is necessarily less than or equal to 1. This shows that we necessarily have $U_{j,j} \leq f_{j,j}^Z$. This relation is also intuitive since $U_{j,j}$ is the probability, starting from state j, to reach state j in one transition while $f_{j,j}^Z$ is the probability, starting from state j, that the first visit to state j occurs in a finite time.

Taking the limit when t tends to ∞ in Relation (3.19), we obtain, using Lemma 3.5.4,

$$f_{j,j}^X = 1 - \frac{1 - f_{j,j}^Z}{1 - U_{j,j}} = \frac{f_{j,j}^Z - U_{j,j}}{1 - U_{j,j}}. \tag{3.20}$$

Suppose that j is recurrent for Z, i.e. that $f_{j,j}^Z = 1$. Then, from (3.20), we have $f_{j,j}^X = 1$, which means that state j is recurrent for X.

Suppose now that j is recurrent for X, i.e. that $f_{j,j}^X = 1$. Then, again from (3.20), we have $f_{j,j}^Z = 1$, which means that state j is recurrent for Z. The rest of the proof follows by contraposition. ∎

It follows from this theorem and from Theorem 2.3.10 that, as in the discrete-time case, a finite continuous-time Markov chain cannot be transient.

Note that this theorem is still valid if we consider the embedded discrete-time Markov Y instead of Z, see for instance [10] or [105].

3.6.2 Irreducible case

If the Markov chain X is irreducible, which means that X is recurrent since the state space is finite, we have, from Theorems 3.5.3 and 3.6.3, that Z is also irreducible and recurrent. In this case, concerning the expectations, we have from (3.15),

$$\mathbb{E}\{\tau_Z(j) \mid Z_0 = j\} = \sum_{n=0}^{\infty} \mathbb{P}\{\tau_Z(j) > n \mid Z_0 = j\}$$

$$= \sum_{n=0}^{\infty} \left((Q^{(j)})^n \mathbb{1} \right)_j$$

$$= \left((I - Q^{(j)})^{-1} \mathbb{1} \right)_j.$$

To see that matrix $I - Q^{(j)}$ is invertible, consider an auxiliary discrete-time Markov chain over the state space $S \cup \{a\}$ where the transition probabilities between states of S are the entries of matrix $Q^{(j)}$. Since Z is irreducible, all states in S are transient for the auxiliary chain. From Lemma 2.9.2, the above series converges.

Using (3.18), Fubini's Theorem, and the monotone convergence theorem, we obtain

$$\mathbb{E}\{\tau_X(j) \mid X_0 = j\} = \int_0^{\infty} \mathbb{P}\{\tau_X(j) > t \mid X_0 = j\} dt$$

$$= \int_0^{\infty} \sum_{n=0}^{\infty} e^{-\lambda t} \frac{(\lambda t)^n}{n!} \sum_{k=0}^{n} \left((Q^{(j)})^k \mathbb{1} \right)_j (U_{j,j})^{n-k} dt$$

$$= \sum_{n=0}^{\infty} \int_0^{\infty} e^{-\lambda t} \frac{(\lambda t)^n}{n!} dt \sum_{k=0}^{n} \left((Q^{(j)})^k \mathbb{1} \right)_j (U_{j,j})^{n-k}$$

$$= \frac{1}{\lambda} \sum_{n=0}^{\infty} \sum_{k=0}^{n} \left((Q^{(j)})^k \mathbb{1} \right)_j (U_{j,j})^{n-k}$$

$$= \frac{1}{\lambda} \sum_{k=0}^{\infty} \left((Q^{(j)})^k \mathbb{1} \right)_j \sum_{n=k}^{\infty} (U_{j,j})^{n-k}$$

$$= \frac{1}{\lambda} \left((I - Q^{(j)})^{-1} \mathbb{1} \right)_j \frac{1}{1 - U_{j,j}}.$$

Since $U_{j,j} = 1 - \lambda_j/\lambda$, we finally obtain

$$\mathbb{E}\{\tau_X(j) \mid X_0 = j\} = \frac{1}{\lambda_j} \mathbb{E}\{\tau_Z(j) \mid Z_0 = j\}. \tag{3.21}$$

From Corollary 2.5.4, we have

$$\mathbb{E}\{\tau_X(j) \mid X_0 = j\} = \frac{1}{\lambda_j \pi_j}, \tag{3.22}$$

where $\pi = (\pi_i, i \in S)$ is the unique positive invariant probability of Z, which is also the limiting distribution of X, from Theorem 3.5.5.

3.7 Ergodic theorem

THEOREM 3.7.1. *Let X be an irreducible Markov chain on the finite state space, S. For every $j \in S$, we have*

$$\lim_{t \longrightarrow \infty} \frac{1}{t} \int_0^t 1_{\{X_s = j\}} ds = \frac{1}{\lambda_j m_j} = \pi_j, \text{ with probability 1,}$$

where $\pi = (\pi_j, j \in S)$ is the limiting distribution of X and $m_j = \mathbb{E}\{\tau_X(j) \mid X_0 = j\}$ is the expected return time to state j.
  Moreover, for any function $r : S \longrightarrow \mathbb{R}$, we have

$$\lim_{t \longrightarrow \infty} \frac{1}{t} \int_0^t r_{X_s} ds = \sum_{j \in S} r_j \pi_j \text{ with probability 1.}$$

Proof.   The fact that $\pi_j = 1/(\lambda_j m_j)$ comes directly from Relation (3.22). Let j be a fixed state of S. We denote by T_j^0 the first instant t such that $X_t = j$, i.e.

$$T_j^0 = \inf\{t \geq 0 \mid X_t = j\}.$$

For $n \geq 1$, we denote by M_j^n the time spent by X during its nth visit to state j, by L_j^n the time spent by X between two successive entrances to state j, and by T_j^n the time of the nth return to state j. More formally, we have for $n \geq 0$,

$$M_j^{n+1} = \inf\{t > T_j^n \mid X_t \neq j\} - T_j^n,$$
$$T_j^{n+1} = \inf\{t > T_j^n + M_j^{n+1} \mid X_t = j\},$$
$$L_j^{n+1} = T_j^{n+1} - T_j^n.$$

It has been shown in Theorem 3.1.7 that the successive sojourns in the states of X are independent and exponentially distributed. This implies that the random variables $L_j^1, L_j^2, \ldots$ are independent and identically distributed with mean $m_j = \mathbb{E}\{\tau(j) \mid X_0 = j\}$. In the same way, the random variables $M_j^1, M_j^2, \ldots$ are independent and exponentially distributed with mean $1/\lambda_j$. By the strong law of large numbers, see Theorem 2.8.3, we have

$$\lim_{n \longrightarrow \infty} \frac{L_j^1 + \cdots + L_j^n}{n} = m_j \text{ with probability 1}$$

and

$$\lim_{n \longrightarrow \infty} \frac{M_j^1 + \cdots + M_j^n}{n} = \frac{1}{\lambda_j} \text{ with probability 1.}$$

Now, for $n \geq 1$ and $T_j^n \leq t < T_j^{n+1}$, we have

$$M_j^1 + \cdots + M_j^n = \int_0^{T_j^n} 1_{\{X_s=j\}} ds \leq \int_0^t 1_{\{X_s=j\}} ds$$

$$\leq \int_0^{T_j^{n+1}} 1_{\{X_s=j\}} ds = M_j^1 + \cdots + M_j^{n+1}$$

and

$$T_j^n = T_j^0 + L_j^1 + \cdots + L_j^n \leq t < T_j^0 + L_j^1 + \cdots + L_j^{n+1} = T_j^{n+1}.$$

We thus obtain

$$\frac{M_j^1 + \cdots + M_j^n}{T_j^0 + L_j^1 + \cdots + L_j^{n+1}} \leq \frac{1}{t} \int_0^t 1_{\{X_s=j\}} ds \leq \frac{M_j^1 + \cdots + M_j^{n+1}}{T_j^0 + L_j^1 + \cdots + L_j^n}.$$

For the upper bound, we have

$$\frac{M_j^1 + \cdots + M_j^{n+1}}{T_j^0 + L_j^1 + \cdots + L_j^n} = \frac{n+1}{n} \frac{\dfrac{M_j^1 + \cdots + M_j^{n+1}}{n+1}}{\dfrac{T_0^j}{n} + \dfrac{L_j^1 + \cdots + L_j^n}{n}}.$$

T_0^j being finite and independent of n, we have $T_0^j/n \longrightarrow 0$ with probability 1 when $n \longrightarrow \infty$. So,

$$\lim_{n \to \infty} \frac{M_j^1 + \cdots + M_j^{n+1}}{T_j^0 + L_j^1 + \cdots + L_j^n} = \frac{1}{\lambda_j m_j} \quad \text{with probability 1.}$$

In the same way, for the lower bound, we have

$$\frac{M_j^1 + \cdots + M_j^n}{T_j^0 + L_j^1 + \cdots + L_j^{n+1}} = \frac{n}{n+1} \frac{\dfrac{M_j^1 + \cdots + M_j^n}{n}}{\dfrac{T_0^j}{n+1} + \dfrac{L_j^1 + \cdots + L_j^{n+1}}{n+1}},$$

that is,

$$\lim_{n \to \infty} \frac{M_j^1 + \cdots + M_j^n}{T_j^0 + L_j^1 + \cdots + L_j^{n+1}} = \frac{1}{\lambda_j m_j} \quad \text{with probability 1.}$$

When $t \longrightarrow \infty$, we have $n \longrightarrow \infty$, since n is the total number of visits to state j during the interval $[0, t]$. Thus

$$\lim_{t \to \infty} \int_0^t 1_{\{X_s=j\}} ds = \frac{1}{\lambda_j m_j} \quad \text{with probability 1,}$$

which completes the first part of the proof.

To prove the second part, we write

$$\frac{1}{t}\int_0^t r_{X_s}\,ds = \sum_{j\in S} r_j\left(\frac{1}{t}\int_0^t 1_{\{X_s=j\}}\,ds\right).$$

From the first part of the proof and since the state space is finite, we have

$$\lim_{t\to\infty}\sum_{j\in S} r_j\left(\frac{1}{t}\int_0^t 1_{\{X_s=j\}}\,ds\right) = \sum_{j\in S} r_j\pi_j,\ \text{with probability 1,}$$

which completes the proof. ∎

3.8 Absorbing Markov chains

DEFINITION 3.8.1. *A continuous-time Markov chain is called absorbing if it has at least one absorbing state, at least one transient state, and no other recurrent state than the absorbing ones.*

Let X be an absorbing Markov chain with infinitesimal generator A over the finite state space, S. The state space can be decomposed as $S = B \cup H$, where B is the set of transient states and H is the set of absorbing states. By definition, B and H are both non-empty. Following this decomposition, matrix A can be written as

$$A = \begin{pmatrix} A_B & A_{BH} \\ 0 & 0 \end{pmatrix},$$

where submatrix A_B contains the transition rates between transient states and A_{BH} contains the transition rates from states of B to states of H. Matrices 0 are here the null matrices with dimension $(|H|, |B|)$ and $|H|$, respectively. Without any loss of generality, we suppose that for every absorbing state, j, there exists a transient state, i, such that $A_{i,j} \neq 0$. This means that matrix A_{BH} has no null column. If this is not the case, it would mean that there exists an absorbing state which remains alone, i.e. not accessible from any other state, which has no interest.

LEMMA 3.8.2. *The matrix A_B is invertible.*

Proof. Let Z be the Markov chain obtained after uniformization of X with respect to the uniformization rate λ, which is such that $\lambda \geq \max\{\lambda_i, i \in B\}$. The transition probability matrix, U, of Z is given, following the same decomposition as for matrix A, by

$$U = \begin{pmatrix} U_B & U_{BH} \\ 0 & I \end{pmatrix},$$

where U_B and U_{BH} are given by

$$U_B = I + A_B/\lambda \quad \text{and} \quad U_{BH} = A_{BH}/\lambda.$$

From Theorem 3.6.3, the states of B are also transient for Z. From Lemma 3.8.2, the matrix $I - U_B$ is invertible and so is matrix $A_B = -\lambda(I - U_B)$.   ∎

The initial probability distribution, α, of X can be decomposed through subsets B and H as

$$\alpha = (\alpha_B \; \alpha_H),$$

where row subvector α_B (resp. α_H) contains the initial probabilities of states of B (resp. H).

For every $j \in H$, we recall that the hitting time of state j is defined by

$$\tau(j) = \inf\{t \geq T_1 \mid X_t = j\},$$

where T_1 is the first jump time of X and $\tau(j) = \infty$ if this set is empty. With this definition, we have, for every $t > 0$,

$$\text{if } X_0 \in B \text{ then } \tau(j) \leq t \iff X_t = j,$$
$$\text{if } X_0 \in H \text{ then } \tau(j) = \infty.$$

We then have, from Relation (3.9), for every $t \geq 0$,

$$\mathbb{P}\{\tau(j) \leq t\} = \sum_{i \in B} \alpha_i \mathbb{P}\{X_t = j \mid X_0 = i\} = \sum_{i \in B} \alpha_i \left(e^{At}\right)_{i,j}.$$

Since matrix A is block triangular, we have, for $n \geq 1$,

$$A^n = \begin{pmatrix} (A_B)^n & (A_B)^{n-1} A_{BH} \\ 0 & 0 \end{pmatrix}, \tag{3.23}$$

which leads to

$$e^{At} = \begin{pmatrix} e^{A_B t} & (e^{A_B t} - I)(A_B)^{-1} A_{BH} \\ 0 & I \end{pmatrix}. \tag{3.24}$$

We then obtain,

$$\mathbb{P}\{\tau(j) \leq t\} = \alpha_B (e^{A_B t} - I)(A_B)^{-1} w_j = -\alpha_B (A_B)^{-1} w_j + \alpha_B (A_B)^{-1} e^{A_B t} w_j,$$

where w_j is the column vector containing the jth column of matrix A_{BH}.

Since the states of B are transient, combining Corollary 2.3.9 and Lemma 3.5.4, we obtain $\lim_{t \to \infty} e^{A_B t} = 0$. We then have, as expected,

$$\mathbb{P}\{\tau(j) < \infty\} = -\alpha_B (A_B)^{-1} w_j. \tag{3.25}$$

As we did in the discrete-time case, the complementary cumulative distribution function of $\tau(j)$ is obtained using the two following relations:

$$\alpha_B \mathbb{1} + \alpha_H \mathbb{1} = 1,$$

$$-(A_B)^{-1} A_{BH} \mathbb{1} = -(A_B)^{-1}(-A_B \mathbb{1}) = \mathbb{1}.$$

Indeed, we have

$$\mathbb{P}\{\tau(j) > t\} = 1 + \alpha_B (A_B)^{-1} w_j - \alpha_B (A_B)^{-1} e^{A_B t} w_j$$

$$= \alpha_B \mathbb{1} + \alpha_H \mathbb{1} + \alpha_B (A_B)^{-1} w_j - \alpha_B (A_B)^{-1} e^{A_B t} w_j$$

$$= -\alpha_B (A_B)^{-1} A_{BH} \mathbb{1} + \alpha_H \mathbb{1} + \alpha_B (A_B)^{-1} w_j - \alpha_B (A_B)^{-1} e^{A_B t} w_j$$

$$= -\alpha_B (A_B)^{-1} (A_{BH} \mathbb{1} - w_j) - \alpha_B (A_B)^{-1} e^{A_B t} w_j + \alpha_H \mathbb{1}.$$

When $|H| \geq 2$, that is when there are at least two absorbing states, the expected time, $\mathbb{E}\{\tau(j)\}$, to hit state $j \in H$ is thus given by

$$\mathbb{E}\{\tau(j)\} = \int_0^\infty \mathbb{P}\{\tau(j) > t\} dt = \infty.$$

Indeed, observe that, if $|H| \geq 2$, we have, for every $j \in H$, $A_{BH} \mathbb{1} - w_j \neq 0$.

Since the non-absorbing states are transient states, absorption occurs in finite time with probability 1 in one of the absorbing states. For every $j \in H$, we denote by p_j the probability of the Markov chain X to be absorbed in state j. We denote by T the time to absorption, i.e.

$$T = \inf\{t \geq T_1 \mid X_t \in H\},$$

where $T = \infty$ if this set is empty. We have $T > 0$ and, for $t > 0$,

$$\text{if } X_0 \in B \text{ then } T \leq t \iff X_t \in H,$$

$$\text{if } X_0 \in H \text{ then } T = \infty.$$

Note that when H has only one element j, we have $T = \tau(j)$.

Using (3.24), we obtain,

$$\mathbb{P}\{T \leq t\} = \sum_{i \in B} \alpha_i \mathbb{P}\{T \leq t \mid X_0 = i\} + \sum_{i \in H} \alpha_i \mathbb{P}\{T \leq t \mid X_0 = i\}$$

$$= \sum_{i \in B} \alpha_i \mathbb{P}\{X_t \in H \mid X_0 = i\}$$

$$= \sum_{i \in B} \alpha_i \sum_{j \in H} \left(e^{At}\right)_{i,j}$$

$$= \alpha_B (e^{A_B t} - I)(A_B)^{-1} A_{BH} \mathbb{1}$$

$$= -\alpha_B (e^{A_B t} - I) \mathbb{1}$$

$$= \alpha_B \mathbb{1} - \alpha_B e^{A_B t} \mathbb{1}$$

and thus,

$$P\{T < \infty\} = \alpha_B \mathbb{1}.$$

This leads to

$$P\{T > t\} = \alpha_H \mathbb{1} + \alpha_B e^{A_B t} \mathbb{1}$$

and to

$$E\{T\} = \int_0^\infty P\{T > t\} dt = \begin{cases} -\alpha_B (A_B)^{-1} \mathbb{1} & \text{if } \alpha_H = 0, \\ \infty & \text{if } \alpha_H \neq 0. \end{cases}$$

As in the discrete-time case, for every $j \in H$, we denote by p_j the probability of the Markov chain X being absorbed in state j, i.e.

$$p_j = P\{X_T = j\}.$$

When $X_0 \in B$, we have, for every $t > 0$,

$$X_T = j \text{ and } T \leq t \iff X_t = j.$$

So, for every $i \in B$, we obtain, from (3.24),

$$P\{X_T = j, T \leq t \mid X_0 = i\} = P\{X_t = j \mid X_0 = i\}$$

$$= \left(e^{At}\right)_{i,j}$$

$$= \left[(e^{A_B t} - I)(A_B)^{-1} w_j\right]_i.$$

Taking the limit when t tends to ∞ and since T is finite when $X_0 \in B$, we have for every $i \in B$,

$$P\{X_T = j \mid X_0 = i\} = -\left[(A_B)^{-1} w_j\right]_i.$$

We then obtain

$$p_j = \sum_{i \in B} \alpha_i P\{X_T = j \mid X_0 = i\} + \sum_{i \in H} \alpha_i P\{X_T = j \mid X_0 = i\}$$

$$= \sum_{i \in B} \alpha_i P\{X_T = j \mid X_0 = i\} + \sum_{i \in H} \alpha_i \mathbb{1}_{\{i = j\}}$$

$$= \alpha_j - \sum_{i \in B} \alpha_i \left[(A_B)^{-1} w_j\right]_i$$

$$= \alpha_j - \alpha_B (A_B)^{-1} w_j.$$

Note that, from (3.25), we have, for every $j \in H$, $p_j = \alpha_j + P\{\tau(j) < \infty\}$.

For every $j \in H$, the expected time to absorption in state j is defined by $\mathbb{E}\{T1_{\{X_T=j\}}\}$. Using the previous results, we have, for every $i \in B$,

$$
\begin{aligned}
\mathbb{E}\{T1_{\{X_T=j\}} \mid X_0 = i\} \\
&= \int_0^\infty \mathbb{P}\{X_T = j, T > t \mid X_0 = i\} dt \\
&= \int_0^\infty [\mathbb{P}\{X_T = j \mid X_0 = i\} - \mathbb{P}\{X_T = j, T \leq t \mid X_0 = i\}] dt \\
&= \int_0^\infty \left[-\left[(A_B)^{-1} w_j\right]_i - \left[(e^{A_B t} - I)(A_B)^{-1} w_j\right]_i \right] dt \\
&= \int_0^\infty \left[-e^{A_B t}(A_B)^{-1} w_j\right]_i dt \\
&= \left[(A_B)^{-2} w_j\right]_i.
\end{aligned}
$$

So we have

$$
\mathbb{E}\{T1_{\{X_T=j\}}\} = \begin{cases} \alpha_B(A_B)^{-2} w_j & \text{if } \alpha_j = 0, \\ \infty & \text{if } \alpha_j \neq 0. \end{cases}
$$

The expected time to absorption given that absorption occurs in state j is given by

$$
\mathbb{E}\{T \mid X_T = j\} = \frac{\mathbb{E}\{T1_{\{X_T=j\}}\}}{p_j} = \begin{cases} \dfrac{\alpha_B(A_B)^{-2} w_j}{-\alpha_B(A_B)^{-1} w_j} & \text{if } \alpha_j = 0 \\[2mm] \infty & \text{if } \alpha_j \neq 0. \end{cases}
$$

4 State aggregation

In this chapter we analyze the conditions under which the aggregated process constructed from a homogeneous Markov chain over a given partition of its state space is also a homogeneous Markov chain. The results of this chapter mainly come from [89] and [91] for irreducible discrete-time Markov chains, in [94] for irreducible continuous-time Markov chains, and [58] for absorbing Markov chains. A necessary condition to obtain a Markovian aggregated process has been obtained in [58]. These works are based on the pioneering works of [50] and [1]. They are themselves the basis of other extensions such as [56], which deals with infinite state spaces, or [57] in which the author derives new results for the lumpability of reducible Markov chains and obtains spectral properties associated with lumpability.

4.1 State aggregation in irreducible DTMC

4.1.1 Introduction and notation

In this first section we consider the aggregated process constructed from an irreducible and homogeneous discrete-time Markov chain (DTMC), X, over a given partition of the state space. We analyze the conditions under which this aggregated process is also Markov homogeneous and we give a characterization of this situation.

The Markov chain $X = \{X_n, n \in \mathbb{N}\}$ is supposed to be irreducible and homogeneous. The state space is supposed to be finite and denoted by $S = \{1, 2, \dots, N\}$. All the vectors used are row vectors except when specified. For instance, the column vector with all its entries equal to 1 is denoted by $\mathbb{1}$ and its dimension is defined by the context.

The stationary distribution π of X is the unique solution to the system

$$\pi = \pi P \quad \text{and} \quad \pi \mathbb{1} = 1,$$

where P is the transition probability matrix of X. Let us denote by $\mathcal{B} = \{B(1), \dots, B(M)\}$ a partition of the state space S in M proper subsets and by $n(i)$ the cardinality of $B(i)$.

We assume the states of S ordered such that

$$B(1) = \{1,\ldots,n(1)\},$$

$$\vdots$$

$$B(m) = \{n(1)+\cdots+n(m-1)+1,\ldots,n(1)+\cdots+n(m)\},$$

$$\vdots$$

$$B(M) = \{n(1)+\cdots+n(M-1)+1,\ldots,N\}.$$

With the given process X we associate the aggregated stochastic process Y with values in $F = \{1,2,\ldots,M\}$, defined by

$$Y_n = m \iff X_n \in B(m) \quad \text{for all } n \geq 0.$$

It is easily checked from this definition and the irreducibility of X that the obtained process Y is also irreducible in the following sense:

$$\forall m \in F, \forall l \in F, \exists n \geq 0 \text{ such that } \mathbb{P}\{Y_n = m \mid Y_0 = l\} > 0.$$

The Markov property of X means that given the state in which X is at the present time, the future and the past of the process are independent. As discussed in the introduction, this is not true for the aggregated process Y in the general case. The question addressed here is under which conditions is Y also a homogeneous Markov chain.

Let us denote by $\mathcal{A}$ the set of all probability vectors with N entries. For every $\alpha \in \mathcal{A}$ and $l \in F$, we denote by $\alpha_{B(l)}$ the restriction of α to the class $B(l)$. The vector $\alpha_{B(l)}$ has $n(l)$ entries and its kth entry is the $(n(1)+\cdots+n(l-1)+k)$th entry of α. The operator $\alpha \mapsto \alpha_{B(l)}$ is denoted by T_l, so we have $T_l(\alpha) = \alpha_{B(l)}$. Conversely, for every probability vector, β, whose dimension is $n(l)$, we define the vector $\gamma = T_l^{-1}(\beta)$, belonging to $\mathcal{A}$, by $\gamma_i = 0$ if $i \notin B(l)$ and $\gamma_i = \beta_{i-j}$ if $i \in B(l)$, where $j = n(1)+\cdots+n(l-1)$.

For $\alpha \in \mathcal{A}$ and $l \in F$, we also define, when $\alpha_{B(l)} \neq 0$, the vector $\alpha^{B(l)}$ of $\mathcal{A}$ by

$$\alpha_i^{B(l)} = \begin{cases} \dfrac{\alpha_i}{\alpha_{B(l)}\mathbb{1}} & \text{if } i \in B(l) \\ 0 & \text{otherwise.} \end{cases}$$

If vector α is interpreted as the distribution of some random variable Z with values in S, then $\alpha_i^{B(l)} = \mathbb{P}\{Z = i \mid Z \in B(l)\}$. Observe that if $\alpha_{B(l)} \neq 0$, we have

$$\alpha^{B(l)} = \frac{T_l^{-1}(\alpha_{B(l)})}{\alpha_{B(l)}\mathbb{1}}. \tag{4.1}$$

The following examples illustrate the notation.

EXAMPLE 4.1.1. Consider a Markov chain, X, with $N = 5$ states and the partition $\mathcal{B} = \{B(1), B(2)\}$ with $B(1) = \{1,2,3\}$ and $B(2) = \{4,5\}$.

When $\alpha = (1/10, 1/10, 1/5, 2/5, 1/5)$, we obtain

$$\alpha_{B(1)} = (1/10, 1/10, 1/5), \qquad \alpha_{B(2)} = (2/5, 1/5),$$
$$\alpha^{B(1)} = (1/4, 1/4, 1/2, 0, 0), \qquad \alpha^{B(2)} = (0, 0, 0, 2/3, 1/3),$$
$$T_1(\alpha^{B(1)}) = (1/4, 1/4, 1/2), \qquad T_1(\alpha^{B(2)}) = (0, 0, 0).$$

If $\beta = (1/4, 3/4)$ then $T_2^{-1}(\beta) = (0, 0, 0, 1/4, 3/4)$.
When $\alpha = (1/2, 1/2, 0, 0, 0)$, we have $\alpha^{B(1)} = (1/2, 1/2, 0, 0, 0)$ and $\alpha^{B(2)}$ is not defined since $\alpha_{B(2)} = (0, 0)$.   ∎

For $\alpha \in \mathcal{A}$, we denote by $X = (\alpha, P)$ the Markov chain with initial probability distribution α and transition probability matrix P. We also denote by $(\cdot, P)$ the set of Markov chains, X, with transition probability matrix P, that is,

$$(\cdot, P) = \{(\alpha, P), \ \alpha \in \mathcal{A}\}.$$

In the same way, we denote the aggregated process Y by $\mathrm{agg}(\alpha, P, \mathcal{B})$ when necessary, to specify that Y is the aggregated process constructed from the Markov chain $X = (\alpha, P)$ and the partition $\mathcal{B}$.

We denote by $(P^n)_{i, B(m)}$ the probability that X_n is in a state of the subset $B(m)$ when starting from state i, i.e.

$$(P^n)_{i, B(m)} = \mathbb{P}\{X_n \in B(m) \mid X_0 = i\} = \sum_{j \in B(m)} (P^n)_{i, j}.$$

When we need to specify the initial probability distribution of the Markov chain X, we will simply write it in the index of the probability measure. For instance, $\mathbb{P}_\beta\{X_n \in B(m)\}$ represents the probability that $X = (\beta, P)$ is in a state of subset $B(m)$ after n transitions, i.e.

$$\mathbb{P}_\beta\{X_n \in B(m)\} = \sum_{j \in B(m)} (\beta P^n)_j = \sum_{i \in S} \beta_i (P^n)_{i, B(m)}.$$

4.1.2   Preliminaries

In this subsection we state the basic tools of our analysis. We begin with some preliminary results and formulas. They basically concern the T operator.

LEMMA 4.1.2.   *Every probability distribution, $\alpha \in \mathcal{A}$, can be written in a unique way as a convex combination of the $\alpha^{B(l)}$:*

$$\alpha = \sum_{l \in F, \alpha_{B(l)} \neq 0} \alpha_{B(l)} \mathbb{1} \alpha^{B(l)}.$$

Proof. Let $\alpha \in \mathcal{A}$. We have

$$\alpha = \sum_{l \in F} T_l^{-1}(\alpha_{B(l)})$$

$$= \sum_{l \in F, \alpha_{B(l)} \neq 0} \alpha_{B(l)} \mathbb{1} \alpha^{B(l)} \text{ from Relation (4.1).}$$

To prove the uniqueness, suppose that α can be written as

$$\alpha = \sum_{l \in F, \alpha_{B(l)} \neq 0} \lambda_l \alpha^{B(l)} \text{ where } 0 \leq \lambda_l \leq 1 \quad \text{and} \quad \sum_{l \in F} \lambda_l = 1.$$

Then for all $m \in F$ such that $\alpha_{B(m)} \neq 0$, we have $\alpha_{B(m)} \mathbb{1} = \lambda_m \alpha^{B(m)} \mathbb{1} = \lambda_m$. ∎

LEMMA 4.1.3. *Let $l \in F$, $x_1, \ldots, x_n$ be n probability vectors of $\mathcal{A}$ and $\lambda_1, \ldots, \lambda_n$ be n nonnegative real numbers such that $\lambda_1 + \cdots + \lambda_n = 1$. Then,*

$$\left(\sum_{k=1}^{n} \lambda_k x_k \right)^{B(l)} = \sum_{\substack{1 \leq k \leq n, \\ (x_k)_{B(l)} \neq 0}} \frac{\lambda_k (x_k)_{B(l)} \mathbb{1}}{\sum_{h=1}^{n} \lambda_h (x_h)_{B(l)} \mathbb{1}} (x_k)^{B(l)},$$

when $\left(\sum_{k=1}^{n} \lambda_k x_k \right)_{B(l)} \neq 0.$

Proof. By definition, we have

$$\left(\sum_{k=1}^{n} \lambda_k x_k \right)^{B(l)} = \frac{T_l^{-1} \left(\left(\sum_{k=1}^{n} \lambda_k x_k \right)_{B(l)} \right)}{\left(\sum_{k=1}^{n} \lambda_k x_k \right)_{B(l)} \mathbb{1}} \text{ from Relation (4.1)}$$

$$= \frac{T_l^{-1} \left(\sum_{k=1}^{n} \lambda_k (x_k)_{B(l)} \right)}{\sum_{k=1}^{n} \lambda_k (x_k)_{B(l)} \mathbb{1}}$$

$$= \frac{\sum_{k=1}^{n} \lambda_k T_l^{-1} ((x_k)_{B(l)})}{\sum_{k=1}^{n} \lambda_k (x_k)_{B(l)} \mathbb{1}}$$

$$= \sum_{\substack{1 \leq k \leq n, \\ (x_k)_{B(l)} \neq 0}} \frac{\lambda_k (x_k)_{B(l)} \mathbb{1}}{\sum_{h=1}^{n} \lambda_h (x_h)_{B(l)} \mathbb{1}} (x_k)^{B(l)} \text{ from Relation (4.1),}$$

which completes the proof. ∎

Now, we state some immediate consequences of the fact that X is a Markov chain.

LEMMA 4.1.4.   *For $l, m \in F$ and $\alpha \in \mathcal{A}$, such that $\alpha_{B(l)} \neq 0$ and $(\alpha P^n)_{B(l)} \neq 0$, we have*

$$\mathbb{P}_\alpha\{X_n \in B(m)\} = \mathbb{P}_{\alpha P^n}\{X_0 \in B(m)\}, \tag{4.2}$$

$$\mathbb{P}_\alpha\{X_n \in B(m) \mid X_0 \in B(l)\} = \mathbb{P}_{\alpha^{B(l)}}\{X_n \in B(m)\}, \tag{4.3}$$

$$\mathbb{P}_\alpha\{X_{n+1} \in B(m) \mid X_n \in B(l)\} = \mathbb{P}_{\alpha P^n}\{X_1 \in B(m) \mid X_0 \in B(l)\}. \tag{4.4}$$

Proof.   First note that

$$\mathbb{P}_\alpha\{X_0 \in B(m)\} = \sum_{j \in B(m)} \alpha_j.$$

Relation (4.2) is then easy to obtain, since

$$\mathbb{P}_\alpha\{X_n \in B(m)\} = \sum_{j \in B(m)} (\alpha P^n)_j = \mathbb{P}_{\alpha P^n}\{X_0 \in B(m)\}.$$

To prove Relation (4.3), we write

$$\mathbb{P}_\alpha\{X_n \in B(m) \mid X_0 \in B(l)\} = \frac{\displaystyle\sum_{i \in B(l)} \alpha_i (P^n)_{i, B(m)}}{\alpha_{B(l)} \mathbb{1}}$$

$$= \sum_{i \in B(l)} \alpha_i^{B(l)} (P^n)_{i, B(m)}$$

$$= \mathbb{P}_{\alpha^{B(l)}}\{X_n \in B(m)\}.$$

For Relation (4.4), we have

$$\mathbb{P}_\alpha\{X_{n+1} \in B(m) \mid X_n \in B(l)\} = \frac{\displaystyle\sum_{i \in B(l)} \mathbb{P}_\alpha\{X_n = i\} P_{i, B(m)}}{\mathbb{P}_\alpha\{X_n \in B(l)\}}$$

$$= \sum_{i \in B(l)} \left(\frac{\mathbb{P}_{\alpha P^n}\{X_0 = i\}}{\mathbb{P}_{\alpha P^n}\{X_0 \in B(l)\}} \right) P_{i, B(m)} \quad \text{from (4.2)}$$

$$= \sum_{i \in B(l)} \left(\frac{(\alpha P^n)_i}{(\alpha P^n)_{B(l)} \mathbb{1}} \right) P_{i, B(m)}$$

$$= \sum_{i \in B(l)} (\alpha P^n)_i^{B(l)} P_{i, B(m)}$$

$$= \mathbb{P}_{(\alpha P^n)^{B(l)}}\{X_1 \in B(m)\}$$

$$= \mathbb{P}_{\alpha P^n}\{X_1 \in B(m) \mid X_0 \in B(l)\} \quad \text{from (4.3)},$$

which completes the proof.   ∎

Let us consider now sequences of elements of the partition $\mathcal{B}$.

DEFINITION 4.1.5. *A sequence* $(C_0, C_1, \ldots, C_j)$ *of elements of* $\mathcal{B}$ *is called a possible sequence for the distribution* $\alpha \in \mathcal{A}$ *if*

$$\mathbb{P}_\alpha\{X_0 \in C_0, \ldots, X_j \in C_j\} > 0.$$

In particular, for $l \in F$, the sequence $(B(l))$ is a possible sequence for α if $\alpha_{B(l)} \neq 0$.

DEFINITION 4.1.6. *Let* $\alpha \in \mathcal{A}$ *and* $(C_0, C_1, \ldots, C_j)$ *be a possible sequence for* α. *We define the distribution* $f(\alpha, C_0, C_1, \ldots, C_j) \in \mathcal{A}$ *recursively by*

$$f(\alpha, C_0) = \alpha^{C_0}$$
$$f(\alpha, C_0, C_1, \ldots, C_k) = (f(\alpha, C_0, C_1, \ldots, C_{k-1})P)^{C_k}, \quad for \; k = 1, \ldots, j.$$

For instance, when $j = 2$, we have $f(\alpha, C_0, C_1, C_2) = ((\alpha^{C_0}P)^{C_1}P)^{C_2}$. In Lemma 4.1.10 below, we show that if $X = (\alpha, P)$ and if sequence $(C_0, \ldots, C_n)$ is possible for α, then vector $f(\alpha, C_0, \ldots, C_n)$ is the distribution of X conditional on the event $\{X_0 \in C_0, X_1 \in C_1, \ldots, X_n \in C_n\}$. Before proving this, we need some technical lemmas.

Let us decompose matrix P into the blocks induced by the partition $\mathcal{B}$, so that the sub-matrix $P_{B(l)B(m)}$ of dimension $n(l) \times n(m)$ contains the one-step transition probabilities from states of $B(l)$ to states of $B(m)$.

LEMMA 4.1.7. *Let* $\alpha \in \mathcal{A}$ *and* $(C_0, \ldots, C_n)$ *be a possible sequence for* α. *The restriction of vector* $\beta = f(\alpha, C_0, \ldots, C_n)$ *to subset* C_n *is*

$$\beta_{C_n} = \frac{\alpha_{C_0}P_{C_0C_1}\ldots P_{C_{n-1}C_n}}{\alpha_{C_0}P_{C_0C_1}\ldots P_{C_{n-1}C_n}\mathbb{1}}.$$

Proof. The result is true for $n = 0$, since $\beta = f(\alpha, C_0) = \alpha^{C_0}$. Suppose that the result is true for a possible sequence containing n subsets of $\mathcal{B}$. This means that if $\gamma = f(\alpha, C_0, \ldots, C_{n-1})$ then

$$\gamma_{C_{n-1}} = \frac{\alpha_{C_0}P_{C_0C_1}\ldots P_{C_{n-2}C_{n-1}}}{\alpha_{C_0}P_{C_0C_1}\ldots P_{C_{n-2}C_{n-1}}\mathbb{1}}.$$

By definition, we have $\beta = (\gamma P)^{C_n}$, and so

$$\beta_{C_n} = \frac{(\gamma P)_{C_n}}{(\gamma P)_{C_n}\mathbb{1}}.$$

But, we have $(\gamma P)_{C_n} = \gamma_{C_{n-1}}P_{C_{n-1}C_n}$ since $\gamma_i = 0$ if $i \notin C_{n-1}$. So, we obtain

$$\beta_{C_n} = \frac{\gamma_{C_{n-1}}P_{C_{n-1}C_n}}{\gamma_{C_{n-1}}P_{C_{n-1}C_n}\mathbb{1}}$$
$$= \frac{\alpha_{C_0}P_{C_0C_1}\ldots P_{C_{n-1}C_n}}{\alpha_{C_0}P_{C_0C_1}\ldots P_{C_{n-1}C_n}\mathbb{1}},$$

which completes the proof. ∎

LEMMA 4.1.8.   *Let $\alpha, \beta \in \mathcal{A}$ and let $(C_0, \ldots, C_n)$ be a possible sequence for both α and β. Let λ and μ be two nonnegative real numbers such that $\lambda + \mu = 1$. We have*

$$f(\lambda \alpha + \mu \beta, C_0, \ldots, C_n) = \frac{\lambda K_\alpha}{K} f(\alpha, C_0, \ldots, C_n) + \frac{\mu K_\beta}{K} f(\beta, C_0, \ldots, C_n),$$

where $K_\alpha = \alpha_{C_0} P_{C_0 C_1} \ldots P_{C_{n-1} C_n} \mathbb{1}$, $K_\beta = \beta_{C_0} P_{C_0 C_1} \ldots P_{C_{n-1} C_n} \mathbb{1}$ and $K = \lambda K_\alpha + \mu K_\beta$.

Proof.   The relation is clearly true for the entries that are not in C_n, since in that case the three terms are equal to 0. To prove the relation for the entries that are in C_n, it is sufficient to consider the corresponding restrictions to C_n. Using Lemma 4.1.7 and the fact that $(\lambda \alpha + \mu \beta)_{C_0} = \lambda \alpha_{C_0} + \mu \beta_{C_0}$, we obtain

$$(f(\lambda \alpha + \mu \beta, C_0, \ldots, C_n))_{C_n} = \frac{(\lambda \alpha + \mu \beta)_{C_0} P_{C_0 C_1} \ldots P_{C_{n-1} C_n}}{(\lambda \alpha + \mu \beta)_{C_0} P_{C_0 C_1} \ldots P_{C_{n-1} C_n} \mathbb{1}}$$

$$= \frac{\lambda \alpha_{C_0} P_{C_0 C_1} \ldots P_{C_{n-1} C_n}}{K} + \frac{\mu \beta_{C_0} P_{C_0 C_1} \ldots P_{C_{n-1} C_n}}{K}$$

$$= \lambda \frac{K_\alpha}{K} (f(\alpha, C_0, \ldots, C_n))_{C_n} + \mu \frac{K_\beta}{K} (f(\beta, C_0, \ldots, C_n))_{C_n},$$

which completes the proof.   ∎

LEMMA 4.1.9.   *For every $k \geq 0$, $n \geq k$, $m \in F$, and any possible sequence $(C_{n-k}, \ldots, C_n)$ for αP^{n-k}, we have*

$$\mathbb{P}_\alpha \{X_{n+1} \in B(m) \mid X_n \in C_n, \ldots, X_{n-k} \in C_{n-k}\} = \mathbb{P}_\beta \{X_1 \in B(m)\},$$

where $\beta = f(\alpha P^{n-k}, C_{n-k}, \ldots, C_n)$.

Proof.   We prove this result by recurrence on the number of subsets contained in the possible sequence. If there is only one subset in this sequence, the result is given by Lemma 4.1.4 using successively Relations (4.4) and (4.3).

Suppose that the result is true for a possible sequence containing k subsets belonging to $\mathcal{B}$. We then have

$$\mathbb{P}_\alpha \{X_{n+1} \in B(m) \mid X_n \in C_n, \ldots, X_{n-k} \in C_{n-k}\}$$

$$= \frac{\mathbb{P}_\alpha \{X_{n+1} \in B(m), X_n \in C_n, \ldots, X_{n-k} \in C_{n-k}\}}{\mathbb{P}_\alpha \{X_n \in C_n, \ldots, X_{n-k} \in C_{n-k}\}}$$

$$= \frac{\sum_{i \in C_n} \mathbb{P}_\alpha \{X_{n+1} \in B(m) \mid X_n = i\} \mathbb{P}_\alpha \{X_n = i \mid X_{n-1} \in C_{n-1}, \ldots, X_{n-k} \in C_{n-k}\}}{\sum_{i \in C_n} \mathbb{P}_\alpha \{X_n = i \mid X_{n-1} \in C_{n-1}, \ldots, X_{n-k} \in C_{n-k}\}}$$

$$= \sum_{i \in C_n} P_{i,B(m)} \left(\frac{\mathbb{P}_\gamma \{X_1 = i\}}{\mathbb{P}_\gamma \{X_1 \in C_n\}} \right) \quad \begin{array}{l} \text{where } \gamma = f(\alpha P^{n-k}, C_{n-k}, \ldots, C_{n-1}) \\ \text{by the recurrence hypothesis} \end{array}$$

$$= \sum_{i \in C_n} P_{i,B(m)} \left(\frac{\mathbb{P}_{\gamma P}\{X_0 = i\}}{\mathbb{P}_{\gamma P}\{X_0 \in C_n\}} \right) \quad \text{by Lemma 4.1.4, Relation (4.2)}$$

$$= \sum_{i \in C_n} (\gamma P)_i^{C_n} P_{i,B(m)}$$

$$= \mathbb{P}_\beta\{X_1 \in B(m)\} \text{ since } (\gamma P)^{C_n} = \beta,$$

where the second equality follows from the Markov property. ∎

The next result gives a probabilistic interpretation of function f.

LEMMA 4.1.10.   *The vector* $\beta = f(\alpha, C_0, \ldots, C_n)$ *is the distribution of* X_n *given that* $\{X_n \in C_n, \ldots, X_0 \in C_0\}$. *That is,*

$$\beta_i = \mathbb{P}_\alpha\{X_n = i \mid X_n \in C_n, \ldots, X_0 \in C_0\}.$$

Proof.   If $i \notin C_n$ the result is evident. Let $i \in C_n$. We have

$$\mathbb{P}_\alpha\{X_n = i \mid X_n \in C_n, \ldots, X_0 \in C_0\}$$

$$= \frac{\mathbb{P}_\alpha\{X_n = i, X_n \in C_n, \ldots, X_0 \in C_0\}}{\mathbb{P}_\alpha\{X_n \in C_n, \ldots, X_0 \in C_0\}}$$

$$= \frac{\mathbb{P}_\alpha\{X_n = i, X_{n-1} \in C_{n-1}, \ldots, X_0 \in C_0\}}{\mathbb{P}_\alpha\{X_n \in C_n, \ldots, X_0 \in C_0\}}$$

$$= \frac{\mathbb{P}_\alpha\{X_n = i \mid X_{n-1} \in C_{n-1}, \ldots, X_0 \in C_0\}\mathbb{P}_\alpha\{X_{n-1} \in C_{n-1}, \ldots, X_0 \in C_0\}}{\mathbb{P}_\alpha\{X_n \in C_n \mid X_{n-1} \in C_{n-1}, \ldots, X_0 \in C_0\}\mathbb{P}_\alpha\{X_{n-1} \in C_{n-1}, \ldots, X_0 \in C_0\}}$$

$$= \frac{\mathbb{P}_\alpha\{X_n = i \mid X_{n-1} \in C_{n-1}, \ldots, X_0 \in C_0\}}{\mathbb{P}_\alpha\{X_n \in C_n \mid X_{n-1} \in C_{n-1}, \ldots, X_0 \in C_0\}}$$

$$= \frac{\mathbb{P}_\delta\{X_1 = i\}}{\mathbb{P}_\delta\{X_1 \in C_n\}} \text{ where } \delta = f(\alpha, C_0, \ldots, C_{n-1}), \text{ from Lemma 4.1.9}$$

$$= (\delta P)_i^{C_n} = \beta_i,$$

which completes the proof. ∎

4.1.3   Strong and weak lumpability

We now analyze the aggregated process $Y = \mathrm{agg}(\alpha, P, \mathcal{B})$. From now on, when we use the notation $f(\alpha, C_1, \ldots, C_j)$, we implicitly suppose that the sequence $(C_1, \ldots, C_j)$ of elements of $\mathcal{B}$ is a possible sequence for α. For every $l \in F$ and $\alpha \in \mathcal{A}$, we denote by $\mathcal{A}(\alpha, B(l))$ the set of all distributions of $\mathcal{A}$ of the form $f(\alpha, C_1, \ldots, C_j, B(l))$, where $C_1, \ldots, C_j$ belong to $\mathcal{B}$. To simplify the notation, we say that the sequence $(C_1, \ldots, C_j)$ is empty when $j = 0$. We thus have

$$\mathcal{A}(\alpha, B(l)) = \{\beta \in \mathcal{A} \mid \exists j \geq 0 \quad \text{and} \quad (C_1, \ldots, C_j) \in \mathcal{B}^j$$

$$\text{such that } \beta = f(\alpha, C_1, \ldots, C_j, B(l))\}.$$

In words, a vector $\beta \in \mathcal{A}(\alpha, B(l))$ is the distribution of X_n for some n, conditional on $X_n \in B(l)$.

It is easy to check that for all $\alpha \in \mathcal{A}$ and for all $l \in F$, the set $\mathcal{A}(\alpha, B(l))$ is not empty. This is an immediate consequence of the irreducibility of process Y. Note also that if the subset $B(l)$ is reduced to only one state, then $\mathcal{A}(\alpha, B(l))$ contains only one distribution since $\mathcal{A}(\alpha, \{i\}) = \{e_i\}$ where e_i is the probability distribution whose ith element is equal to 1. With this notation, a first characterization of the fact that Y is a homogeneous Markov chain is given by the following theorem that has been obtained by Kemeny and Snell [50] when X is an irreducible and aperiodic Markov chain. We show that this theorem is valid for irreducible Markov chains that are not necessarily aperiodic.

THEOREM 4.1.11.   *The process* $Y = \mathrm{agg}(\alpha, P, \mathcal{B})$ *is a homogeneous Markov chain if and only if* $\forall l, m \in F$, *the probability* $\mathbb{P}_\beta\{X_1 \in B(m)\}$ *has the same value for all* $\beta \in \mathcal{A}(\alpha, B(l))$. *This common value is the transition probability from state* l *to state* m *for the Markov chain* Y.

Proof.   The proof is based on Lemma 4.1.9. This lemma allows us to write that for all $l, m \in F$, for all $n \geq 0$, and for all sequences $(C_0, \ldots, C_{n-1}, B(l)) \in \mathcal{B}^{n+1}$ possible for α, we have

$$\mathbb{P}_\alpha\{X_{n+1} \in B(m) \mid X_n \in B(l), X_{n-1} \in C_{n-1}, \ldots, X_0 \in C_0\} = \mathbb{P}_\beta\{X_1 \in B(m)\},$$

where $\beta = f(\alpha, C_0, \ldots, C_{n-1}, B(l))$.

Suppose that the probability $\mathbb{P}_\beta\{X_1 \in B(m)\}$ is the same for all $\beta \in \mathcal{A}(\alpha, B(l))$. By choosing the particular distribution $\beta = \alpha^{B(l)}$ and using Relation (4.3) of Lemma 4.1.4, we obtain

$$\mathbb{P}_\beta\{X_1 \in B(m)\} = \mathbb{P}_{\alpha^{B(l)}}\{X_1 \in B(m)\} = \mathbb{P}_\alpha\{X_1 \in B(m) \mid X_0 \in B(l)\},$$

which proves that Y is a homogeneous Markov chain.

Conversely, suppose that Y is a homogeneous Markov chain and let β be any distribution of $\mathcal{A}(\alpha, B(l))$. Vector β can then be written $\beta = f(\alpha, C_0, \ldots, C_{n-1}, B(l))$, where $(C_0, \ldots, C_{n-1}) \in \mathcal{B}^n$ is a possible sequence for α. Using Lemma 4.1.9, we have

$$\mathbb{P}_\beta\{X_1 \in B(m)\} = \mathbb{P}_\alpha\{X_{n+1} \in B(m) \mid X_n \in B(l), X_{n-1} \in C_{n-1}, \ldots, X_0 \in C_0\}$$
$$= \mathbb{P}_\alpha\{X_{n+1} \in B(m) \mid X_n \in B(l)\} \text{ since } Y \text{ is a Markov chain.}$$

This last quantity does not depend on n since Y is homogeneous. Thus, $\mathbb{P}_\beta\{X_1 \in B(m)\}$ has the same value for all $\beta \in \mathcal{A}(\alpha, B(l))$.   ∎

A particular case is given by the following corollary.

COROLLARY 4.1.12.   *Let* $\alpha \in \mathcal{A}$. *If for every* $l, m \in F$ *such that* $\alpha_{B(l)} \neq 0$, $\alpha_{B(m)} \neq 0$ *and* $(\alpha^{B(l)}P)_{B(m)} \neq 0$ *we have* $(\alpha^{B(l)}P)^{B(m)} = \alpha^{B(m)}$, *then* $Y = \mathrm{agg}(\alpha, P, \mathcal{B})$ *is a homogeneous Markov chain.*

Proof.   The condition $(\alpha^{B(l)}P)^{B(m)} = \alpha^{B(m)}$ implies that $\mathcal{A}(\alpha, B(m)) = \{\alpha^{B(m)}\}$. The result then follows from Theorem 4.1.11.   ∎

We are interested in the analysis of the set of all initial probability distributions of X leading to a homogeneous Markov chain for the process $Y = \mathrm{agg}(\alpha, P, \mathcal{B})$. We denote this set by $\mathcal{A}_{\mathcal{M}}$, that is

$$\mathcal{A}_{\mathcal{M}} = \{\alpha \in \mathcal{A} \mid Y = \mathrm{agg}(\alpha, P, \mathcal{B}) \text{ is a homogeneous Markov chain}\}.$$

Let us consider the following example taken from [50].

EXAMPLE 4.1.13. Consider the state space $S = \{1, 2, 3\}$ and the partition of S defined by $\mathcal{B} = \{B(1), B(2)\}$, where $B(1) = \{1\}$ and $B(2) = \{2, 3\}$. We define the set of Markov chains $(\cdot, P)$ with state space S and transition probability matrix P given by

$$P = \left(\begin{array}{c|cc} 1/4 & 1/4 & 1/2 \\ \hline 0 & 1/6 & 5/6 \\ 7/8 & 1/8 & 0 \end{array} \right).$$

Consider the set U defined by

$$U = \{\alpha \in \mathcal{A} \mid \alpha = (1 - 3a, a, 2a), \ a \in [0, 1/3]\}.$$

It is easily checked that for $\alpha \in U$, we have

- if $a \neq 1/3$, $\alpha^{B(1)} = (1, 0, 0) \in U$,
- if $a \neq 0$, $\alpha^{B(2)} = (0, 1/3, 2/3) \in U$,
- $\alpha P = ((1 + 4a)/4, (3 - 4a)/12, (3 - 4a)/6) \in U$.

For every $\alpha \in U$, we have

$$\mathcal{A}(\alpha, B(1)) = \{(1, 0, 0)\} \quad \text{and} \quad \mathcal{A}(\alpha, B(2)) = \{(0, 1/3, 2/3)\}.$$

The condition of Theorem 4.1.11 is thus satisfied for every initial distribution in U. This means that $U \subseteq \mathcal{A}_{\mathcal{M}}$.

Co nsider now the vector $\alpha = (0, 0, 1)$. We define the two vectors $\beta = f(\alpha, B(2), B(2))$ and $\gamma = f(\beta, B(2), B(2))$, which both belong to $\mathcal{A}(\alpha, B(2))$. We have

$$\beta = (\alpha^{B(2)} P)^{B(2)} = (0, 1, 0) \quad \text{and} \quad \gamma = (\beta^{B(2)} P)^{B(2)} = (0, 1/6, 5/6).$$

But we have

$$\mathbb{P}_\beta\{X_1 \in B(1)\} = 0 \quad \text{and} \quad \mathbb{P}_\gamma\{X_1 \in B(1)\} = 35/48,$$

which means, from Theorem 4.1.11, that $(0, 0, 1) \notin \mathcal{A}_{\mathcal{M}}$, i.e. $Y = \mathrm{agg}((0, 0, 1), P, \mathcal{B})$ is not a homogeneous Markov chain. ∎

We now give some properties of the set $\mathcal{A}_{\mathcal{M}}$.

THEOREM 4.1.14.   *Let $\alpha \in \mathcal{A}_\mathcal{M}$. Then*

- *for every $h \in F$ such that $\alpha_{B(h)} \neq 0$, $\alpha^{B(h)} \in \mathcal{A}_\mathcal{M}$,*
- *for every $k \in \mathbb{N}$, $\alpha P^k \in \mathcal{A}_\mathcal{M}$,*
- *for every $n \in \mathbb{N}^*$, $\dfrac{1}{n}\displaystyle\sum_{k=1}^{n}\alpha P^k \in \mathcal{A}_\mathcal{M}$.*

Proof.   Let $\alpha \in \mathcal{A}_\mathcal{M}$.

For the first point, let $h \in F$ such that $\alpha_{B(h)} \neq 0$. From Lemma 4.1.9, we have

$$\mathbb{P}_{\alpha^{B(h)}}\{X_{n+1} \in B(m) \mid X_n \in B(l), X_{n-1} \in C_{n-1}, \ldots, X_0 \in B(h)\}$$
$$= \mathbb{P}_\beta\{X_1 \in B(m)\},$$

where $\beta = f(\alpha^{B(h)}, B(h), C_1, \ldots, C_{n-1}, B(l)) = f(\alpha, B(h), C_1, \ldots, C_{n-1}, B(l))$. So, we have

$$\mathbb{P}_\beta\{X_1 \in B(m)\}$$
$$= \mathbb{P}_\alpha\{X_{n+1} \in B(m) \mid X_n \in B(l), X_{n-1} \in C_{n-1}, \ldots, X_0 \in B(h)\},$$

which has the same value for every $\beta \in \mathcal{A}(\alpha^{B(h)}, B(l))$. This proves, from Theorem 4.1.11, that $\alpha^{B(h)} \in \mathcal{A}_\mathcal{M}$.

To prove the second point, let $k \in \mathbb{N}$. From Lemma 4.1.9, we have

$$\mathbb{P}_\alpha\{X_{n+k+1} \in B(m) \mid X_{n+k} \in B(l), X_{n+k-1} \in C_{n-1}, \ldots, X_k \in C_0\}$$
$$= \mathbb{P}_{\alpha P^k}\{X_{n+1} \in B(m) \mid X_n \in B(l), X_{n-1} \in C_{n-1}, \ldots, X_0 \in C_0\}.$$

Since $\alpha \in \mathcal{A}_\mathcal{M}$, the left-hand side of this equality is equal to $\mathbb{P}_\alpha\{X_{k+1} \in B(m) \mid X_k \in B(l)\}$, which is equal, from Relation (4.4) in Lemma 4.1.4, to $\mathbb{P}_{\alpha P^k}\{X_1 \in B(m) \mid X_0 \in B(l)\}$. This means, again from Theorem 4.1.11, that $\alpha P^k \in \mathcal{A}_\mathcal{M}$ for every $k \in \mathbb{N}$.

Consider now the third point. Let γ_n be defined by

$$\gamma_n = \frac{1}{n}\sum_{k=1}^{n}\alpha P^k.$$

From Lemma 4.1.9, we have

$$\mathbb{P}_{\gamma_n}\{X_{n+1} \in B(m) \mid X_n \in B(l), X_{n-1} \in C_{n-1}, \ldots, X_0 \in C_0\}$$
$$= \mathbb{P}_{\gamma_n'}\{X_1 \in B(m)\},$$

where

$$\gamma_n' = f(\gamma_n, C_0, \ldots, C_{n-1}, B(l)).$$

If we define $\alpha_k' = f(\alpha P^k, C_0, \ldots, C_{n-1}, B(l))$, Lemma 4.1.8 gives

$$\gamma_n' = \frac{1}{n}\sum_{k=1}^{n}\frac{K_k}{K}\alpha_k',$$

where $K_k = (\alpha P^k)_{C_0} P_{C_0 C_1} \ldots P_{C_{n-1} B(l)} \mathbb{1}$ and $K = \dfrac{1}{n} \sum\limits_{k=1}^{n} K_k$. So, we obtain

$$\mathbb{P}_{\gamma_n'}\{X_1 \in B(m)\} = \frac{1}{n} \sum_{k=1}^{n} \frac{K_k}{K} \mathbb{P}_{\alpha_k'}\{X_1 \in B(m)\},$$

and

$$\mathbb{P}_{\alpha_k'}\{X_1 \in B(m)\} = \mathbb{P}_{\alpha P^k}\{X_{n+1} \in B(m) \mid X_n \in B(l), \ldots, X_0 \in C_0\}$$
$$= \mathbb{P}_{\alpha}\{X_{n+k+1} \in B(m) \mid X_{n+k} \in B(l), \ldots, X_k \in C_0\}$$
$$= \mathbb{P}_{\alpha}\{X_1 \in B(m) \mid X_0 \in B(l)\} \text{ (since } \alpha \in \mathcal{A}_{\mathcal{M}}).$$

We finally obtain

$$\mathbb{P}_{\gamma_n'}\{X_1 \in B(m)\} = \frac{1}{n} \sum_{k=1}^{n} \frac{K_k}{K} \mathbb{P}_{\alpha}\{X_1 \in B(m) \mid X_0 \in B(l)\}$$
$$= \mathbb{P}_{\alpha}\{X_1 \in B(m) \mid X_0 \in B(l)\}.$$

Thus the probability $\mathbb{P}_{\gamma_n'}\{X_1 \in B(m)\}$ has the same value for every $\gamma_n' \in \mathcal{A}(\gamma_n, B(l))$, which implies that $\gamma_n \in \mathcal{A}_{\mathcal{M}}$, for all $n \geq 1$. ∎

LEMMA 4.1.15. *For every $m \in F$, for every $k \geq 1$, $n \geq k$, and for every possible sequence $(C_{n-k}, \ldots, C_{n-1})$, the function g from $\mathcal{A}$ to $[0, 1]$ defined by*

$$g : v \mapsto \mathbb{P}_v\{X_n \in B(m) \mid X_{n-1} \in C_{n-1}, \ldots, X_{n-k} \in C_{n-k}\}$$

is continuous on $\mathcal{A}$.

Proof. We have, for every $v \in \mathcal{A}$,

$$g(v) = \sum_{i \in S, \, v_i > 0} v_i \mathbb{P}_v\{X_n \in B(m) \mid X_{n-1} \in C_{n-1}, \ldots, X_{n-k} \in C_{n-k}, X_0 = i\}.$$

The quantity $\mathbb{P}_v\{X_n \in B(m) \mid X_{n-1} \in C_{n-1}, \ldots, X_{n-k} \in C_{n-k}, X_0 = i\}$ is constant for all $v \in \mathcal{A}$ such that $v_i > 0$. So, we can write, for every $v, v' \in \mathcal{A}$,

$$|g(v') - g(v)| = \left| \sum_{i \in S, \, v_i > 0, \, v_i' > 0} (v_i' - v_i) \right.$$
$$\left. \times \mathbb{P}\{X_n \in B(m) \mid X_{n-1} \in C_{n-1}, \ldots, X_{n-k} \in C_{n-k}, X_0 = i\} \right|$$

$$\leq \sum_{i \in S, \, v_i > 0, \, v_i' > 0} |v_i' - v_i|$$

$$\times \mathbb{P}\{X_n \in B(m) \mid X_{n-1} \in C_{n-1}, \ldots, X_{n-k} \in C_{n-k}, X_0 = i\}$$

$$\leq \sum_{i \in S} |v_i' - v_i|,$$

which implies the continuity of g on $\mathcal{A}$.   ∎

The following theorem proves the uniqueness of the transition probability matrix of the aggregated homogeneous Markov chain. It is proved in [50] for irreducible and aperiodic Markov chains. To obtain the result when X is only irreducible, we need the previous lemma.

When $Y = \mathrm{agg}(\alpha, P, \mathcal{B})$ is a homogeneous Markov chain, we denote by $\hat{P}$ its one-step transition probability matrix and by $\hat{P}_{l,m}$ the transition probability of Y from state l to state m, for every $l, m \in F$.

THEOREM 4.1.16.   *Let $\alpha \in \mathcal{A}_{\mathcal{M}}$. The transition probability matrix $\hat{P}$ of the homogeneous Markov chain $Y = \mathrm{agg}(\alpha, P, \mathcal{B})$ is the same for all $\alpha \in \mathcal{A}_{\mathcal{M}}$. Moreover, the stationary distribution π of X satisfies $\pi \in \mathcal{A}_{\mathcal{M}}$.*

Proof.   Let $\alpha \in \mathcal{A}_{\mathcal{M}}$. From Theorem 4.1.14, we have, for all $n \geq 1$,

$$v_n = \frac{1}{n} \sum_{k=1}^{n} \alpha P^k \in \mathcal{A}_{\mathcal{M}}.$$

For every k such that $(\alpha P^k)_{B(l)} \neq 0$, we have

$$\hat{P}_{l,m} = \mathbb{P}_\alpha \{X_1 \in B(m) \mid X_0 \in B(l)\}$$

$$= \mathbb{P}_\alpha \{X_{k+1} \in B(m) \mid X_k \in B(l)\}$$

$$= \mathbb{P}_{\alpha P^k} \{X_1 \in B(m) \mid X_0 \in B(l)\}.$$

Let n be large enough so that $\sum_{k=1}^{n} (\alpha P^k)_{B(l)} \mathbb{1} \neq 0$. Such an integer, n, exists since X is irreducible. For $k = 1, \ldots, n$, let us define

$$\gamma_k = \frac{(\alpha P^k)_{B(l)} \mathbb{1}}{\displaystyle\sum_{h=1}^{n} (\alpha P^h)_{B(l)} \mathbb{1}}.$$

$\hat{P}_{l,m}$ is independent of k, so it can be written as

$$\hat{P}_{l,m} = \sum_{k=1}^{n} \gamma_k \hat{P}_{l,m}$$

$$= \sum_{1 \leq k \leq n, \, (\alpha P^k)_{B(l)} \neq 0} \gamma_k \mathbb{P}_{\alpha P^k} \{X_1 \in B(m) \mid X_0 \in B(l)\}$$

$$= \sum_{1 \le k \le n,\, (\alpha P^k)_{B(l)} \ne 0} \gamma_k \mathbb{P}_{(\alpha P^k)^{B(l)}} \{X_1 \in B(m)\}$$

$$= \mathbb{P}_\Gamma \{X_1 \in B(m)\} \text{ where } \Gamma = \sum_{1 \le k \le n,\, (\alpha P^k)_{B(l)} \ne 0} \gamma_k (\alpha P^k)^{B(l)},$$

the last equality coming from Lemma 4.1.8.

Using Lemma 4.1.3, in which we set $\lambda_k = 1/n$ and $x_k = \alpha P^k$, Γ can be written as

$$\Gamma = v_n^{B(l)},$$

where

$$v_n = \frac{1}{n} \sum_{k=1}^{n} \alpha P^k.$$

We finally obtain

$$\hat{P}_{l,m} = \mathbb{P}_{v_n^{B(l)}} \{X_1 \in B(m)\}$$
$$= \mathbb{P}_{v_n} \{X_1 \in B(m) \mid X_0 \in B(l)\}.$$

When $n \to \infty$, we have $v_n \to \pi$ by Theorem 2.8.4, and thanks to Lemma 4.1.15, we obtain

$$\hat{P}_{l,m} = \mathbb{P}_\pi \{X_1 \in B(m) \mid X_0 \in B(l)\},$$

which does not depend on α.

Note that we have shown that if $\mathcal{A}_\mathcal{M} \ne \emptyset$ then $\pi \in \mathcal{A}_\mathcal{M}$. ∎

It has been shown in Example 4.1.13 that it is possible to have the situation in which some initial distributions lead to the Markov property for Y and some others don't. This motivates the following definitions.

DEFINITION 4.1.17.

- We say that the Markov chain $X = (\cdot, P)$ is weakly lumpable *with respect to the partition* $\mathcal{B}$ if the set $\mathcal{A}_\mathcal{M}$ is not empty.
- We say that the Markov chain $X = (\cdot, P)$ is strongly lumpable *with respect to the partition* $\mathcal{B}$ if $\mathcal{A}_\mathcal{M} = \mathcal{A}$.

The characterization of strong lumpability is straightforward and is given by the following theorem.

THEOREM 4.1.18. *The Markov chain $X = (\cdot, P)$ is strongly lumpable with respect to the partition $\mathcal{B}$ if and only if for every $l, m \in F$, the probability $P_{i,B(m)}$ has the same value* *for every $i \in B(l)$. This common value is the transition probability from state l to state m for the aggregated homogeneous Markov chain Y.*

Proof. Suppose that for every $l, m \in F$, the probability $P_{i,B(m)}$ has the same value for every $i \in B(l)$. Let $\alpha \in \mathcal{A}$ and $\beta \in \mathcal{A}(\alpha, B(l))$. By definition, β can be written as $\beta = f(\alpha, C_0, \dots, C_{n-1}, B(l))$ for some $n \geq 1$, and from Lemma 4.1.9, we have

$$\mathbb{P}_\beta\{X_1 \in B(m)\} = \mathbb{P}_\alpha\{X_{n+1} \in B(m) \mid X_n \in B(l), X_{n-1} \in C_{n-1}, \dots, X_0 \in C_0\}$$

$$= \frac{\displaystyle\sum_{i \in B(l)} P_{i,B(m)} \mathbb{P}_\alpha\{X_n = i \mid X_{n-1} \in C_{n-1}, \dots, X_0 \in C_0\}}{\displaystyle\sum_{i \in B(l)} \mathbb{P}_\alpha\{X_n = i \mid X_{n-1} \in C_{n-1}, \dots, X_0 \in C_0\}}$$

$$= P_{i,B(m)} \text{ since } P_{i,B(m)} \text{ has the same value for every } i \in B(l).$$

This shows, from Theorem 4.1.11, that Y is a homogeneous Markov chain for every initial probability distribution $\alpha \in \mathcal{A}$.

Conversely, suppose that for every initial distribution $\alpha \in \mathcal{A}$, $Y = \mathrm{agg}(\alpha, P, \mathcal{B})$ is a homogeneous Markov chain. Take $i \in B(l)$ and $\alpha = e_i$ where the ith entry of e_i is equal to 1 and the others are equal to 0. Applying Theorem 4.1.11, we obtain that $\mathbb{P}_\beta\{X_1 \in B\}$ has the same value for every $\beta \in \mathcal{A}(e_i, B(l))$ and is equal to $\mathbb{P}_{e_i}\{X_1 \in B(m) \mid X_0 \in B(l)\} = P_{i,B(m)}$. From Theorem 4.1.16, this value is independent of the initial distribution, that is, independent of i. ∎

Another important property of the set $\mathcal{A}_\mathcal{M}$ is stated in the following theorem.

THEOREM 4.1.19. *The set $\mathcal{A}_\mathcal{M}$ is a closed convex set.*

Proof. It follows immediately from the continuity Lemma 4.1.15 that $\mathcal{A}_\mathcal{M}$ is a closed set. To prove its convexity, consider α and $\beta \in \mathcal{A}_\mathcal{M}$ and let $\gamma = \lambda\alpha + \mu\beta$ where λ and μ are nonnegative real numbers with $\lambda + \mu = 1$.
From Lemma 4.1.9, we have

$$\mathbb{P}_\gamma\{X_{n+1} \in B(m) \mid X_n \in B(l), X_{n-1} \in C_{n-1}, \dots, X_0 \in C_0\} = \mathbb{P}_{\gamma'}\{X_1 \in B(m)\},$$

where $\gamma' = f(\gamma, C_0, \dots, C_{n-1}, B(l))$.
Let $\alpha' = f(\alpha, C_0, \dots, C_{n-1}, B(l))$ and $\beta' = f(\beta, C_0, \dots, C_{n-1}, B(l))$. Lemma 4.1.8 then gives

$$\gamma' = \lambda \frac{K_\alpha}{K} \alpha' + \mu \frac{K_\beta}{K} \beta',$$

where $K_\alpha = \alpha_{C_0} P_{C_0 C_1} \ldots P_{C_{n-1} B(l)} \mathbb{1}$, $K_\beta = \beta_{C_0} P_{C_0 C_1} \ldots P_{C_{n-1} B(l)} \mathbb{1}$ and $K = \lambda K_\alpha + \mu K_\beta$. Finally, we obtain

$$\mathbb{P}_{\gamma'}\{X_1 \in B(m)\} = \mathbb{P}_{(\lambda K_\alpha \alpha' + \mu K_\beta \beta')/K}\{X_1 \in B(m)\}$$

$$= \lambda \frac{K_\alpha}{K} \mathbb{P}_{\alpha'}\{X_1 \in B(m)\} + \mu \frac{K_\beta}{K} \mathbb{P}_{\beta'}\{X_1 \in B(m)\}$$

$$= \lambda \frac{K_\alpha}{K} \hat{P}_{l,m} + \mu \frac{K_\beta}{K} \hat{P}_{l,m} \text{ from Theorem 4.1.16}$$

$$= \hat{P}_{l,m},$$

which proves that $\gamma \in \mathcal{A}_\mathcal{M}$ and so $\mathcal{A}_\mathcal{M}$ is a convex set. ∎

Let us introduce the notation $\mathcal{A}_\pi$ for the convex hull generated by the vectors $\pi^{B(l)}$, $l \in F$, that is,

$$\mathcal{A}_\pi = \left\{ \sum_{l \in F} \lambda_l \pi^{B(l)}, \lambda_l \geq 0 \quad \text{and} \quad \sum_{l \in F} \lambda_l = 1 \right\}.$$

COROLLARY 4.1.20. *If the set $\mathcal{A}_\mathcal{M}$ is not empty then $\mathcal{A}_\pi \subseteq \mathcal{A}_\mathcal{M}$.*

Proof. If $\mathcal{A}_\mathcal{M}$ is not empty then, from Theorem 4.1.16, the stationary distribution π belongs to $\mathcal{A}_\mathcal{M}$ and from Theorem 4.1.14 the distributions $\pi^{B(l)}$ belong to $\mathcal{A}_\mathcal{M}$. The proof is achieved using Theorem 4.1.19. ∎

4.1.4 Characterization of weak lumpability

To characterize the set $\mathcal{A}_\mathcal{M}$ of all initial probability distributions of X leading to an aggregated homogeneous Markov chain Y, we need some supplementary notation and definitions.

- For $l \in F$, we denote by $\tilde{P}^{(l)}$ the $n(l) \times M$ matrix whose (i, m) entry is given by $\mathbb{P}\{X_1 \in B(m) \mid X_0 = n(1) + \cdots + n(l-1) + i\}$, i.e.

$$\tilde{P}^{(l)}_{i,m} = P_{n(1)+\cdots+n(l-1)+i, B(m)}, \quad 1 \leq i \leq n(l), \; m \in F.$$

From Theorem 4.1.16, we deduce the relation $\hat{P}^{(l)} = \left(T_l(\pi^{B(l)})\right) \tilde{P}^{(l)}$, where $\hat{P}^{(l)}$ is the row vector containing the lth row of $\hat{P}$. In the following, we always define the matrix $\hat{P}$ in this way, even if the aggregated process Y is not a homogeneous Markov chain.

- For $l \in F$, we denote by σ_l the linear system $x_l \tilde{P}^{(l)} = \hat{P}^{(l)}$. This system has M equations and $n(l)$ unknowns. If x_l is a solution to σ_l then necessarily $x_l \mathbb{1} = 1$. By construction, for every $l \in F$, σ_l has at least one solution, which is $T_l(\pi^{B(l)})$.

- From the solutions to these M linear systems, we construct the set $\mathcal{A}^*$ defined by

$$\mathcal{A}^* = \{\alpha \in \mathcal{A} \mid T_l(\alpha^{B(l)}) \text{ is a solution to } \sigma_l \text{ for every } l \in F \text{ s.t. } \alpha_{B(l)} \neq 0\}.$$

Observe that $\pi \in \mathcal{A}^*$.

- For every $l \in F$, we denote by $\mathcal{A}_l^*$ the set of nonnegative solutions to the system σ_l, i.e.

$$\mathcal{A}_l^* = \{x \in (\mathbb{R}^+)^{n(l)} \mid x\tilde{P}^{(l)} = \hat{P}^{(l)}\}.$$

Observe that $T_l(\pi^{B(l)}) \in \mathcal{A}_l^*$.

- We say that a subset $\mathcal{U}$ of $\mathcal{A}$ is *stable* by P if $\forall x \in \mathcal{U}$, the vector $xP \in \mathcal{U}$.

Let us now give some properties of the set $\mathcal{A}^*$. First, recall that a convex polyhedral of $\mathbb{R}^N$ is a subset of $\mathbb{R}^N$ of the form

$$\left\{x \in \mathbb{R}^N \mid \exists H \geq 1, \text{ matrix } A \in \mathbb{R}^N \times \mathbb{R}^H \text{ and vector } b \in \mathbb{R}^H \text{ s.t. } xA \leq b\right\}.$$

A polytope of $\mathbb{R}^N$ is a bounded convex polyhedral of $\mathbb{R}^N$; see for instance [84]. The set $\mathcal{A}_l^*$ of nonnegative solutions to system σ_l is then a polytope. Let us define $T_l^{-1}(\mathcal{A}_l^*)$ by

$$T_l^{-1}(\mathcal{A}_l^*) = \{T_l^{-1}(x), \, x \in \mathcal{A}_l^*\}.$$

Observe that the set $T_l^{-1}(\mathcal{A}_l^*)$ is also a polytope.

LEMMA 4.1.21.   *The set $\mathcal{A}^*$ is the convex hull generated from the sets $T_l^{-1}(\mathcal{A}_l^*)$, i.e.*

$$\mathcal{A}^* = \sum_{l \in F} \lambda_l T_l^{-1}(\mathcal{A}_l^*), \text{ where } \lambda_l \geq 0 \quad \text{and} \quad \sum_{l \in F} \lambda_l = 1.$$

In particular, $\mathcal{A}^*$ is also a polytope.

Proof.   Let $x_1 \in \mathcal{A}_1^*, x_2 \in \mathcal{A}_2^*, \ldots, x_M \in \mathcal{A}_M^*$. Let $\lambda_1, \ldots, \lambda_M$ be M nonnegative real numbers such that $\lambda_1 + \cdots + \lambda_M = 1$. By definition of $\mathcal{A}^*$, it is clear that $x = \sum_{l \in F} \lambda_l T_l^{-1}(x_l) \in \mathcal{A}^*$, since $\forall l \in F$, $x^{B(l)} = T_l^{-1}(x_l)$.

Conversely, let $x \in \mathcal{A}^*$. From Lemma 4.1.2, vector x can be written in a unique way as

$$x = \sum_{l \in F} \lambda_l x^{B(l)} \quad \text{with } \lambda_l = x_{B(l)} \mathbb{1}$$

$$= \sum_{l \in F} \lambda_l T_l^{-1}(T_l(x^{B(l)})).$$

The result follows since the sets $\mathcal{A}_l^*$ are polytopes.   ∎

A first important result concerning weak lumpability is given by the following theorem.

THEOREM 4.1.22. $\mathcal{A}^*$ *is stable by* $P \Longleftrightarrow \mathcal{A}_M = \mathcal{A}^*$.

Proof.   From Lemma 4.1.14, the set $\mathcal{A}_M$ is stable by P, which means that the implication from the right to the left is true. To prove the converse, observe first that Theorem 4.1.11 can be written as

$Y = \text{agg}(\alpha, P, \mathcal{B})$ is a homogeneous Markov chain $\Longleftrightarrow \forall l \in F$, $T_l(\beta)$ is a solution to σ_l for every $\beta \in \mathcal{A}(\alpha, B(l))$.

By taking in particular, $\beta = \alpha^{B(l)}$, we obtain that $\mathcal{A}_\mathcal{M} \subseteq \mathcal{A}^*$.

Let $\alpha \in \mathcal{A}^*$ and suppose that $\mathcal{A}^*$ is stable by P. Consider a sequence $(C_1, \ldots, B(l))$ possible for α. We then have $\alpha^{C_1} \in \mathcal{A}^*$ by definition of $\mathcal{A}^*$ and $\alpha^{C_1}P \in \mathcal{A}^*$ by hypothesis. From the definition of $\mathcal{A}(\alpha, B(l))$, we obtain by induction that $\mathcal{A}(\alpha, B(l)) \subseteq \mathcal{A}^*$ for every $l \in F$, which is equivalent to the fact that $Y = \text{agg}(\alpha, P, \mathcal{B})$ is a homogeneous Markov chain. So, $\alpha \in \mathcal{A}_\mathcal{M}$. ∎

To obtain a characterization of the set $\mathcal{A}_\mathcal{M}$, we introduce the following subsets of $\mathcal{A}$. For every $j \geq 1$, we define the set $\mathcal{A}^j$ as

$$\mathcal{A}^j = \{\alpha \in \mathcal{A}^* \mid \forall \beta = f(\alpha, B(i_1)), \ldots, B(i_k)) \text{ with } 1 \leq k \leq j,\ \beta \in \mathcal{A}^*\}.$$

A first remark about these sets is that

$$\mathcal{A}^j = \{\alpha \in \mathcal{A} \mid \forall \beta = f(\alpha, B(i_1)), \ldots, B(i_k)) \text{ with } 1 \leq k \leq j,\ \beta \in \mathcal{A}^*\}.$$

By definition of $\mathcal{A}^*$ and since $\alpha^{B(l)} = f(\alpha, B(l))$, we have

$$\mathcal{A}^1 = \{\alpha \in \mathcal{A}^* \mid \forall l \in F \text{ such that } \alpha_{B(l)} \neq 0,\ \alpha^{B(l)} \in \mathcal{A}^*\} = \mathcal{A}^*$$

and, by definition of $\mathcal{A}^j$, we have

$$\forall j \geq 1,\ \mathcal{A}^{j+1} \subseteq \mathcal{A}^j.$$

The following theorem gives a first expression of the set $\mathcal{A}_\mathcal{M}$ by means of the sets $\mathcal{A}^j$.

THEOREM 4.1.23.

$$\mathcal{A}_\mathcal{M} = \bigcap_{j \geq 1} \mathcal{A}^j.$$

Proof. As in the proof of Theorem 4.1.22, we use the following equivalence, which is a rewriting of Theorem 4.1.11,

$Y = \text{agg}(\alpha, P, \mathcal{B})$ is a homogeneous Markov chain $\Longleftrightarrow \forall l \in F$, $T_l(\beta)$ is a solution to σ_l for every $\beta \in \mathcal{A}(\alpha, B(l))$.

This equivalence can also be written as

$$\mathcal{A}_\mathcal{M} = \{\alpha \in \mathcal{A}^* \mid \forall l \in F,\ \mathcal{A}(\alpha, B(l)) \subseteq \mathcal{A}^*\}.$$

It follows, by definition of $\mathcal{A}^j$, that $\forall j \geq 1$, $\mathcal{A}_\mathcal{M} \subseteq \mathcal{A}^j$.

Conversely, if $\forall j \geq 1$, $\alpha \in \mathcal{A}^j$ then, by definition, $\forall l \in F$, $\mathcal{A}(\alpha, B(l)) \subseteq \mathcal{A}^*$, which means that $\alpha \in \mathcal{A}_\mathcal{M}$. ∎

An evident necessary condition to have weak lumpability, i.e. to have $\mathcal{A}_{\mathcal{M}} \neq \emptyset$, is that $\forall j \geq 1$, $\mathcal{A}^j \neq \emptyset$.

We now give some properties of the sets $\mathcal{A}^j$. We have seen that $\mathcal{A}^1 = \mathcal{A}^*$ and $\mathcal{A}^{j+1} \subseteq \mathcal{A}^j$, for every $j \geq 1$. The following lemma gives an expression of the set $\mathcal{A}^{j+1}$ as a function of $\mathcal{A}^j$.

LEMMA 4.1.24.   *For every $j \geq 1$,*

$$\mathcal{A}^{j+1} = \{\alpha \in \mathcal{A}^j \mid \alpha^{B(l)} P \in \mathcal{A}^j \text{ for every } l \in F \text{ such that } \alpha_{B(l)} \neq 0\}.$$

Proof.   Let $\alpha \in \mathcal{A}^{j+1}$, let $l \in F$ such that $\alpha_{B(l)} \neq 0$ and define the vector β by $\beta = \alpha^{B(l)} P$. If $(B(i_1), \ldots, B(i_k))$ is a possible sequence for β with $1 \leq k \leq j$, then

$$f(\beta, B(i_1), \ldots, B(i_k)) = f(\alpha, B(l), B(i_1), \ldots, B(i_k)) \in \mathcal{A}^*$$

because $\alpha \in \mathcal{A}^{j+1}$ and $k \leq j$. Thus, the vector $\beta = \alpha^{B(l)} P$ belongs to $\mathcal{A}^j$.

Conversely, let $\alpha \in \mathcal{A}^j$ such that $\alpha^{B(l)} P \in \mathcal{A}^j$ for every $l \in F$ such that $\alpha_{B(l)} \neq 0$, and let $(B(i_1), \ldots, B(i_k))$ be a possible sequence for α with $1 \leq k \leq j + 1$. We have

$$f(\alpha, B(i_1), \ldots, B(i_k)) = f(\alpha^{B(i_1)} P, B(i_2), \ldots, B(i_k)) \in \mathcal{A}^*,$$

since by hypothesis, $\alpha^{B(i_1)} P \in \mathcal{A}^j$ and $k \leq j + 1$.   ∎

Using this result, we prove the convexity of the sets $\mathcal{A}^j$.

LEMMA 4.1.25.   *For every $j \geq 1$, the set $\mathcal{A}^j$ is a convex set.*

Proof.   We prove this property by recurrence on integer j. We have proved in Lemma 4.1.21 the convexity of the set $\mathcal{A}^1$. Assume that $\mathcal{A}^j$ is a convex set. Let α and β be two elements of $\mathcal{A}^{j+1}$ and let λ and μ be nonnegative real numbers such that $\lambda + \mu = 1$. We define $\gamma = \lambda \alpha + \mu \beta$. By the recurrence hypothesis and since $\mathcal{A}^{j+1} \subseteq \mathcal{A}^j$, we have $\gamma \in \mathcal{A}^j$. Then, for every $l \in F$ such that $\gamma_{B(l)} \neq 0$, it is clear, by applying Lemma 4.1.3, that $\gamma^{B(l)} P \in \mathcal{A}^j$, which implies that $\gamma \in \mathcal{A}^{j+1}$ from Lemma 4.1.24.   ∎

LEMMA 4.1.26.   *For every $j \geq 1$, we have*

$$\alpha \in \mathcal{A}^j \iff \alpha^{B(l)} \in \mathcal{A}^j \text{ for every } l \in F \text{ such that } \alpha_{B(l)} \neq 0.$$

Proof.   Let $\alpha \in \mathcal{A}^j$ and $l \in F$ such that $\alpha_{B(l)} \neq 0$. For every $k \leq j$ and every sequence $(B(l), B(i_2), \ldots, B(i_k))$ possible for α, we have

$$f(\alpha^{B(l)}, B(l), B(i_2), \ldots, B(i_k)) = f(\alpha, B(l), B(i_2), \ldots, B(i_k)) \in \mathcal{A}^*,$$

since $k \leq j$ and $\alpha \in \mathcal{A}^j$. So, $\alpha^{B(l)} \in \mathcal{A}^j$.

The converse is immediate, since every element $\alpha \in \mathcal{A}$ can be written in a unique way, from Lemma 4.1.2, as

$$\alpha = \sum_{l \in F} \alpha_{B(l)} \mathbb{1} \alpha^{B(l)},$$

and since, from Lemma 4.1.25, $\mathcal{A}^j$ is a convex set. ∎

A sufficient condition to have weak lumpability, i.e. to have $\mathcal{A}_\mathcal{M} \neq \emptyset$, is given by the following result.

COROLLARY 4.1.27. *If $\exists j \geq 1$ such that $\mathcal{A}^{j+1} = \mathcal{A}^j$ then $\mathcal{A}_\mathcal{M} = \mathcal{A}^{j+k}$, $\forall k \geq 0$.*

Proof. If $\mathcal{A}^j = \emptyset$, the proof is trivial. Let $\mathcal{U} = \mathcal{A}^j = \mathcal{A}^{j+1}$ and $\alpha \in \mathcal{U}$. From Lemma 4.1.24, we have

$$\alpha \in \mathcal{U} \Longrightarrow \alpha^{B(l)}P \in \mathcal{U}, \; \forall l \in F \text{ such that } \alpha_{B(l)} \neq 0.$$

By recursively applying this implication, we obtain that if $\alpha \in \mathcal{U}$ then $\forall l \in F$, $\mathcal{A}(\alpha, B(l)) \subseteq \mathcal{U}$, which implies that $\alpha \in \mathcal{A}_\mathcal{M}$. Finally, Theorem 4.1.23 leads to the result. ∎

Let us give now a generalization of Theorem 4.1.22.

THEOREM 4.1.28. *$\mathcal{A}^j$ is stable by $P \Longleftrightarrow \mathcal{A}_\mathcal{M} = \mathcal{A}^j$.*

Proof. From Lemma 4.1.14, the set $\mathcal{A}_\mathcal{M}$ is stable by P, which means that the implication from the right to the left is true. To prove the converse, observe again that Theorem 4.1.11 can be written as

$Y = \text{agg}(\alpha, P, \mathcal{B})$ is a homogeneous Markov chain $\Longleftrightarrow \forall l \in F$, $T_l(\beta)$ is a solution to σ_l for every $\beta \in \mathcal{A}(\alpha, B(l))$.

Let $\alpha \in \mathcal{A}^j$ and suppose that $\mathcal{A}^j$ is stable by P. Consider a sequence $(C_1, \ldots, B(l))$ possible for α. We then have, from Lemma 4.1.26, that $\alpha^{C_1} \in \mathcal{A}^j$ and $\alpha^{C_1}P \in \mathcal{A}^j$ by hypothesis. From the definition of $\mathcal{A}(\alpha, B(l))$, we obtain by induction that $\mathcal{A}(\alpha, B(l)) \subseteq \mathcal{A}^j$ for every $l \in F$, which is equivalent to the fact that $Y = \text{agg}(\alpha, P, \mathcal{B})$ is a homogeneous Markov chain. So, $\alpha \in \mathcal{A}_\mathcal{M}$. ∎

Let us now introduce the following notation. For every $j \geq 1$ and for every $l \in F$,

- $\mathcal{A}_l^j = \{T_l(\alpha^{B(l)}) \mid \alpha \in \mathcal{A}^j \text{ and } \alpha_{B(l)} \neq 0\}$,
- $T_l^{-1}(\mathcal{A}_l^j) = \{T_l^{-1}(x) \mid x \in \mathcal{A}_l^j\}$.

We deduce from these definitions that $\forall j \geq 1$, $\forall l \in F$,

- $\mathcal{A}_l^* = \mathcal{A}_l^1$,
- $\mathcal{A}_l^{j+1} \subseteq \mathcal{A}_l^j$,
- $T_l^{-1}(\mathcal{A}_l^j) = \{\alpha^{B(l)} \mid \alpha \in \mathcal{A}^j \text{ and } \alpha_{B(l)} \neq 0\} \subseteq \mathcal{A}^j$.

We then have the following recurrence on the sequence $(\mathcal{A}_l^j)_{j \geq 1}$.

LEMMA 4.1.29. *For every $l \in F$ and $j \geq 1$,*

$$\mathcal{A}_l^{j+1} = \{x \in \mathcal{A}_l^j \mid (T_l^{-1}(x))P \in \mathcal{A}^j\}.$$

Proof. Let $x \in A_l^{j+1}$. By definition, there exists $\alpha \in A^{j+1}$, with $\alpha_{B(l)} \neq 0$, such that $x = T_l(\alpha^{B(l)})$. So, we have $x \in A_l^j$ and $(T_l^{-1}(x))P = (T_l^{-1}(T_l(\alpha^{B(l)})))P = \alpha^{B(l)}P$. Let $k \leq j$ and $(B(i_1), \ldots, B(i_k))$ be a possible sequence for $\alpha^{B(l)}P$. We have

$$f(\alpha^{B(l)}P, B(i_1), \ldots, B(i_k)) = f(\alpha, B(l), B(i_1), \ldots, B(i_k)) \in A^*$$

since $k \leq j$ and $\alpha \in A^{j+1}$. So, $\alpha^{B(l)}P \in A^j$.

Conversely, let $x \in A_l^j$ such that $(T_l^{-1}(x))P \in A^j$. The vector x belongs to A_l^j, so there exists $\alpha \in A^j$, with $\alpha_{B(l)} \neq 0$, such that $x = T_l(\alpha^{B(l)})$. From Lemma 4.1.26, we can choose $\alpha = T_l^{-1}(x)$. We thus obtain $(T_l^{-1}(x))P = \alpha P = \alpha^{B(l)}P \in A^j$ which means, from Lemma 4.1.24, that $\alpha \in A^{j+1}$ and thus that $x \in A_l^{j+1}$. ∎

LEMMA 4.1.30. *For every $j \geq 1$, the set A^j is the convex hull generated by the sets $T_l^{-1}(A_l^j)$, $l \in F$, i.e.*

$$A^j = \sum_{l \in F} \lambda_l T_l^{-1}(A_l^j), \text{ where } \lambda_l \geq 0 \quad \text{and} \quad \sum_{l \in F} \lambda_l = 1.$$

Proof. Let $x_1 \in A_1^j, x_2 \in A_2^j, \ldots, x_M \in A_M^j$. Let $\lambda_1, \ldots, \lambda_M$ be M nonnegative real numbers such that $\lambda_1 + \cdots + \lambda_M = 1$. We have seen that $T_l^{-1}(A_l^j) \subseteq A^j$ for every $l \in F$ and, from Lemma 4.1.25, that A^j is a convex set. So,

$$x = \sum_{l \in F} \lambda_l T_l^{-1}(x_l) \in A^j.$$

Conversely, let $x \in A^j$. The vector x can be written, from Lemma 4.1.2, in a unique way as

$$x = \sum_{l \in F, x_{B(l)} \neq 0} \lambda_l x^{B(l)} \text{ where } \lambda_l = x_{B(l)} \mathbb{1}.$$

The result then follows by applying Lemma 4.1.26. ∎

LEMMA 4.1.31. *For every $j \geq 1$, for every $l \in F$, A_l^j and A^j are polytopes.*

Proof. The proof is by induction. We have seen in Lemma 4.1.21 that the result is true for $j = 1$. We have also seen in Lemma 4.1.29 that

$$A_l^{j+1} = \{x \in A_l^j \mid (T_l^{-1}(x))P \in A^j\}.$$

Assume the result is true until order j. By Lemma 4.1.30, A^j is a polytope. Let $l \in F$ and let Φ_l be the following application:

$$\Phi_l: \quad \mathbb{R}^{n(l)} \longrightarrow \quad \mathbb{R}^N$$
$$x \longmapsto (T_l^{-1}(x))P.$$

Φ_l is a linear application and so the set $\Phi_l^{-1}(\mathcal{A}^j) = \{x \in \mathbb{R}^{n(l)} \mid (T_l^{-1}(x))P \in \mathcal{A}^j\}$ is a polytope as shown in [84, pp. 18 and 174]. Finally, $\mathcal{A}_l^{j+1} = \mathcal{A}_l^j \cap \Phi_l^{-1}(\mathcal{A}^j)$ is also a polytope by the recurrence hypothesis. We then deduce from Lemma 4.1.30 that for every $j \geq 1$, the set $\mathcal{A}^j$ is a polytope. ∎

We now give the final characterization of the set, $\mathcal{A}_\mathcal{M}$. This characterization has the form of an algorithm which determines the set $\mathcal{A}_\mathcal{M}$ after a finite number of iterations. More precisely, we prove that $\mathcal{A}_\mathcal{M} = \mathcal{A}^N$, where N is the number of states of the Markov chain X. To do this, we need the following technical lemmas.

LEMMA 4.1.32. *Let α be a distribution of $\mathcal{A}^{j+1}$ and let β be a distribution of $\mathcal{A}^j$ not belonging to $\mathcal{A}^{j+1}$. Then there is no point of the segment $(\alpha, \beta]$ belonging to $\mathcal{A}^{j+1}$.*

Proof. Let γ be a point of the segment $(\alpha, \beta]$, that is, let $\lambda \in [0, 1)$ such that

$$\gamma = \lambda\alpha + (1 - \lambda)\beta.$$

Assume that $\gamma \in \mathcal{A}^{j+1}$. We show that we must necessarily have $\beta \in \mathcal{A}^{j+1}$, which contradicts the hypothesis. To prove that $\beta \in \mathcal{A}^{j+1}$, it is sufficient, from Lemma 4.1.24, to prove that $\forall l \in F$ such that $\beta_{B(l)} \neq 0$, $\beta^{B(l)}P \in \mathcal{A}^j$, or equivalently, that $\forall l \in F$ such that $\beta_{B(l)} \neq 0$ and $\forall$ sequence s possible for $\beta^{B(l)}P$ whose length is less than or equal to j, $f(\beta^{B(l)}P, s) \in \mathcal{A}^1$. Let $l \in F$ such that $\beta_{B(l)} \neq 0$. This implies, in particular, that $\gamma_{B(l)} \neq 0$. Let s be a possible sequence for $\beta^{B(l)}P$ whose length is less than or equal to j.

If $\alpha_{B(l)} = 0$ then $\gamma^{B(l)} = \beta^{B(l)}$, and so $f(\beta^{B(l)}P, s) = f(\gamma^{B(l)}P, s) \in \mathcal{A}^1$.

If $\alpha_{B(l)} \neq 0$ and if s is not possible for $\alpha^{B(l)}P$ then, from Lemma 4.1.7, we have $f(\beta^{B(l)}P, s) = f(\gamma^{B(l)}P, s) \in \mathcal{A}^1$.

Finally, if $\alpha_{B(l)} \neq 0$ and if s is possible for $\alpha^{B(l)}P$ then s is also possible for $\gamma^{B(l)}P$ and then $f(\gamma^{B(l)}P, s) \in \mathcal{A}^1$. Applying Lemma 4.1.8 we have

$$f(\gamma^{B(l)}P, s) = f(\gamma, B(l), s)$$
$$= \lambda \frac{K_\alpha}{K} f(\alpha, B(l), s) + (1 - \lambda)\frac{K_\beta}{K}f(\beta, B(l), s)$$
$$= \lambda \frac{K_\alpha}{K} f(\alpha^{B(l)}P, s) + (1 - \lambda)\frac{K_\beta}{K}f(\beta^{B(l)}P, s),$$

where K_α and K_β are given in Lemma 4.1.8 and $K = \lambda K_\alpha + (1 - \lambda)K_\beta$. But now, $f(\gamma^{B(l)}P, s) \in \mathcal{A}^1$ and $f(\alpha^{B(l)}P, s) \in \mathcal{A}^1$. If we denote by $B(m)$ the last element of sequence s, we have

$$(1 - \lambda)\frac{K_\beta}{K}T_m(f(\beta^{B(l)}P, s))\tilde{P}^{(m)}$$
$$= T_m(f(\gamma^{B(l)}P, s))\tilde{P}^{(m)} - \lambda\frac{K_\alpha}{K}T_m(f(\alpha^{B(l)}P, s))\tilde{P}^{(m)}$$

$$= \hat{P}^{(m)} - \lambda \frac{K_\alpha}{K} \hat{P}^{(m)}$$

$$= (1 - \lambda) \frac{K_\beta}{K} \hat{P}^{(m)}.$$

As $\lambda \neq 1$ and $K_\beta \neq 0$, we obtain $T_m(f(\beta^{B(l)}P, s))\tilde{P}^{(m)} = \hat{P}^{(m)}$ which means that $f(\beta^{B(l)}P, s) \in \mathcal{A}^1$. ∎

Remark. It is easy to check that Lemma 4.1.32 holds for $j = 0$ if we define $\mathcal{A}^0 = \mathcal{A}$.

Let α and β be two distinct distributions of $\mathcal{A}^j$. We denote by $D(\alpha, \beta)$ the whole line of $\mathbb{R}^N$ containing α and β.

LEMMA 4.1.33.   *If α and β are two distinct points of $\mathcal{A}^{j+1}$ then $D(\alpha, \beta) \cap \mathcal{A}^j \subseteq \mathcal{A}^{j+1}$.*

Proof. We have seen that $\mathcal{A}^{j+1}$ is a convex set, so $[\alpha, \beta] \subseteq \mathcal{A}^{j+1}$. Suppose that there exists a point $\gamma \notin [\alpha, \beta]$ such that $\gamma \in D(\alpha, \beta) \cap \mathcal{A}^j$ and $\gamma \notin \mathcal{A}^{j+1}$. Assume that the point $\beta \in (\alpha, \gamma]$. From Lemma 4.1.32, we conclude that there is no point of $(\alpha, \gamma]$ in $\mathcal{A}^{j+1}$. But by hypothesis, $\beta \in \mathcal{A}^{j+1}$, which is a contradiction. ∎

We denote by intrel($\mathcal{C}$) the interior of the set $\mathcal{C}$ relative to the smallest affine subspace S containing $\mathcal{C}$ and, in the same way, we denote by boundrel($\mathcal{C}$) the boundary of $\mathcal{C}$ relative to S.

LEMMA 4.1.34.   *For every $j \geq 1$, if $\mathcal{A}^{j+1}$ is not reduced to a single point, then*

$$\text{boundrel}(\mathcal{A}^{j+1}) \subseteq \text{boundrel}(\mathcal{A}^j).$$

Proof. The result is trivial if $\mathcal{A}^{j+1} = \mathcal{A}^j$ or if $\mathcal{A}^{j+1} = \emptyset$. Assume that $\emptyset \neq \mathcal{A}^{j+1} \subset \mathcal{A}^j$. Let $\alpha \in \text{boundrel}(\mathcal{A}^{j+1})$ and suppose that $\alpha \notin \text{boundrel}(\mathcal{A}^j)$. By hypothesis, there exists a point $\beta \in \mathcal{A}^{j+1}$. Applying Lemma 4.1.32, we deduce that $D(\alpha, \beta) \cap \mathcal{A}^j \subseteq \mathcal{A}^{j+1}$, so $D(\alpha, \beta) \cap \mathcal{A}^j$ is a segment $[a, b]$ with $a, b \in \text{boundrel}(\mathcal{A}^j) \cap \text{boundrel}(\mathcal{A}^{j+1})$. It necessarily follows that $\alpha \in (a, b)$, which contradicts the hypothesis. ∎

LEMMA 4.1.35.   *Let α be a point of $\mathcal{A}^{j+1}$ and β be a point of $\mathcal{A}^j$ not in $\mathcal{A}^{j+1}$. Then there is no point of the line $D(\alpha, \beta)$, different to α, belonging to $\mathcal{A}^{j+1}$.*

Proof. If such a point γ exists, then from Lemma 4.1.33, we would necessarily have $\beta \in \mathcal{A}^{j+1}$ which contradicts the hypothesis. ∎

The dimension of a convex set $\mathcal{C}$ of $\mathbb{R}^N$ is the dimension of the smallest affine subset containing $\mathcal{C}$. We shall denote it by $\dim(\mathcal{C})$. We then have the following lemma which allows us to determine the set $\mathcal{A}_\mathcal{M}$ in a finite number of iterations.

LEMMA 4.1.36.   *Let $j \geq 1$ such that $\mathcal{A}^{j+1} \neq \emptyset$. If $\mathcal{A}^j \neq \mathcal{A}^{j+1}$ then $\dim(\mathcal{A}^{j+1}) < \dim(\mathcal{A}^j)$.*

Proof. We have $\dim(\mathcal{A}^{j+1}) \leq \dim(\mathcal{A}^j)$ since $\mathcal{A}^{j+1} \subseteq \mathcal{A}^j$. Suppose that these two dimensions are equal. This common dimension cannot be equal to zero by hypothesis and by the properties of the sets, $\mathcal{A}^j$. So, there exists a point $x \in \text{intrel}(\mathcal{A}^{j+1})$. Let y be a point of

$\mathcal{A}^j \setminus \mathcal{A}^{j+1}$. Given that $\dim(\mathcal{A}^{j+1}) = \dim(\mathcal{A}^j)$, there exists $z \in \mathrm{intrel}(\mathcal{A}^{j+1})$ such that $z \neq x$ and $z \in D(x,y)$. From Lemma 4.1.33, we have

$$[D(x,z) \cap \mathcal{A}^j] \subseteq \mathcal{A}^{j+1},$$

which is in contradiction with the fact that $y \notin \mathcal{A}^{j+1}$. ∎

Remark. The previous results also hold if we consider the subsets $\mathcal{A}^j_l$ instead of the subsets $\mathcal{A}^j$. Let us give the corresponding properties without proof.

- Let x be a point of $\mathcal{A}^{j+1}_l$ and y a point of $\mathcal{A}^j_l$ not in $\mathcal{A}^{j+1}_l$, then there is no point of the line $D(x,y)$ different to x and belonging to $\mathcal{A}^{j+1}_l$.
- If x and y are two different points of $\mathcal{A}^{j+1}_l$, then $D(x,y) \cap \mathcal{A}^j_l \subseteq \mathcal{A}^{j+1}_l$.
- If $\mathcal{A}^{j+1}_l$ is not reduced to a single point, then $\mathrm{boundrel}(\mathcal{A}^{j+1}_l) \subseteq \mathrm{boundrel}(\mathcal{A}^j_l)$.
- If $\mathcal{A}^{j+1}_l \neq \emptyset$ and $\mathcal{A}^{j+1}_l \neq \mathcal{A}^j_l$, then $\dim(\mathcal{A}^{j+1}_l) < \dim(\mathcal{A}^j_l)$.

We are now ready to prove the main result of the discrete-time case. Recall that N denotes the number of states of Markov chain X.

THEOREM 4.1.37. $\mathcal{A}_\mathcal{M} = \mathcal{A}^N$.

Proof. If $\mathcal{A}^1 = \mathcal{A}$, then since $\mathcal{A}$ is stable by P, we have $\mathcal{A}_\mathcal{M} = \mathcal{A}^1$ from Lemma 4.1.22 and we also have $\mathcal{A}_\mathcal{M} = \mathcal{A}^j$ for every j, from Lemma 4.1.23.

Otherwise, consider the sequence $\mathcal{A}^1, \ldots, \mathcal{A}^N$. If two consecutive elements of this sequence are equal, from Corollary 4.1.27, we have $\mathcal{A}_\mathcal{M} = \mathcal{A}^N$. If all its elements are different, then from Lemma 4.1.36, the dimensions of these sets are strictly decreasing. We thus have $\dim(\mathcal{A}^1) < \dim(\mathcal{A}) = N - 1$ and so $\dim(\mathcal{A}^j) < N - j$ for $j = 1, 2, \ldots, N-1$, and $\mathcal{A}^N = \emptyset$. Theorem 4.1.23 then completes the proof. ∎

An algorithm to determine $\mathcal{A}_\mathcal{M}$

To determine the set $\mathcal{A}_\mathcal{M}$, we may proceed as follows. Assume that the chain X is not strongly lumpable with respect to the partition $\mathcal{B}$. The first step consists of calculating the set $\mathcal{A}^1$ and determining if it is stable by P. If $\mathcal{A}^1$ is stable by P, Theorem 4.1.22 tells us that $\mathcal{A}_\mathcal{M} = \mathcal{A}^1$. If $\mathcal{A}^1$ is not stable by P, we first check if $\forall l \in F$ the vector $\pi^{B(l)}P$ is in $\mathcal{A}^1$. If $\exists l \in F$ such that $\pi^{B(l)}P \notin \mathcal{A}^1$, then $\pi \notin \mathcal{A}^2$ and $\mathcal{A}_\mathcal{M} = \emptyset$, which is a consequence of Theorems 4.1.16 and 4.1.23. If $\pi \in \mathcal{A}^2$, then we calculate this set and we repeat the previous operations.

We denote by Ψ the following procedure.

$$\Psi : \mathcal{P}(A) \longrightarrow \mathcal{P}(A)$$
$$\mathcal{U} \subseteq A \longmapsto \{\alpha \in \mathcal{U} \mid \forall l \in F \text{ such that } \alpha_{B(l)} \neq 0 \text{ we have } \alpha^{B(l)}P \in \mathcal{U}\},$$

where $\mathcal{P}(A)$ denotes the set of all subsets of A. With this notation, we have, from Lemma 4.1.24, $\mathcal{A}^{j+1} = \Psi(\mathcal{A}^j)$. The algorithm then has the following form.

if X is strongly lumpable **then** $\mathcal{A}_\mathcal{M} = \mathcal{A}$; **stop**
else $\mathcal{U} := \mathcal{A}^1$;
     loop
         if $\mathcal{U}$ is stable by P **goto** notempty **endif** ;
         if $\exists l \in F$ such that $\pi^{B(l)}P \notin \mathcal{U}$ **goto** empty **endif** ;
         $\mathcal{U} := \Psi(\mathcal{U})$
     endloop
     outputs
         output notempty : $\mathcal{A}_\mathcal{M} := \mathcal{U}$
         output empty : $\mathcal{A}_\mathcal{M} := \emptyset$
     endoutputs
endif

If we are interested in the minimal representation of the polytope $\mathcal{A}_\mathcal{M}$, that is the set of its vertices, then the incremental Chernikova's algorithm [64] is well suited to compute these vertices with the above recursive definition of $\mathcal{A}^N$.

4.2   State aggregation in absorbing DTMC

We consider a homogeneous absorbing Markov chain X evolving in discrete time.

- The partition $\mathcal{B} = \{B(0), B(1), \ldots, B(M)\}$ of the state space $S = \{1, \ldots, N\}$ is such that $B(0)$ contains only the absorbing states. The cardinality of the class $B(l)$ is denoted by $n(l)$.
- For simplicity, we assume, without any loss of generality, that the class $B(0)$ contains only one absorbing state, which is denoted by 1. We denote by T the set $\{2, \ldots, N\}$ and we suppose that the states of T are all transient.
- With the given process X, we associate the aggregated process Y with state space $F = \{0, 1, 2, \ldots, M\}$ defined by

$$Y_n = l \Longrightarrow X_n \in B(l), \quad \text{for any } n \in \mathbb{N}.$$

- If v is a vector, then the notation $v > 0$ means that each component of v is positive. The column vector e_i denotes the ith vector of the canonical basis of $\mathbb{R}^N$. We denote by I the identity matrix and by $\text{diag}(v)$ the diagonal matrix with entries v_i, the dimensions being defined by the context.
- We denote by $\mathcal{A}^T$ the subset of distributions of $\mathcal{A}$ with support T, i.e.

$$\mathcal{A}^T = \left\{ \alpha \in \mathcal{A} \,\middle|\, \sum_{i \in T} \alpha_i = 1 \right\}.$$

4.2.1 Quasi-stationary distribution

The stationary distribution associated with the irreducible case is a central notion for weak lumpability characterization. In the absorbing case, this distribution is trivial and without interest. The vector playing the equivalent role in the context of absorbing chains is the quasi-stationary distribution [26], [101]. In this section, we recall the definitions and the main properties of this distribution. In particular we discuss the periodic case which is not considered in [26], [101].

The transition probability matrix, P, can be decomposed as

$$P = \begin{pmatrix} 1 & 0 \\ q & Q \end{pmatrix},$$

where matrix Q is irreducible. In the same way, the initial probability distribution, α, can be written as $\alpha = (\alpha_{\{1\}}, \alpha_T)$. For any $n \in \mathbb{N}^*$, we define the probability vector, $\pi(n) = (\pi_j(n), j \in T)$ by

$$\pi_j(n) = \mathbb{P}_\alpha\{X_n = j \mid X_n \in T\} = \frac{\alpha_T Q^n e_j}{\alpha_T Q^n \mathbb{1}}, \quad \text{if } Q \text{ is aperiodic,} \tag{4.5}$$

$$\pi_j(n) = \frac{\sum_{k=1}^n \alpha_T r^k Q^k e_j}{\sum_{k=1}^n \alpha_T r^k Q^k \mathbb{1}}, \quad \text{if } Q \text{ is periodic,} \tag{4.6}$$

where $r = 1/\rho$, ρ is the spectral radius of Q and $v > 0$ is the unique left probability eigenvector of Q associated with ρ. Let us denote by w the unique right probability eigenvector of Q associated with ρ. Denoting $W = \text{diag}(w)$, we define

$$\bar{Q} = \frac{1}{\rho} W^{-1} Q W. \tag{4.7}$$

Matrix $\bar{Q}$ is stochastic and irreducible. It is aperiodic (resp. periodic) if and only if matrix Q is aperiodic (resp. periodic). Putting expression (4.7) in definitions (4.5) and (4.6) and using standard arguments about the asymptotic behavior of Markov chains seen in Chapter 2, we obtain the basic property of the *quasi-stationary distribution* v of the absorbing Markov chain X:

$$\lim_{n \to \infty} \pi(n) = v > 0, \quad \text{for any initial distribution } \alpha \in \mathcal{A}.$$

See Subsection 1.2.2 for an example.

4.2.2 Weak lumpability

For any $\alpha \in \mathcal{A}$ such that $\alpha_T \mathbb{1} \neq 0$ and any $l \in F$, $\mathcal{A}(\alpha, B(l))$ is not empty, which follows from the irreducibility of Y on $F \setminus \{0\}$ and the fact that $q \neq 0$. As in Theorem 4.1.11, a necessary and sufficient condition for Y to be a homogeneous Markov chain can be exhibited after noting that any possible sequence for α whose last element is $B(l)$, $l \neq 0$, never contains $B(0)$ and that for any $\beta \in \mathcal{A}(\alpha, B(0))$, we have $\mathbb{P}_\beta\{X_1 \in B(0)\} = 1$ and $\mathbb{P}_\beta\{X_1 \in B(m)\} = 0$ for $m \neq 0$. We deduce the following result.

THEOREM 4.2.1.   *The chain $Y = \mathrm{agg}(\alpha, P, \mathcal{B})$ is a homogeneous Markov chain if and only if $\forall l \in F \setminus \{0\}$, $\forall m \in F$, the probability $\mathbb{P}_\beta\{X_1 \in B(m)\}$ is the same for every $\beta \in \mathcal{A}(\alpha, B(l))$. This common value is the transition probability of the chain Y to move from state l to state m.*

Let us define $\mathcal{A}_{\mathcal{M}}$ as in the irreducible case, and let $\mathcal{A}^T_{\mathcal{M}}$ denote the subset of $\mathcal{A}_{\mathcal{M}}$ composed of the distributions having support T. We thus have

$$\mathcal{A}_{\mathcal{M}} = \lambda_1\{(1, 0, \ldots, 0)\} + \lambda_T \mathcal{A}^T_{\mathcal{M}},$$

where $\lambda_T, \lambda_1 \geq 0$ and $\lambda_T + \lambda_1 = 1$. Therefore, we restrict the analysis to the set $\mathcal{A}^T_{\mathcal{M}}$. When necessary, we specify the dependency of these sets with respect to matrix P, denoting them by $\mathcal{A}_{\mathcal{M}}(P)$ and $\mathcal{A}^T_{\mathcal{M}}(P)$.

The set $\mathcal{A}^T_{\mathcal{M}}$ has analogous properties to those obtained for $\mathcal{A}_{\mathcal{M}}$ in the irreducible case. The following property can be proved as in Theorem 4.1.14.

If $\alpha \in \mathcal{A}^T_{\mathcal{M}}$, then, for every $n \geq 1$, we have

$$\left(0, \frac{\alpha_T Q^n}{\alpha_T Q^n \mathbb{1}}\right) \in \mathcal{A}^T_{\mathcal{M}} \quad \text{and} \quad \left(0, \frac{\sum_{k=1}^n \alpha_T Q^k r^k}{\sum_{k=1}^n \alpha_T Q^k r^k \mathbb{1}}\right) \in \mathcal{A}^T_{\mathcal{M}}.$$

Using the asymptotic interpretation of the quasi-stationary distribution and Lemma 4.1.15, the previous relations lead to the following result.

$$\text{If } \mathcal{A}^T_{\mathcal{M}} \neq \emptyset \quad \text{then } (0, v) \in \mathcal{A}^T_{\mathcal{M}}.$$

We then obtain, using Theorem 4.2.1, a result similar to Theorem 4.1.16, which tells us that the transition probability matrix, denoted by $\hat{P}$, of the homogeneous Markov chain $Y = \mathrm{agg}(\alpha, P, \mathcal{B})$, is the same for every $\alpha \in \mathcal{A}^T_{\mathcal{M}}$ and is given, for all $l, m \in F$, $l \neq 0$, by

$$\hat{P}_{l,m} = \sum_{i \in B(l)} v_i^{B(l)} P_{i,B(m)}, \hat{P}_{0,m} = 0 \quad \text{and} \quad \hat{P}_{0,0} = 1. \tag{4.8}$$

DEFINITION 4.2.2.   *The family of homogeneous Markov chains, $(., P)$, is weakly lumpable with respect to the partition $\mathcal{B}$ if $\mathcal{A}^T_{\mathcal{M}} \neq \emptyset$. For any $\alpha \in \mathcal{A}_{\mathcal{M}}$, the aggregated chain $Y = \mathrm{agg}(\alpha, P, \mathcal{B})$ is a homogeneous Markov chain and its transition probability matrix $\hat{P}$ is given by (4.8).*

4.2.3 Link with the irreducible case

From the decomposition of the transition probability matrix, P, in Subsection 4.2.1, we define a new matrix R as

$$R = \begin{pmatrix} 0 & v \\ q & Q \end{pmatrix},$$

where v is the quasi-stationary probability vector associated with P. This matrix, R, is the transition probability matrix of an irreducible Markov chain and we have the following theorem.

THEOREM 4.2.3. *If $\mathcal{A}_{\mathcal{M}}(R)$ denotes the set of distributions α such that $Y = \mathrm{agg}(\alpha, R, \mathcal{B})$ is a homogeneous Markov chain, then we have $\mathcal{A}_{\mathcal{M}}(P) = \mathcal{A}_{\mathcal{M}}(R)$. Moreover, the transition probability matrix, $\hat{R}$, of Y is given for every $l, m \in F$, $l \neq 0$, by*

$$\hat{R}_{l,m} = \hat{P}_{l,m}, \quad \hat{R}_{0,m} = v_{B(m)}\mathbb{1} \quad and \quad \hat{R}_{0,0} = 1,$$

where matrix $\hat{P}$ is given by (4.8).

Proof. We denote the function of Definition 4.1.6 by f_P (resp. f_R) when related to matrix P (resp. R). In a similar way, X (resp. $X^{(v)}$) denotes a Markov chain with transition probability matrix P (resp. R). If $\alpha \in \mathcal{A}_{\mathcal{M}}(R)$, then we have by definition of weak lumpability that for any $l, m \in F$ and any $\beta' = f_R(\alpha, C_0, \ldots, B(l))$,

$$\hat{R}_{l,m} = \mathbb{P}_{\beta'}\{X_1^{(v)} \in B(m)\} = \sum_{i \in B(l)} \beta'_i R_{i,B(m)}.$$

We must show that $\alpha \in \mathcal{A}_{\mathcal{M}}(P)$. The construction of matrix R implies that for any $l, m \in F$, $l \neq 0$ and $\beta = f_P(\alpha, C_0, \ldots, B(l))$, we have $\beta = \beta'$

$$\mathbb{P}_{\beta}\{X_1 \in B(m)\} = \sum_{i \in B(l)} \beta'_i R_{i,B(m)}$$

$$= \hat{R}_{l,m}.$$

The characterization condition of Theorem 4.2.1 is then satisfied.

Conversely, suppose that $\alpha \in \mathcal{A}_{\mathcal{M}}(P)$. If $l = 0$, then any vector $\beta' = f_R(\alpha, C_0, \ldots, B(l))$ reduces to $(1, 0, \ldots, 0)$. Therefore, we have

$$\mathbb{P}_{\beta'}\{X_1^{(v)} \in B(m)\} = \begin{cases} ((1,0,\ldots,0)R)_{B(m)}\mathbb{1} = v_{B(m)}\mathbb{1} & \text{if } m \neq 0, \\ 0 & \text{if } m = 0, \end{cases}$$

and this probability, which depends only on m, is therefore $\hat{R}_{0,m}$.

Suppose that the expression of β' does not contain the class $B(0)$. In this case, $\beta' = f_P(\alpha, C_0, \ldots, B(l))$ because only matrix Q is concerned. We deduce that

$$\mathbb{P}_{\beta'}\{X_1^{(v)} \in B(m)\} = \hat{P}_{l,m}.$$

Finally, suppose that $l \neq 0$ and that in the definition of vector β' the set $B(0)$ appears at least once. This vector can be written as

$$\beta' = f_R(\alpha, C_0, \ldots, C_{j-1}, B(0), C_{j+1}, \ldots, C_n, B(l)),$$

where j is the greatest integer between 0 and n such that the sequence $(C_{j+1}, \ldots, C_n, B(l))$ does not contain $B(0)$, with the convention that if $j = n$ then the sequence is reduced to $B(l)$. Using the recursive definition of f, β' can also be expressed as

$$f_R\left(f_R(\alpha, \ldots, C_{j-1}, B(0))R, C_{j+1}, \ldots, C_n, B(l)\right) = f_R\left((0, v), C_{j+1}, \ldots, C_n, B(l)\right).$$

Therefore, we return to the previous situation and the proof is completed.   ∎

This last theorem allows us to use the finite algorithm described for irreducible Markov chains.

4.3   State aggregation in CTMC

Let $X = \{X_t,\ t \in \mathbb{R}^+\}$ be a homogeneous Markov chain with state space S. We show in this section that the continuous-time case can be reduced to the discrete-time one using the uniformization technique described in Chapter 3. This can be done as in [94] where some intermediate steps are given. We give here a more concise proof of this result which holds in the irreducible and absorbing cases. The process X is given by its infinitesimal generator A and we denote by $\mathrm{agg}(\alpha, A, \mathcal{B})$ the aggregated process of X with respect to the partition $\mathcal{B}$. The set of all initial distributions of X leading to a homogeneous Markov chain for $\mathrm{agg}(\alpha, A, \mathcal{B})$ is denoted by $\mathcal{C}_\mathcal{M}$, that is,

$$\mathcal{C}_\mathcal{M} = \{\alpha \mid \mathrm{agg}(\alpha, A, \mathcal{B}) \text{ is a homogeneous Markov chain}\}.$$

We consider a Poisson process $N = \{N_t,\ t \in \mathbb{R}_+\}$ with rate λ, such that

$$\lambda \geq \max\{-A_{i,i},\ i \in S\}.$$

Let $Z = \{Z_n,\ n \in \mathbb{N}\}$ be a discrete-time homogeneous Markov chain, independent of the Poisson process N, on the state space S, and with transition probability matrix P given by $P = I + A/\lambda$. It was shown in Chapter 3 that the two processes X and $\{Z_{N_t},\ t \in \mathbb{R}_+\}$ are stochastically equivalent. Using this property, we prove the following result.

THEOREM 4.3.1.   *Process* $\mathrm{agg}(\alpha, P, \mathcal{B})$ *is a homogeneous Markov chain if and only if* $\mathrm{agg}(\alpha, A, \mathcal{B})$ *is a homogeneous Markov chain. So, we have* $\mathcal{C}_\mathcal{M} = \mathcal{A}_\mathcal{M}$.

Proof.   For all $k \in \mathbb{N}$, $B_0, \ldots, B_k \in \mathcal{B}$, $0 < t_1 < \cdots < t_k$ and $0 < n_1 < \cdots < n_k$, we define

$$F_X(k) = \mathbb{P}_\alpha\{X_{t_k} \in B_k, \ldots, X_{t_1} \in B_1, X_0 \in B_0\},$$

$$F_Z(k) = \mathbb{P}_\alpha\{Z_{n_k} \in B_k, \ldots, Z_{n_1} \in B_1, Z_0 \in B_0\},$$

$$\text{and}\quad F_N(k) = \mathbb{P}\{N_{t_k} = n_k, \ldots, N_{t_1} = n_1\}.$$

Since N is a Poisson process with rate λ, we have

$$F_N(k) = F_N(k-1)\mathbb{P}\{N_{t_k-t_{k-1}} = n_k - n_{k-1}\}. \tag{4.9}$$

As the processes Z and N are independent, we have

$$F_X(k) = \sum_{n_1=0}^{+\infty} \sum_{n_2=n_1}^{+\infty} \cdots \sum_{n_{k-1}=n_{k-2}}^{+\infty} \sum_{n_k=n_{k-1}}^{+\infty} F_Z(k)F_N(k). \tag{4.10}$$

Assume that $\text{agg}(\alpha, P, \mathcal{B})$ is Markov homogeneous. This implies that

$$F_Z(k) = F_Z(k-1)\mathbb{P}_\alpha\{Z_{n_k-n_{k-1}} \in B_k \mid Z_0 \in B_{k-1}\}. \tag{4.11}$$

We have to show that

$$F_X(k) = F_X(k-1)\mathbb{P}_\alpha\{X_{t_k-t_{k-1}} \in B_k \mid X_0 \in B_{k-1}\}.$$

Replacing $F_Z(k)$ and $F_N(k)$ in (4.10) by their respective expressions (4.9) and (4.11), we obtain

$$F_X(k) = \sum_{n_1 \geq 0} \cdots \sum_{n_k \geq n_{k-1}} F_Z(k-1)\mathbb{P}_\alpha\{Z_{n_k-n_{k-1}} \in B_k \mid Z_0 \in B_{k-1}\}$$

$$\times F_N(k-1)\mathbb{P}\{N_{t_k-t_{k-1}} = n_k - n_{k-1}\}$$

$$= \sum_{n_1 \geq 0} \cdots \sum_{n_{k-1} \geq n_{k-2}} F_Z(k-1)F_N(k-1)$$

$$\times \sum_{n_k \geq n_{k-1}} \mathbb{P}_\alpha\{Z_{n_k-n_{k-1}} \in B_k \mid Z_0 \in B_{k-1}\}\mathbb{P}\{N_{t_k-t_{k-1}} = n_k - n_{k-1}\}$$

$$= F_X(k-1) \sum_{n_k \geq n_{k-1}} \mathbb{P}_\alpha\{Z_{n_k-n_{k-1}} \in B_k \mid Z_0 \in B_{k-1}\}\mathbb{P}\{N_{t_k-t_{k-1}} = n_k - n_{k-1}\},$$

i.e.

$$F_X(k) = F_X(k-1)\mathbb{P}_\alpha\{X_{t_k-t_{k-1}} \in B_k \mid X_0 \in B_{k-1}\}. \tag{4.12}$$

Conversely, if Relation (4.12) holds then we can rewrite it using (4.10) and (4.9) as

$$F_X(k) = F_X(k-1) \sum_{n_k \geq n_{k-1}} \mathbb{P}_\alpha\{Z_{n_k-n_{k-1}} \in B_k \mid Z_0 \in B_{k-1}\}\mathbb{P}\{N_{t_k-t_{k-1}} = n_k - n_{k-1}\}$$

$$= \sum_{n_1=0}^{+\infty} \cdots \sum_{n_k=n_{k-1}}^{+\infty} F_N(k)F_Z(k-1)\mathbb{P}_\alpha\{Z_{n_k-n_{k-1}} \in B_k \mid Z_0 \in B_{k-1}\}.$$

Replacing $F_X(k)$ by its expression (4.10), we obtain

$$\sum_{n_1=0}^{+\infty} \cdots \sum_{n_k=n_{k-1}}^{+\infty} F_N(k)\left[F_Z(k) - F_Z(k-1)\mathbb{P}_\alpha\{Z_{n_k-n_{k-1}} \in B_k \mid Z_0 \in B_{k-1}\}\right] = 0.$$

This relation is valid for all real numbers $0 < t_1 < \cdots < t_n$ and since the terms $F_N(k)$ are positive, we deduce that for all $n_k > \cdots > n_1 > 0$,

$$F_Z(k) = F_Z(k-1)\mathbb{P}_\alpha\{Z_{n_k - n_{k-1}} \in B_k \mid Z_0 \in B_{k-1}\},$$

and so $\mathrm{agg}(\alpha, P, \mathcal{B})$ is a homogeneous Markov chain.   ∎

In [94] a different scheme is followed. Before obtaining the main result, it is shown that the transition rate matrix, $\hat{A}$, of $\mathrm{agg}(\alpha, A, \mathcal{B})$ is the same for any α such that $\mathrm{agg}(\alpha, A, \mathcal{B})$ is a homogeneous Markov chain. This can be also deduced from Theorem 4.3.1, which leads to $\hat{P} = I + \hat{A}/\lambda$.

5 Sojourn times in subsets of states

In this chapter we study the successive times a Markov chain, X, spends in a proper subset, B, of its state space, S. We write $S = \{1, 2, \ldots, N\}$, where $N \geq 2$ and, to simplify the notation, $B = \{1, 2, \ldots, L\}$ with $1 \leq L < N$. We derive the distribution of these random variables, and, in the case of irreducible models, their asymptotic behavior. When the model is absorbing, we study the accumulated sojourn time in B before absorption. We also study some relationships between sojourn times and lumping, the subject of Chapter 4.

5.1 Successive sojourn times in irreducible DTMC

In this first section, X is an irreducible DTMC with transition probability matrix P and initial probability distribution α. Let us denote by x its stationary probability distribution, that is, $x = xP$, $x > 0$ and $x\mathbb{1} = 1$, where $\mathbb{1}$ is a column vector of dimension given by the context and with all entries equal to 1.

The partition $\{B, B^c\}$ of S induces a decomposition of P into four submatrices and a decomposition of the initial distribution, α, and the stationary distribution, x, into two subvectors:

$$P = \begin{pmatrix} P_B & P_{BB^c} \\ P_{B^c B} & P_{B^c} \end{pmatrix}, \qquad \alpha = (\alpha_B \quad \alpha_{B^c}), \qquad x = (x_B \quad x_{B^c}).$$

LEMMA 5.1.1. *The matrix $I - P_B$ is invertible.*

Proof. Consider a Markov chain on the state space $B \cup \{0\}$, with transition probability matrix, P', given, for every $i, j \in B$, by

$$
\begin{aligned}
P'(i, j) &= P_B(i, j), \\
P'(0, j) &= 0, \\
P'(i, 0) &= 1 - \sum_{\ell \in B} P_B(i, \ell), \\
P'(0, 0) &= 1.
\end{aligned}
$$

Matrix P being irreducible, this chain has exactly two classes: B, which is transient, and $\{0\}$ which is absorbing. Therefore, for every $i, j \in B$, we have

$$\lim_{k \to \infty} (P_B)^k (i, j) = 0,$$

which implies, using the same argument as in Lemma 2.9.2, that $I - P_B$ is invertible. ∎

In the same way, matrix $I - P_{B^c}$ is invertible.

LEMMA 5.1.2. *The vector x_B is a solution to the linear system in the L-dimensioned vector ϕ: $\phi = \phi U_B$, $\phi > 0$, where $U_B = P_{BB^c}(I - P_{B^c})^{-1}P_{B^cB}(I - P_B)^{-1}$.*

Proof. The result follows immediately from the decomposition of the system, $x = xP$:

$$x_B = x_B P_B + x_{B^c} P_{B^c B},$$

$$x_{B^c} = x_B P_{BB^c} + x_{B^c} P_{B^c},$$

by replacing in the first equation the value of vector x_{B^c} obtained from the second one.

∎

In the same way, we have:

$$x_{B^c} = x_{B^c} U_{B^c}, \quad \text{where } U_{B^c} = P_{B^c B}(I - P_B)^{-1}P_{BB^c}(I - P_{B^c})^{-1}.$$

DEFINITION 5.1.3. *We call "sojourn of X in B" every sequence $X_m, X_{m+1}, \ldots, X_{m+k}$ where $k \geq 1$, $X_m, X_{m+1}, \ldots, X_{m+k-1} \in B$, $X_{m+k} \notin B$ and if $m > 0$, $X_{m-1} \notin B$. This sojourn begins at time m and finishes at time $m + k$; it lasts k, its first state is X_m (the "entering" state) and its last state is X_{m+k-1} (the "exit" state).*

For any $n \geq 1$, we denote by V_n the random variable "state of B in which the nth sojourn of X in B begins". Since X is irreducible, there is an infinite number of sojourns of X in B with probability 1.

By definition, $\{V_n, n \geq 1\}$ is a process living in the state space, B, and it is obviously a homogeneous Markov chain. Let G be the (L, L) transition probability matrix of this chain and v_n its probability distribution at time n, that is $v_n = (\mathbb{P}\{V_n = 1\}, \ldots, \mathbb{P}\{V_n = L\})$. We have $v_n = v_1 G^{n-1}$. The Markov chain $(V_n)_{n \geq 1}$ is thus characterized by G and v_1, which are given in the following theorem.

THEOREM 5.1.4. *Matrix G and vector v_1 are given by the following expressions.*

(i) $G = (I - P_B)^{-1}P_{BB^c}(I - P_{B^c})^{-1}P_{B^c B} = (I - P_B)^{-1}U_B(I - P_B)$.
(ii) $v_1 = \alpha_B + \alpha_{B^c}(I - P_{B^c})^{-1}P_{B^c B}$.

Proof. (i) Let $i \in B^c$ and $j \in B$. Define $H(i,j) = \mathbb{P}\{V_1 = j \mid X_0 = i\}$ and let H be the $(N - L, L)$ matrix with entries $H(i,j)$. Conditioning on the state of the chain at time 1 and using the strong Markov property, we have, for every $i \in B^c$ and $j \in B$,

$$\mathbb{P}\{V_1 = j \mid X_0 = i\} = P(i,j) + \sum_{k \in B^c} P(i,k)\mathbb{P}\{V_1 = j \mid X_1 = k, X_0 = i\}$$

$$= P(i,j) + \sum_{k \in B^c} P(i,k)\mathbb{P}\{V_1 = j \mid X_0 = k\},$$

and for every $i \in B$ and $j \in B$,

$$
\begin{aligned}
G(i,j) &= \mathbb{P}\{V_2 = j \mid V_1 = i\} = \mathbb{P}\{V_2 = j \mid X_0 = i\} \\
&= \sum_{k \in E} P(i,k)\mathbb{P}\{V_2 = j \mid X_0 = i, X_1 = k\} \\
&= \sum_{k \in B} P(i,k)\mathbb{P}\{V_2 = j \mid X_0 = k\} + \sum_{k \in B^c} P(i,k)\mathbb{P}\{V_1 = j \mid X_0 = k\},
\end{aligned}
$$

which gives, in matrix notation

$$
H = P_{B^c B} + P_{B^c} H,
$$
$$
G = P_B G + P_{BB^c} H.
$$

Therefore,

$$
H = (I - P_{B^c})^{-1} P_{B^c B}
$$

and

$$
G = (I - P_B)^{-1} P_{BB^c}(I - P_{B^c})^{-1} P_{B^c B}.
$$

(*ii*) Let $j \in B$. We have

$$
\mathbb{P}\{V_1 = j\} = \mathbb{P}\{X_0 = j\} + \sum_{i \in B^c} \mathbb{P}\{V_1 = j \mid X_0 = i\}\, \mathbb{P}\{X_0 = i\},
$$

which can be written in matrix notation as

$$
v_1 = \alpha_B + \alpha_{B^c} H,
$$

and this concludes the proof. ∎

The Markov chain $\{V_n,\, n \geq 1\}$ contains only one recurrent class, the subset B' of the states of B directly accessible from B^c:

$$
B' = \{ j \in B \mid \exists\, i \in B^c \text{ with } P(i,j) > 0 \}.
$$

Without any loss of generality, we assume that $B' = \{1, \ldots, L'\}$ where $1 \leq L' \leq L$. We then denote by B'' the set $B \setminus B'$. The partition $\{B', B''\}$ induces the following decomposition on matrices G and H,

$$
G = \begin{pmatrix} G' & 0 \\ G'' & 0 \end{pmatrix}, \qquad H = (H'\ \ 0).
$$

In the same way, the partition $\{B', B'', B^c\}$ induces the following decomposition on P,

$$
P = \begin{pmatrix}
P_{B'} & P_{B'B''} & P_{B'B^c} \\
P_{B''B'} & P_{B''} & P_{B''B^c} \\
P_{B^c B'} & 0 & P_{B^c}
\end{pmatrix}.
$$

Now, as for G, we have the following expression for G'.

THEOREM 5.1.5.

$$G' = (I - P_{B'} - P_{B'B''}(I - P_{B''})^{-1}P_{B''B'})^{-1}$$
$$(P_{B'B^c} + P_{B'B''}(I - P_{B''})^{-1}P_{B''B^c})(I - P_{B^c})^{-1}P_{B^cB'}.$$

Proof. The proof is as in (i) of Theorem 5.1.4. We give only the linear system satisfied by matrices G', G'', and H' (derived as the linear system satisfied by matrices H and G of Theorem 5.1.4). We have

$$G' = P_{B'}G' + P_{B'B''}G'' + P_{B'B^c}H',$$

$$G'' = P_{B''B'}G' + P_{B''}G'' + P_{B''B^c}H',$$

$$H' = P_{B^cB'}G' + P_{B^c}H'.$$

For instance, the first two equations provide a new one without matrix G'', and this equation together with the last one allows us to infer the expression given for G' in the theorem. ∎

Note that the expression given in the previous theorem can considerably reduce the time needed to compute G' when $L' < L$. Indeed, instead of inverting a matrix of size L when computing G from the formula of Theorem 5.1.4, we have to invert two matrices of sizes L' and $L - L'$.

We denote now by $H_{B,n}$, for $n \geq 1$, the random variable taking values in $\mathbb{N}^*$ and defined by the time spent by X during its nth sojourn in B. We have the following explicit expression for the distribution of $H_{B,n}$.

THEOREM 5.1.6.   *For every $n, k \in \mathbb{N}^*$, we have*

$$\mathbb{P}\{H_{B,n} = k\} = v_n P_B^{k-1}(I - P_B)\mathbb{1}.$$

Proof. We start by deriving the distribution of the length of the first sojourn of X in B, which is $H_{B,1}$. Let us condition on the state in which the sojourn in B begins. We obtain, for every $i \in B$,

$$\mathbb{P}\{H_{B,1} = 1 \mid V_1 = i\} = \sum_{j \in B^c} P(i,j).$$

Let $k > 1$ and $i \in B$. We have

$$\mathbb{P}\{H_{B,1} = k \mid V_1 = i\} = \sum_{j \in B} P(i,j)\mathbb{P}\{H_{B,1} = k - 1 \mid V_1 = j\}.$$

If we define the column vector

$$h_k = (\mathbb{P}\{H_{B,1} = k \mid V_1 = 1\}, \ldots, \mathbb{P}\{H_{B,1} = k \mid V_1 = L\}),$$

we can rewrite these two last relations as

$$h_1 = P_{BB^c}\mathbb{1} = (I - P_B)\mathbb{1}$$

and, for $k > 1$,

$$h_k = P_B h_{k-1},$$

that is, for every $k \geq 1$,

$$h_k - P_B^{k-1}(I - P_B)\mathbb{1}$$

and thus

$$\mathbb{P}\{H_{B,1} = k\} = v_1 P_B^{k-1}(I - P_B)\mathbb{1}.$$

If we now consider the nth sojourn of X in B, we have

$$\mathbb{P}\{H_{B,n} = k\} = \sum_{i \in B} \mathbb{P}\{H_{B,n} = k \mid V_n = i\}\,\mathbb{P}\{V_n = i\}$$

$$= \sum_{i \in B} \mathbb{P}\{H_{B,1} = k \mid V_1 = i\}\,\mathbb{P}\{V_n = i\}$$

$$= v_n h_k,$$

which is the explicit expression given in the statement. ∎

Let us now compute the moments of the random variable $H_{B,n}$, for $n \geq 1$. An elementary matrix calculus gives

$$\mathbb{E}\{H_{B,n}\} = v_n \left(\sum_{k=1}^{+\infty} k P_B^{k-1} \right)(I - P_B)\mathbb{1}$$

$$= v_n(I - P_B)^{-2}(I - P_B)\mathbb{1}$$

$$= v_n(I - P_B)^{-1}\mathbb{1}.$$

For higher-order moments, it is more comfortable to work with factorial moments instead of standard ones. Recall that the k-order factorial moment of a discrete random variable, Y, which is denoted by $\mathbb{FM}_k\{Y\}$, is defined, for $k \geq 1$, by

$$\mathbb{FM}_k\{Y\} = \mathbb{E}\{Y(Y-1)\ldots(Y-k+1)\}.$$

The following property holds: if $\mathbb{E}\{Y^j\} = \mathbb{E}\{Z^j\}$ for $j = 1, 2, \ldots, k-1$ then

$$\mathbb{FM}_k\{Y\} = \mathbb{FM}_k\{Z\} \iff \mathbb{E}\{Y^k\} = \mathbb{E}\{Z^k\}.$$

Note also that $\mathbb{FM}_1\{Y\} = \mathbb{E}\{Y\}$.

Then, by a classic matrix computation as we did for the mean value of $H_{B,n}$, we have

$$\mathbb{FM}_k\{H_{B,n}\} = k! v_n P_B^{k-1}(I - P_B)^{-k}\mathbb{1}. \tag{5.1}$$

Let us consider now the asymptotic behavior of the nth sojourn time of X in B.

COROLLARY 5.1.7. *For any $k \geq 1$, the sequence $\left(\mathbb{P}\{H_{B,n} = k\}\right)_{n \geq 1}$ converges in the Cesàro sense as $n \longrightarrow \infty$ to the limit $v P_B^{k-1}(I - P_B)\mathbb{1}$, where the L-dimensional row* *vector v is the unique solution to the system $\phi = \phi G$, $\phi \geq 0$ and $\phi \mathbb{1} = 1$, which is given by*

$$v = \frac{x_{B^c} P_{B^c B}}{x_{B^c} P_{B^c B} \mathbb{1}}.$$

The convergence is simple for any initial distribution of X if and only if G' is aperiodic.

Proof. The proof is an elementary consequence of general properties of Markov chains, since the sequence depends on n only through v_n and $(V_n)_{n \geq 1}$ is a finite homogeneous Markov chain with only one recurrent class B'. Note that $v = (v'\ 0)$ according to the partition $\{B', B''\}$ of B. If we consider the block decomposition of G with respect to that partition, a simple recurrence on integer n allows us to write:

$$G^n = \begin{pmatrix} G'^n & 0 \\ G'' G'^{n-1} & 0 \end{pmatrix}.$$

Sequence (G'^n) converges in the Cesàro sense to $\mathbb{1} v'$ and in the same way, $G'' G'^{n-1}$ converges in the Cesàro sense to $\mathbb{1} v'$ because $G'' \mathbb{1} = \mathbb{1}$ (recall that the dimension of vector $\mathbb{1}$ is given by the context). The expression for v is easily checked using Lemma 5.1.2. The second part of the proof is an immediate corollary of the convergence properties of Markov chains. ∎

Let us define the random variable $H_{B,\infty}$ with values in $\mathbb{N}^*$ by making its distribution equal to the previous limit,

$$\mathbb{P}\{H_{B,\infty} = k\} = v P_B^{k-1}(I - P_B)\mathbb{1}, \quad \text{for any } k \geq 1.$$

Then, by taking limits in the Cesàro sense in Expression (5.1), giving the factorial moments of $H_{B,n}$, we obtain the limit given in the following corollary which needs no proof.

COROLLARY 5.1.8. *The sequence $\left(\mathbb{FM}_k\{H_{B,n}\}\right)_{n \geq 1}$ converges in the Cesàro sense and we have*

$$\lim_{n \to \infty} \frac{1}{n} \sum_{l=1}^{n} \mathbb{FM}_k\{H_{B,l}\} = \mathbb{FM}_k\{H_{B,\infty}\} = k! v P_B^{k-1}(I - P_B)^{-k}\mathbb{1}.$$

The convergence is simple for any initial distribution of X if and only if the matrix G' is aperiodic.

There is no relation between the periodicity of P and the periodicity of G'. That is, the four situations obtained by combining the two properties *periodicity* and *aperiodicity* of each matrix are possible as we show below in the four examples of Figures 5.1 to 5.4. In the four cases, the state space is $S = \{1, 2, 3, 4\}$, $B = \{1, 2\}$, and there is an arrow between two states if and only if the corresponding transition probability is positive.

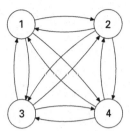

Figure 5.1 P and G' are both aperiodic

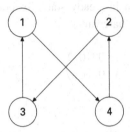

Figure 5.2 P is periodic (period 4) and G' is periodic (period 2)

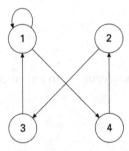

Figure 5.3 P is aperiodic and G' is periodic (period 2)

Example. Let us illustrate these results with an example. We consider a homogeneous Markov chain with state space $\{1,2,3,4\}$ given by the following transition probability matrix

$$P = \left(\begin{array}{cc|cc} 0 & 1/4 & 1/4 & 1/2 \\ 1/2 & 0 & 1/4 & 1/4 \\ \hline 1/4 & 3/4 & 0 & 0 \\ 1/2 & 1/2 & 0 & 0 \end{array} \right).$$

The selected subset of states is $B = \{1,2\}$. The stationary distribution is

$$x = \frac{1}{13}(4,4,2,3)$$

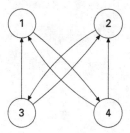

Figure 5.4 P is periodic (period 2) and G' is aperiodic

and we take $\alpha = x$, that is, we consider the chain in steady state. To compute the distribution of $H_{B,n}$ let us give the following intermediate values:

$$H = P_{B^cB} = \frac{1}{4}\begin{pmatrix} 1 & 3 \\ 2 & 2 \end{pmatrix} \quad \text{since } P_{B^c} = \begin{pmatrix} 0 & 0 \\ 0 & 0 \end{pmatrix},$$

$$v_1 = \frac{1}{13}(6,7),$$

$$G = \frac{1}{56}\begin{pmatrix} 23 & 33 \\ 22 & 34 \end{pmatrix},$$

$$v_2 = v_1 G = \frac{1}{182}(73, 109),$$

$$v = \frac{x_{B^c}P_{B^cB}}{x_{B^c}P_{B^cB}\mathbb{1}} = \frac{1}{5}(2,3),$$

$$(P_B)^{2m} = \left(\frac{1}{8}\right)^m I \quad \text{for } m \geq 0, \text{ where } I \text{ is the identity matrix of order 2,}$$

$$(P_B)^{2m+1} = \frac{1}{4}\left(\frac{1}{8}\right)^m \begin{pmatrix} 0 & 1 \\ 2 & 0 \end{pmatrix} \quad \text{for } m \geq 0.$$

A classic matrix computation gives

$$G^n = \frac{1}{5}\begin{pmatrix} 2+3\times 56^{-n} & 3-3\times 56^{-n} \\ 2-2\times 56^{-n} & 3+2\times 56^{-n} \end{pmatrix}.$$

Then, after some algebra, we obtain

$$\mathbb{P}\{H_{B,n} = 2h\} = v_1 G^{n-1} P_B^{2h-1}(I - P_B)\mathbb{1}$$

$$= \left(\frac{1}{8}\right)^h \frac{143 \times 56^n - 448}{65 \times 56^n} \quad \text{for } h \geq 1,$$

$$\mathbb{P}\{H_{B,n} = 2h+1\} = v_1 G^{n-1} P_B^{2h}(I - P_B)\mathbb{1}$$

$$= \frac{1}{4}\left(\frac{1}{8}\right)^h \frac{156 \times 56^n + 224}{65 \times 56^n} \quad \text{for } h \geq 0.$$

5.2 Successive sojourn times in irreducible CTMC

Let $X = \{X_t, \ t \geq 0\}$ be an irreducible homogeneous Markov chain on the state space $S = \{1,\ldots,N\}$, with $N \geq 2$. Let A be the infinitesimal generator of X and define, for every $i \in S$, $\lambda_i = -A(i,i)$, which is the output rate of state i. We denote by Λ the diagonal matrix whose ith entry is λ_i. We denote also by α and π the initial and the stationary distributions of X, respectively. The transition probability matrix, P, of the embedded discrete-time Markov chain at the instants of transition of X is given by $P = I + \Lambda^{-1}A$ and its stationary distribution is $x = \pi \Lambda/(\pi \Lambda \mathbb{1})$.

As in the discrete-time case, we consider a proper subset B of S and we conserve the notation for B, B', and B'', for the decompositions of P (and similarly of A) with respect to the partitions $\{B, B^c\}$ or $\{B', B'', B^c\}$, as well as for the initial distribution α. A sojourn of X in B is now a sequence $X_{t_m},\ldots,X_{t_{m+k}}$, $k \geq 1$, where the t_i are transition instants, $X_{t_m},\ldots,X_{t_{m+k-1}} \in B$, $X_{t_{m+k}} \notin B$ and if $m > 0$, $X_{t_{m-1}} \notin B$. This sojourn begins at time t_m and finishes at time t_{m+k}. It lasts $t_{m+k} - t_m$.

Let $H_{B,n}$ be the random variable representing the time spent during the nth sojourn of X in B. The hypothesis of irreducibility guarantees the existence of an infinite number of sojourns of X in B with probability 1. Defining V_n as in the discrete-time case for the embedded discrete-time Markov chain, we have the following result.

THEOREM 5.2.1.

$$\mathbb{P}\{H_{B,n} \leq t\} = 1 - v_n e^{A_B t} \mathbb{1}$$

where $v_n = v_1 G^{n-1}$, with $v_1 = \alpha_B - \alpha_{B^c} A_{B^c}^{-1} A_{B^cB}$ and $G = A_B^{-1} A_{BB^c} A_{B^c}^{-1} A_{B^cB}$.

Proof. Let us construct a homogeneous Markov chain, $Y = \{Y_t, t \geq 0\}$, on the state space $B \cup \{a\}$, where a is an absorbing state and matrix A' is its infinitesimal generator, given by

$$A' = \begin{pmatrix} A_B & A_{Ba} \\ 0 & 0 \end{pmatrix}.$$

A_{Ba} is the column vector defined by $A_{Ba}(i) = \sum_{j \in B^c} A(i,j)$, $i \in B$.

Therefore, we have, for every $i \in B$,

$$\mathbb{P}_i\{H_{B,1} \leq t\} = \mathbb{P}_i\{Y_t = a\} = 1 - \sum_{j \in B} e^{A_B t}(i,j),$$

where $\mathbb{P}_i\{.\} = \mathbb{P}_i\{. \mid X_0 = i\}$.

For every $i \in B^c$, the following relation holds:

$$\mathbb{P}_i\{H_{B,1} \leq t\} = \sum_{j \in E} P(i,j)\mathbb{P}_j\{H_{B,1} \leq t\}.$$

These two relations give

$$\mathbb{P}\{H_{B,1} \leq t\} = 1 - (\alpha_B + \alpha_{B^c}(I - P_{B^c})^{-1}P_{B^cB})e^{A_B t}\mathbb{1},$$

i.e.,

$$\mathbb{P}\{H_{B,1} \leq t\} = 1 - v_1 e^{A_B t} \mathbb{1}.$$

For any $i \in B$ and for any $n \geq 1$, we can write

$$\mathbb{P}\{H_{B,n} \leq t \mid V_n = i\} = \mathbb{P}_i\{H_{B,1} \leq t\}.$$

We deduce that

$$\mathbb{P}\{H_{B,n} \leq t\} = \sum_{i \in B} \mathbb{P}\{H_{B,n} \leq t \mid V_n = i\} \mathbb{P}\{V_n = i\}$$

$$= \sum_{i \in B} \mathbb{P}\{V_n = i\} \mathbb{P}_i\{H_{B,1} \leq t\}$$

$$= v_n(\mathbb{1} - e^{A_B t} \mathbb{1})$$

$$= 1 - v_n e^{A_B t} \mathbb{1},$$

which is the explicit form of the distribution of $H_{B,n}$. The specific representations given here, in terms of the blocks of A, come immediately from the relation $P = I + \Lambda^{-1} A$. ∎

The moments of $H_{B,n}$ are then easily derived. We have, for every $n \geq 1$ and $k \geq 1$,

$$\mathbb{E}\{H_{B,n}^k\} = (-1)^k k! v_n A_B^{-k} \mathbb{1}.$$

The analog of Corollary 5.1.7 is the following result.

COROLLARY 5.2.2. *For all $t \geq 0$, the sequence $\left(\mathbb{P}\{H_{B,n} \leq t\}\right)_{n \geq 1}$ converges as $n \longrightarrow \infty$ in the Cesàro sense to $1 - v e^{A_B t} \mathbb{1}$, where the L-dimensioned row vector v is the stationary distribution of the Markov chain $\{V_n, n \geq 1\}$ which has been given in Corollary 5.1.7 (in terms of A and π, we have $v = \pi_B A_{BB^c} / (\pi_B A_{BB^c} \mathbb{1})$). The convergence is simple for any initial distribution of X if and only if G' is aperiodic.*

Proof. Note that $A_B = -\Lambda_B(I - P_B)$ and that the canonical embedded Markov chain with transition probability matrix P is irreducible. This implies that matrix $(I - P_B)^{-1}$ exists (Lemma 5.1.1) and so A_B^{-1} exists as well. The proof of the corollary is then as in Corollary 5.1.7. ∎

As in the discrete-time case, we define the positive real-valued random variable $H_{B,\infty}$ by giving its distribution: $\mathbb{P}\{H_{B,\infty} \leq t\} = 1 - v e^{A_B t} \mathbb{1}$. We can then derive the limits in the Cesàro sense of the k-order moments of $H_{B,n}$ and easily verify the following relation.

COROLLARY 5.2.3. *For any $k \geq 1$, the sequence $\left(\mathbb{E}\{H_{B,n}^k\}\right)_{n \geq 1}$ converges in the Cesàro sense and we have*

$$\lim_{n \to \infty} \frac{1}{n} \sum_{l=1}^{n} \mathbb{E}\{H_{B,l}^k\} = \mathbb{E}\{H_{B,\infty}^k\} = (-1)^k k! v A_B^{-k} \mathbb{1}.$$

The convergence is simple for any initial distribution of X if and only if the matrix G' is aperiodic.

Remark: The results in continuous time can also be derived using uniformization, which allows to reduce them to the discrete-time case. In this section, we chose direct proofs instead.

5.3 Pseudo-aggregation

The reader is advised to read Chapter 4 before going through this section.

5.3.1 The pseudo-aggregated process

We consider first the discrete-time case. Let $X = \{X_n, n \geq 0\}$ be a homogeneous irreducible Markov chain transition probability matrix P, and equilibrium or stationary probability distribution x. We construct the pseudo-aggregated homogeneous Markov chain $Z = \{Z_n, n \geq 0\}$ from X with respect to the partition $\mathcal{B} = \{B(1), B(2), \ldots, B(M)\}$ by defining its transition probability matrix, $\hat{P}$. If $F = \{1, 2, \ldots, M\}$, then matrix $\hat{P}$ is defined by

$$\forall i, j \in F, \ \hat{P}(i,j) = \sum_{k \in B(i)} \frac{x(k)}{\sum_{h \in B(i)} x(h)} \sum_{l \in B(j)} P(k,l).$$

If we denote by α the initial distribution of X, then the initial distribution of Z is $T\alpha$ where the operator T is defined by

$$T\alpha = \left(\sum_{\ell \in B(1)} \alpha(i), \ldots, \sum_{\ell \in B(M)} \alpha_i \right).$$

If the initial distribution, α, leads to a homogeneous Markov chain for Y, then Z and Y define the same (in distribution) homogeneous Markov chain. The stationary distribution of Z is Tx as the following lemma shows.

LEMMA 5.3.1. *If $z = Tx$ then we have $z\hat{P} = z$, $z > 0$ and $z\mathbb{1} = 1$.*

Proof. For $1 \leq m \leq M$, the mth entry of $z\hat{P}$ is given by

$$\left(z\hat{P} \right)(m) = \sum_{l=1}^{M} z(l)\hat{P}(l,m)$$

$$= \sum_{l=1}^{M} \left(\sum_{i \in B(l)} x(i) \sum_{j \in B(m)} P(i,j) \right)$$

$$= \sum_{i=1}^{N} x(i) \sum_{j\in B(m)} P(i,j)$$

$$= \sum_{j\in B(m)} \left(\sum_{i=1}^{N} x(i)P(i,j) \right)$$

$$= \sum_{j\in B(m)} x(j) = z(m).$$

The remainder of the proof is immediate.   ∎

The construction is analogous in the continuous-time case. Let $X = \{X_t, \ t \geq 0\}$ be a homogeneous irreducible Markov chain evolving in continuous time with transition rate matrix A. Let π denote the equilibrium probability distribution of X, that is $\pi A = 0$, $\pi > 0$ and $\pi \mathbb{1} = 1$. As in the discrete-time case, we construct the pseudo-aggregated process $Z = \{Z_t, \ t \geq 0\}$ from X with respect to the partition, $\mathcal{B}$, by giving its transition rate matrix, $\hat{A}$:

$$\forall i,j \in F, \hat{A}(i,j) = \sum_{k\in B(i)} \frac{\pi(k)}{\sum_{h\in B(i)} \pi(h)} \sum_{l\in B(j)} A(k,l).$$

We have the same result for the stationary distribution of Z as the one in the discrete-time case:

LEMMA 5.3.2.   *If $z = T\pi$ then we have $z\hat{A} = 0$, $z > 0$ and $z\mathbb{1} = 1$.*

Proof. Since it is immediate that $z > 0$ and $z\mathbb{1} = 1$, let us simply verify, as in Lemma 5.3.1, that $z\hat{A} = 0$. For $1 \leq m \leq M$, the mth entry of $z\hat{A}$ is given by

$$\left(z\hat{A} \right)(m) = \sum_{l=1}^{M} z(l)\hat{A}(l,m)$$

$$= \sum_{l=1}^{M} \left(\sum_{i\in B(l)} \pi(i) \sum_{j\in B(m)} A(i,j) \right)$$

$$= \sum_{i=1}^{N} \pi(i) \sum_{j\in B(m)} A(i,j)$$

$$= \sum_{j\in B(m)} \left(\sum_{i=1}^{N} \pi(i)A(i,j) \right)$$

$$= 0,$$

which concludes the proof.   ∎

In both discrete- and continuous-time cases, the following property holds.

LEMMA 5.3.3. *The pseudo-aggregated process constructed from X with respect to the partition $\mathcal{B}$ and the pseudo-aggregated process obtained after M successive aggregations of X with respect to each B(i), $i \in F$, in any order, are equivalent.*

The proof is an immediate consequence of the construction of the pseudo-aggregated processes. It is for this reason that in the following we consider only the situation where $\mathcal{B}$ contains only one subset having more than one state. That is, we assume, as in the previous sections, that $B = \{1, \ldots, L\}$ where $1 < L < N$, and that $\mathcal{B} = \{B, \{L+1\}, \ldots, \{N\}\}$, with $N \geq 3$. The state space of the pseudo-aggregated process, Z, is denoted by $F = \{b, L+1, \ldots, N\}$. We also denote by B^c the complementary subset $\{L+1, \ldots, N\}$ and by α the initial distribution of X.

5.3.2   Pseudo-aggregation and sojourn times: discrete-time case

We now consider the pseudo-aggregated homogeneous Markov chain, Z, constructed from X with respect to the partition $\mathcal{B} = \{B, \{L+1\}, \ldots, \{N\}\}$ of S. Although the stationary distribution of X over the sets of $\mathcal{B}$ is equal to the stationary distribution of Z (Lemma 5.3.1), the distribution of the nth sojourn time of X in B is, in general, different from the corresponding distribution of the nth holding time of Z in b, which is independent of n and is denoted by H_b. This last (geometric) distribution is given by

$$\mathbb{P}\{H_b = k\} = \frac{x_B(I - P_B)\mathbb{1}}{x_B \mathbb{1}} \left(\frac{x_B P_B \mathbb{1}}{x_B \mathbb{1}} \right)^{k-1}, \quad k \geq 1,$$

which is to be compared with the given expression of $\mathbb{P}\{H_{B,n} = k\}$.

Observe that if there is no internal transition between different states inside B and if the geometric holding time distributions of X in the individual states of B have the same parameter, we have that the distributions of $H_{B,n}$, $H_{B,\infty}$, and H_b are the same. That is, if $P_B = \beta I$ with $0 \leq \beta < 1$, we have

$$\mathbb{P}\{H_{B,n} = k\} = \mathbb{P}\{H_{B,\infty} = k\} = \mathbb{P}\{H_b = k\} = (1 - \beta)\beta^{k-1}, \quad k \geq 1.$$

More generally, if X is a strongly lumpable Markov chain (see Chapter 4) over the partition $\mathcal{B}$ (i.e. if the aggregated chain Y is also Markov homogeneous for any initial distribution of X) we have $P_B \mathbb{1} = \beta \mathbb{1}$, $(I - P_B)\mathbb{1} = (1 - \beta)\mathbb{1}$, where $\beta \in [0, 1)$. In this case, $P_B^{k-1}(I - P_B)\mathbb{1} = (1 - \beta)\beta^{k-1}$ for $k \geq 1$, and the distributions of $H_{B,n}$, $H_{B,\infty}$, and H_b are identical.

We now investigate the relation between the moments (factorial moments) of $H_{B,n}$ and H_b. We give an algebraic necessary and sufficient condition for the equality between the limits in the Cesàro sense of the k-order moments of the random variable, $H_{B,n}$, and the k-order moments of H_b.

THEOREM 5.3.4.   *For any $k \geq 1$,*

$$\left[\lim_{n \to +\infty} \frac{1}{n} \sum_{l=1}^{n} \mathbb{FM}_k\{H_{B,l}\} = \mathbb{FM}_k\{H_b\} \right]$$

$$\Longleftrightarrow$$

$$\left[x_B \left(P_B(I - P_B)^{-1} \right)^{k-1} \mathbb{1} = x_B \mathbb{1} \left(\frac{x_B P_B \mathbb{1}}{x_B(I - P_B)\mathbb{1}} \right)^{k-1} \right].$$

Proof.   For the geometric distribution of H_b, we have:

$$\mathbb{FM}_k\{H_b\} = \frac{k! \, (x_B P_B \mathbb{1})^{k-1} x_B \mathbb{1}}{(x_B(I - P_B)\mathbb{1})^k}.$$

Let us fix $k \geq 1$. Since

$$v = \frac{x_{B^c} P_{B^c B}}{x_{B^c} P_{B^c B} \mathbb{1}} = \frac{x_B(I - P_B)}{x_B(I - P_B)\mathbb{1}},$$

we have

$$\left[\lim_{n \to +\infty} \frac{1}{n} \sum_{l=1}^{n} \mathbb{FM}_k\{H_{B,l}\} = \mathbb{FM}_k\{H_b\} \right]$$

$$\Longleftrightarrow \left[v P_B^{k-1}(I - P_B)^{-k}\mathbb{1} = x_B \mathbb{1} \frac{(x_B P_B \mathbb{1})^{k-1}}{(x_B(I - P_B)\mathbb{1})^k} \right]$$

$$\Longleftrightarrow \left[x_B(I - P_B) P_B^{k-1}(I - P_B)^{-k}\mathbb{1} = x_B \mathbb{1} \left(\frac{x_B P_B \mathbb{1}}{x_B(I - P_B)\mathbb{1}} \right)^{k-1} \right]$$

$$\Longleftrightarrow \left[x_B \left(P_B(I - P_B)^{-1} \right)^{k-1} \mathbb{1} = x_B \mathbb{1} \left(\frac{x_B P_B \mathbb{1}}{x_B(I - P_B)\mathbb{1}} \right)^{k-1} \right],$$

since

$$P_B(I - P_B)^{-1} = \sum_{j \geq 1} P_B^j = (I - P_B)^{-1} P_B,$$

which completes the proof.   ∎

In the important case of $k = 1$, the above condition is always satisfied. The following corollary states this result, whose proof is just a matter of verification.

COROLLARY 5.3.5.

$$\lim_{n \to +\infty} \frac{1}{n} \sum_{l=1}^{n} \mathbb{E}\{H_{B,l}\} = \mathbb{E}\{H_b\}.$$

Consider the numerical example given at the end of Section 5.2, where the distribution of $H_{B,n}$ is derived. When $n \to \infty$, the probabilities $\mathbb{P}\{H_{B,n} = h\}$ converge simply

(G' is aperiodic). The limits are

$$\mathbb{P}\{H_{B,\infty} = 2h\} = \frac{11}{5}\left(\frac{1}{8}\right)^h, \quad \text{for } h \geq 1,$$

$$\mathbb{P}\{H_{B,\infty} = 2h+1\} = \frac{3}{5}\left(\frac{1}{8}\right)^h, \quad \text{for } h \geq 0.$$

In the pseudo-aggregated chain, Z, the corresponding holding time distribution is

$$\mathbb{P}\{H_b = h\} = \frac{5}{8}\left(\frac{1}{8}\right)^{h-1}, \quad \text{for } h \geq 1$$

and we can verify that $\mathbb{E}\{H_{B,\infty}\} = \mathbb{E}\{H_b\} = 8/5$.

When we consider higher-order moments, the equality is not necessarily valid as the following counterexample shows. We compute here the standard moments of the involved random variables $H_{B,n}$, and H_b.

$S = \{1,2,3\}$, $B = \{1,2\}$, and

$$P = \left(\begin{array}{cc|c} 0 & 1 & 0 \\ 1/2 & 0 & 1/2 \\ \hline 1 & 0 & 0 \end{array}\right).$$

The stationary probability vector is $x = (2/5, 2/5, 1/5)$.
The expression giving the second moment of $H_{B,n}$ is

$$\mathbb{E}\{H_{B,n}^2\} = v_n(I + P_B)(I - P_B)^{-2}\mathbb{1}.$$

Taking limits in the Cesàro sense, we obtain

$$\lim_{n \to +\infty} \frac{1}{n}\sum_{l=1}^{n}\mathbb{E}\{H_{B,l}^2\} = v(I + P_B)(I - P_B)^{-2}\mathbb{1} = \frac{x_B(I + P_B)(I - P_B)^{-1}\mathbb{1}}{x_B(I - P_B)\mathbb{1}},$$

which is equal to 24 in the example.
In the pseudo-aggregated process, we have

$$\mathbb{E}\{H_b^2\} = \frac{2 - \rho}{\rho^2} \quad \text{where } \rho = \frac{x_B(I - P_B)\mathbb{1}}{x_B\mathbb{1}},$$

which is equal to 28 in the example.

5.3.3   Pseudo-aggregation and sojourn times: continuous-time case

In continuous time we have analogous results for the relation between properties of sojourn times of the given process X and the corresponding holding times of the pseudo-aggregated process, Z.

Denote by H_b the holding time random variable for process Z. For every $t \geq 0$, we have

$$\mathbb{P}\{H_b \leq t\} = 1 - e^{-\mu t} \quad \text{where } \mu = -\frac{x_B \Lambda_B^{-1} A_B \mathbb{1}}{x_B \Lambda_B^{-1} \mathbb{1}}.$$

The continuous-time version of Theorem 5.3.4 is then given by the following.

THEOREM 5.3.6. *For any $k \geq 1$,*

$$\lim_{n \to +\infty} \frac{1}{n} \sum_{l=1}^{n} \mathbb{E}\{H_{B,l}^k\} = \mathbb{E}\{H_b^k\}$$

$$\Longleftrightarrow \left[(x_B(I - P_B)\mathbb{1})^{k-1} x_B \Lambda_B^{-1}((I - P_B)^{-1} \Lambda_B^{-1})^{k-1} \mathbb{1} = (x_B \Lambda_B^{-1} \mathbb{1})^k \right].$$

Proof. The k-order moment of the holding time, H_b, of Z in b can be written as

$$\mathbb{E}\{H_b^k\} = \frac{k!}{\mu^k} = k! \left(-\frac{x_B \Lambda_B^{-1} \mathbb{1}}{x_B \Lambda_B^{-1} A_B \mathbb{1}} \right)^k;$$

then, we have

$$\left[\lim_{n \to +\infty} \frac{1}{n} \sum_{l=1}^{n} \mathbb{E}\{H_{B,l}^k\} = \mathbb{E}\{H_b^k\} \right]$$

$$\Longleftrightarrow \left[v \left((I - P_B)^{-1} \Lambda_B^{-1} \right)^k \mathbb{1} = \left(\frac{x_B \Lambda_B^{-1} \mathbb{1}}{x_B(I - P_B)\mathbb{1}} \right)^k \right]$$

$$\Longleftrightarrow \left[x_B(I - P_B)\mathbb{1} v \left((I - P_B)^{-1} \Lambda_B^{-1} \right)^k \mathbb{1} = \frac{\left(x_B \Lambda_B^{-1} \mathbb{1} \right)^k}{(x_B(I - P_B)\mathbb{1})^{k-1}} \right]$$

$$\Longleftrightarrow \left[x_B(I - P_B)\mathbb{1} v (I - P_B)^{-1} \Lambda_B^{-1} \left((I - P_B)^{-1} \Lambda_B^{-1} \right)^{k-1} \mathbb{1} = \frac{\left(x_B \Lambda_B^{-1} \mathbb{1} \right)^k}{(x_B(I - P_B)\mathbb{1})^{k-1}} \right]$$

$$\Longleftrightarrow \left[x_B \Lambda_B^{-1} \left((I - P_B)^{-1} \Lambda_B^{-1} \right)^{k-1} \mathbb{1} = \frac{\left(x_B \Lambda_B^{-1} \mathbb{1} \right)^k}{(x_B(I - P_B)\mathbb{1})^{k-1}} \right]$$

$$\Longleftrightarrow \left[(x_B(I - P_B)\mathbb{1})^{k-1} x_B \Lambda_B^{-1} \left((I - P_B)^{-1} \Lambda_B^{-1} \right)^{k-1} \mathbb{1} = (x_B \Lambda_B^{-1} \mathbb{1})^k \right],$$

which concludes the proof. ∎

As in the discrete-time case, the given condition is always satisfied for first-order moments. The following result corresponds to Corollary 5.3.5 with identical immediate verification.

COROLLARY 5.3.7.

$$\lim_{n \to +\infty} \frac{1}{n} \sum_{l=1}^{n} \mathbb{E}\{H_{B,l}\} = \mathbb{E}\{H_{B,\infty}\} = \mathbb{E}\{H_b\}.$$

5.4 The case of absorbing Markov chains

We consider here that states $1, 2, \ldots, N-1$ are transient and state N is absorbing, with $N \geq 2$. In this setting, the nth sojourn of X in B may not exist. The process will visit set B a random number (finite with probability 1) of times, and will then be absorbed. In this section we derive the distribution of the total time X spends in set B.

We denote by C the subset of states $C = B^c \setminus \{N\}$.

5.4.1 Discrete time

The block-decomposition of P and α induced by the partition $\{B, C, \{N\}\}$ is

$$P = \begin{pmatrix} P_B & P_{BC} & p \\ P_{CB} & P_C & q \\ 0 & 0 & 1 \end{pmatrix}, \qquad \alpha = (\begin{array}{ccc} \alpha_B & \alpha_C & 0 \end{array}).$$

We define the random variable V_n as the state by which the nth sojourn of X in B starts, if it exists, and by value 0 otherwise. We also define $v_n(i) = \mathbb{P}\{V_n = i\}$, for $i \in B$, and we denote by v_n the vector indexed by B whose ith component is $v_n(i)$. The same developments as in the first section of this chapter give the following result.

THEOREM 5.4.1. *The nth sojourn of X in B exists with probability $v_n \mathbb{1}$, where $v_n = v_1 G^{n-1}$ with $v_1 = \alpha_B + \alpha_C (I - P_C)^{-1} P_{CB}$ and $G = (I - P_B)^{-1} P_{BC} (I - P_C)^{-1} P_{CB}$.*

Denoting the length of the nth sojourn of X in B by $H_{B,n}$, where this length is equal to zero if X visits B strictly less than n times, then

$$\mathbb{P}\{H_{B,n} = k\} = v_n P_B^k (I - P_B).$$

Proof. The proof is exactly as in the irreducible case at the beginning of this chapter. The only difference is that now $v_n \mathbb{1}$ can be < 1, but the derivations are identical, so we omit them. ∎

From this, we can compute other related quantities in a straightforward manner. If N_B denotes the number of visits that X makes to the subset B, we have $\mathbb{P}\{N_B = 0\} = 1 - v_1 \mathbb{1}$, and, for $k \geq 1$,

$$\mathbb{P}\{N_B = k\} = v_k \mathbb{1} - v_{k+1} \mathbb{1} = v_k (I - G) \mathbb{1}.$$

The distribution of $H_{B,n}$ conditional on $H_{B,n} > 0$ is

$$\mathbb{P}\{H_{B,n} = k \mid H_{B,n} > 0\} = \frac{v_n P_B^k (I - P_B) \mathbb{1}}{v_n \mathbb{1}}.$$

We now move to the total sojourn time of X in B before absorption, a random variable denoted here by W, i.e.

$$W = \sum_{n \geq 1} H_{B,n}.$$

THEOREM 5.4.2.   *We have*

$$\mathbb{P}\{W=0\}=1-v_1\mathbb{1},$$
$$\mathbb{P}\{W=k\}=v_1M^{k-1}(I-M)\mathbb{1},\quad for\ k\geq1,$$

where $M=P_B+P_{BC}(I-P_C)^{-1}P_{CB}$.

Proof.   For $k\geq1$, let us denote $w_i(k)=\mathbb{P}_i\{W=k\}$ for any state $i\neq N$.
First, for $k\geq2$,

$$w_i(k)=\sum_j P(i,j)w_j(k-1)\quad\text{if }i\in B,$$

and

$$w_i(k)=\sum_j P(i,j)w_j(k)\quad\text{if }i\in C.$$

If $w_B(k)$ is a column vector indexed on B where the component associated with i is $w_i(k)$, and analogously for $w_C(k)$, we then have, for $k\geq2$,

$$w_B(k)=P_Bw_B(k-1)+P_{BC}w_C(k-1),$$
$$w_C(k)=P_{CB}w_B(k)+P_Cw_C(k).$$

From the second relation, we obtain

$$w_C(k)=(I-P_C)^{-1}P_{CB}w_B(k),$$

which, replaced in the first relation, gives

$$w_B(k)=Mw_B(k-1)=M^{k-1}w_B(1).$$

Now, for $k=1$, we have

$$w_B(1)=P_{BC}w_C(0)+p.$$

For $i\in C$,

$$w_i(0)=\sum_{j\in C}P(i,j)w_i(0)+P(i,N),$$

that is

$$w_C(0)=P_Cw_C(0)+q,$$

so

$$w_C(0)=(I-P_C)^{-1}q.$$

We thus obtain

$$w_B(1)=P_{BC}w_C(0)+p$$
$$=P_{BC}(I-P_C)^{-1}q+p.$$

Using the fact that P is stochastic, we have $p = (I - P_B)\mathbb{1} - P_{BC}\mathbb{1}$ and $q = (I - P_C)\mathbb{1} - P_{CB}\mathbb{1}$. Replacing and simplifying, we obtain

$$w_B(1) = (I - M)\mathbb{1}.$$

For the distribution of W, we then have, for $k \geq 1$,

$$\begin{aligned}
\mathbb{P}\{W = k\} &= \alpha_B w_B(k) + \alpha_C w_C(k) \\
&= [\alpha_B + \alpha_C(I - P_C)^{-1} P_{CB}] w_B(k) \\
&= v_1 M^{k-1} w_B(1) \\
&= v_1 M^{k-1}(I - M)\mathbb{1},
\end{aligned}$$

which completes the proof. ∎

From this, we have, as in the first section,

$$\mathbb{E}\{W\} = v_1(I - M)^{-1}\mathbb{1}$$

and

$$\mathbb{FM}_k\{W\} = k! v_1 M^{k-1}(I - M)^{-k}\mathbb{1}.$$

5.4.2 Continuous time

The case of continuous-time models is easily solved by reducing it to the discrete-time case through uniformization. The main steps are as follows. If A is the transition rate matrix of X, the partition $\{B, C, \{N\}\}$ induces the block-decomposition

$$A = \begin{pmatrix} A_B & A_{BC} & c \\ A_{CB} & A_C & d \\ 0 & 0 & 0 \end{pmatrix}.$$

Let us denote by η the uniformization rate used, and by P the corresponding stochastic matrix $P = I + A/\eta$.

Let N_B denote the total number of visits that the uniformized process Z makes to states in B. Using the results in the discrete-time case, we have

$$\mathbb{P}\{N_B = 0\} = 1 - v_1\mathbb{1}$$

and, for $k \geq 1$,

$$\mathbb{P}\{N_B = k\} = v_k\mathbb{1} - v_{k+1}\mathbb{1} = v_k(I - G)\mathbb{1},$$

where $G = A_B^{-1} A_{BC} A_C^{-1} A_{CB}$, $v_1 = \alpha_B - \alpha_C A_C^{-1} A_{CB}$ and $v_k = v_1 G^{k-1}$.

Let W be the total sojourn time of X in B before absorption, that is,

$$W = \sum_{n \geq 1} H_{B,n}.$$

If $\tilde{W}(s)$ is the Laplace transform of W in the complex variable s, we have

$$\tilde{W}(s) = \sum_{k \geq 0} \mathbb{P}\{N_B = k\} \left(\frac{\eta}{\eta + s} \right)^k.$$

The derivation of $\tilde{W}(s)$ and then of the distribution of W is a matter of straightforward matrix computations, using the distribution of N_B that is given above. The result is given here as the last theorem in the chapter.

THEOREM 5.4.3. *The distribution of*

$$W = \sum_{n \geq 1} H_{B,n}$$

is given by

$$\mathbb{P}\{W \leq t\} = 1 - v_1 e^{Mt} \mathbb{1},$$

where

$$M = A_B - A_{BC} A_C^{-1} A_{CB}.$$

5.4.3 An illustrative example

We illustrate the previous results by means of a simple model taken from the distributed algorithms area. Consider a set of processes organized in a unidirectional ring and assume that there is a special message (the token) used to control some distributed application. For instance, one can think of a network of processors sharing a common resource (a peripheral device) in mutual exclusion. A processor may use the resource if and only if it possesses the token. Of course, some kind of mechanism is provided to avoid monopoly of the resource by one or a few processors. We are interested in the behavior of such a system from a fault-tolerant point of view. We assume that the token is vulnerable only when traveling from a node (here, node = process = site) to the next one on the ring. The modeling assumptions are as follows. The different sojourn times of the token in each site are i.i.d. random variables exponentially distributed with common parameter λ_r (the mean *resource-token speed*). When passing from one process to the next, the token may be lost (unsuccessful transmission, perturbations, etc.) with probability p. If the transmission is successful, the time of the transfer is negligible (that is, zero in the model). The results of each transmission are independent events. The lifetime of the token is then exponentially distributed with mean $1/(\lambda_r p)$.

Concerning the problem of detecting the token loss and regenerating it, a first solution was proposed in [53] using local clocks. A different approach was considered in [68], where a second token was used to increase the lifetime of the system. The two tokens carry some additional information and leave a trace of their visits when leaving the sites. When one of them is lost, there always exists a node in the ring such that if it is reached by the remaining token, the detection and regeneration take place at that moment in that site. In [82] the previous solution was improved and generalized to support any number of tokens. It must be observed that these solutions need no local clock management in the nodes. Here, we consider this last algorithm when there are two tokens in the ring. The only functional aspect we need to know here is that the site in which a detection and regeneration can take place follows exactly the last one visited by the lost unit (for technical details concerning the algorithms, see [82]).

We assume that the stochastic behavior of the second unit (the *system-token*) is the same as the first one, with mean speed λ_s. Furthermore, while there are two units on the ring their respective evolutions are independent. For illustrative purposes here, let us consider only the case of three sites in the ring. The lifetime of such a system is the absorbing time of the homogeneous Markov chain exhibited in Figure 5.5, where q is defined by $q = 1 - p$.

At any instant, the distance from token x to token y (when they are both "alive") is defined as the number of lines that token x has to pass through to join y (remember that the ring is unidirectional). When one of the tokens is lost, the distance between them is the number of lines that the remaining unit has to pass through to regenerate the lost one.

States D_i, $i = 0, 1, 2$, correspond to the case where two tokens are alive, where i equals the distance from the resource-token to the system-token. States R_i (resp. S_i) correspond to the resource-token alive (resp. the system-token alive) and the other unit lost, i representing the distance between them, $i = 1, 2, 3$. The arcs without endpoints represent the transitions to the absorbing state (system completely down). We define the operational states as the states in which the resource-token is alive. Denoting by B the set of operational states, we have $B = \{D_0, D_1, D_2, R_1, R_2, R_3\}$. The set of nonoperational states is $C = \{S_1, S_2, S_3\}$.

Assume that the two parameters, p and λ_r, are input data and that the user wants to choose the value of the speed of the system-token to satisfy the following "informal" constraints: the lifetime of the system must be "high" and the system must spend "as much time as possible" in the operational states.

Due to the particular structure of matrix A_B, we get the following simple expression

$$\mathbb{P}\{H_{B,n} \leq t\} = 1 - g^{n-1}e^{-\lambda_r p t}, \quad \text{for } n \geq 1,$$

where $g = G(1,1)$. If we want the distribution of the nth sojourn time in B, given that it exists, we have

$$\mathbb{P}\{H_{B,n} \leq t \mid H_{B,n} > 0\} = 1 - e^{-\lambda_r p t}, \quad \text{for } n \geq 1.$$

In this simple structure, the total number, V_B, of visits to states in B has a geometric distribution, that is

$$\mathbb{P}\{V_B = k\} = g^{k-1}(1-g), \quad \text{for } k \geq 1.$$

The mean cumulative operational time until absorption is

$$\mathbb{E}\{W\} = \frac{\mathbb{E}\{V_B\}}{\lambda_r p} = \frac{1}{\lambda_r p(1-g)}.$$

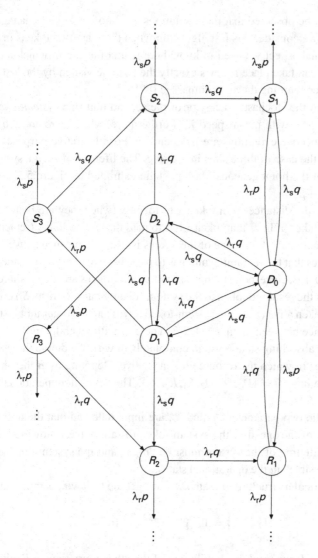

Figure 5.5   Markov model for three nodes

Let us assign numerical values to the input parameters. We take $\lambda_r = 1.0$ (say 1.0/sec) and $p = 10^{-3}$. It can be easily verified that if we denote by LT the lifetime of the system (that is, the absorption time of X), then functions $\lambda_s \mapsto \mathbb{E}\{LT\}$ and $\lambda_s \mapsto \mathbb{P}\{H_{C,n} > t \mid H_{C,n} > 0\}$ (for any fixed value of t) are decreasing.

Also, $\mathbb{P}\{H_{C,n} > t \mid H_{C,n} > 0\}$ does not depend on n in this particular case. Indeed, we have

$$\mathbb{P}\{H_{C,n} > t \mid H_{C,n} > 0\} = \frac{v_1' e^{A_C t} \mathbb{1}}{v_1' \mathbb{1}},$$

where $v_1' = -(1,0,0)A_B^{-1}A_{BC}$.

To formalize the optimization problem, we look for values of λ_s such that the mean lifetime of the system is greater than a given value T_{min} and the conditional probability $\mathbb{P}\{ H_{C,n} > t \mid H_{C,n} > 0 \}$ is less than a given level ε. For instance, let us set $T_{min} = 259\,200$ sec (three days). Since $\mathbb{E}\{LT\}$ decreases with λ_s, we have

$$\mathbb{E}\{LT\} > 259\,200 \text{ sec} \iff \lambda_s \in (0, 0.9387).$$

Assume that the other specification is $t = 5.0$ with a 20% level ($\varepsilon = 0.2$). This leads to

$$\mathbb{P}\{H_{C,n} > 5 \mid H_{C,n} > 0\} < 0.2 \iff \lambda_s > 0.7940.$$

The solution interval is thus

$$\lambda_s \in (0.7940, 0.9387)$$

and the mean lifetime belongs to the interval $(259\,200, 280\,130)$.

Notes

This chapter is based on [90]. Section 5.4 comes from [88]. The example given in Subsection 5.4.3 is taken from [92], where sojourn times are positioned under a more applied point of view in the dependability area. Two other related papers by the same authors are [93] and [96]. In [93], conditions are given under which the successive sojourn times in a set are i.i.d., in order to exploit the property for interval availability evaluation. In [96] and [88] some of the results described in this chapter are extended to semi-Markov reward models. See also the monograph [25] which is entirely based on the material of this chapter.

6 Occupation times of subsets of states – interval availability

Let $X = \{X_u, u \geq 0\}$ be a homogeneous Markov chain on a finite state space, S. The process X is characterized by its infinitesimal generator, A, and its initial probability distribution, α. We denote by $Z = \{Z_n, n \geq 0\}$ the uniformized Markov chain, described in Chapter 3, associated with X, with the same initial distribution, α. Its transition probability matrix, P, is related to the matrix A by the relation $P = I + A/\lambda$, where I is the identity matrix and λ satisfies $\lambda \geq \max\{-A_{i,i}, i \in S\}$. The rate, λ, is the rate of the Poisson process $\{N_u, u \geq 0\}$, independent of Z, which counts the number of transitions of process $\{Z_{N_u}, u \geq 0\}$. As shown in Chapter 3, the processes $\{Z_{N_u}\}$ and X are stochastically equivalent. We consider a partition $S = U \cup D$, $U \cap D = \emptyset$, of the state space, S, and we study the occupation time of the subset U.

6.1 The discrete-time case

We consider the Markov chain $Z = \{Z_n, n \geq 0\}$, and we define the random variable, V_n, as the total number of states of U visited during the n first transitions of Z, that is

$$V_n = \sum_{k=0}^{n} 1_{\{Z_k \in U\}}.$$

The following theorem gives the backward equations for the distribution of the pair (V_n, Z_n).

THEOREM 6.1.1. *For $1 \leq k \leq n$ and $j \in S$, we have, for every $i \in U$,*

$$\mathbb{P}\{V_n \leq k, Z_n = j \mid Z_0 = i\} = \sum_{l \in S} P_{i,l} \mathbb{P}\{V_{n-1} \leq k-1, Z_{n-1} = j \mid Z_0 = l\}$$

and, for every $i \in D$,

$$\mathbb{P}\{V_n \leq k, Z_n = j \mid Z_0 = i\} = \sum_{l \in S} P_{i,l} \mathbb{P}\{V_{n-1} \leq k, Z_{n-1} = j \mid Z_0 = l\}.$$

Proof. For $1 \leq k \leq n$ and $j \in S$, we have

$$
\mathbb{P}\{V_n \leq k, Z_n = j \mid Z_0 = i\}
$$
$$
= \sum_{l \in S} P_{i,l} \mathbb{P}\{V_n \leq k, Z_n = j \mid Z_1 = l, Z_0 = i\}
$$
$$
= \sum_{l \in S} P_{i,l} \mathbb{P}\left\{ \sum_{h=1}^{n} 1_{\{Z_h \in U\}} \leq k - 1_{\{i \in U\}}, Z_n = j \mid Z_1 = l, Z_0 = i \right\}
$$
$$
= \sum_{l \in S} P_{i,l} \mathbb{P}\left\{ \sum_{h=1}^{n} 1_{\{Z_h \in U\}} \leq k - 1_{\{i \in U\}}, Z_n = j \mid Z_1 = l \right\}
$$
$$
= \sum_{l \in S} P_{i,l} \mathbb{P}\left\{ \sum_{h=0}^{n-1} 1_{\{Z_h \in U\}} \leq k - 1_{\{i \in U\}}, Z_{n-1} = j \mid Z_0 = l \right\}
$$
$$
= \sum_{l \in S} P_{i,l} \mathbb{P}\{V_{n-1} \leq k - 1_{\{i \in U\}}, Z_{n-1} = j \mid Z_0 = l\}.
$$

The third equality follows from the Markov property of Z and the fourth follows from the homogeneity of Z. ∎

The following theorem gives the forward equations for the probabilities associated with the pair (V_n, Z_n).

THEOREM 6.1.2.   *For $1 \leq k \leq n$ and $i \in S$, we have, for every $j \in U$,*

$$
\mathbb{P}\{V_n \leq k, Z_n = j \mid Z_0 = i\} = \sum_{l \in S} \mathbb{P}\{V_{n-1} \leq k - 1, Z_{n-1} = l \mid Z_0 = l\} P_{l,j}
$$

and, for every $j \in D$,

$$
\mathbb{P}\{V_n \leq k, Z_n = j \mid Z_0 = i\} = \sum_{l \in S} \mathbb{P}\{V_{n-1} \leq k, Z_{n-1} = l \mid Z_0 = l\} P_{l,j}.
$$

Proof. For $1 \leq k \leq n$ and $i \in S$, we have

$$
\mathbb{P}\{V_n \leq k, Z_n = j, Z_0 = i\}
$$
$$
= \mathbb{P}\{V_{n-1} \leq k - 1_{\{j \in U\}}, Z_n = j, Z_0 = i\}
$$
$$
= \sum_{l \in S} \mathbb{P}\{V_{n-1} \leq k - 1_{\{j \in U\}}, Z_n = j, Z_{n-1} = l, Z_0 = i\}
$$
$$
= \sum_{l \in S} \mathbb{P}\{V_{n-1} \leq k - 1_{\{j \in U\}}, Z_0 = i \mid Z_n = j, Z_{n-1} = l\} \mathbb{P}\{Z_n = j, Z_{n-1} = l\}
$$
$$
= \sum_{l \in S} \mathbb{P}\{V_{n-1} \leq k - 1_{\{j \in U\}}, Z_0 = i \mid Z_{n-1} = l\} \mathbb{P}\{Z_n = j, Z_{n-1} = l\}
$$
$$
= \sum_{l \in S} \mathbb{P}\{V_{n-1} \leq k - 1_{\{j \in U\}}, Z_{n-1} = l, Z_0 = i\} \mathbb{P}\{Z_n = j \mid Z_{n-1} = l\}
$$
$$
= \sum_{l \in S} \mathbb{P}\{V_{n-1} \leq k - 1_{\{j \in U\}}, Z_{n-1} = l, Z_0 = i\} P_{l,j}.
$$

The fourth equality follows from the Markov property of Z and the last one follows from the homogeneity of Z. We thus obtain the desired relation by conditioning with respect to Z_0.   ∎

For $n \geq 0$ and $k \geq 0$, we introduce the matrix $F(n,k) = (F_{i,j}(n,k))$ defined by

$$F_{i,j}(n,k) = \mathbb{P}\{V_n \leq k, Z_n = j \mid Z_0 = i\}.$$

The results of Theorems 6.1.1 and 6.1.2 can be easily expressed in matrix notation. We decompose matrices P and $F(n,k)$ with respect to the partition $\{U,D\}$ as

$$P = \begin{pmatrix} P_U & P_{UD} \\ P_{DU} & P_D \end{pmatrix} \quad \text{and} \quad F(n,k) = \begin{pmatrix} F_U(n,k) & F_{UD}(n,k) \\ F_{DU}(n,k) & F_D(n,k) \end{pmatrix}.$$

The result of Theorem 6.1.1 can now be written as

$$\begin{pmatrix} F_U(n,k) & F_{UD}(n,k) \end{pmatrix} = \begin{pmatrix} P_U & P_{UD} \end{pmatrix} F(n-1,k-1)$$
$$\begin{pmatrix} F_{DU}(n,k) & F_D(n,k) \end{pmatrix} = \begin{pmatrix} P_{DU} & P_D \end{pmatrix} F(n-1,k)$$

or as

$$F(n,k) = \begin{pmatrix} P_U & P_{UD} \\ 0 & 0 \end{pmatrix} F(n-1,k-1) + \begin{pmatrix} 0 & 0 \\ P_{DU} & P_D \end{pmatrix} F(n-1,k).$$

In the same way, Theorem 6.1.2 can be written as

$$\begin{pmatrix} F_U(n,k) \\ F_{DU}(n,k) \end{pmatrix} = F(n-1,k-1) \begin{pmatrix} P_U \\ P_{DU} \end{pmatrix}$$

$$\begin{pmatrix} F_{UD}(n,k) \\ F_D(n,k) \end{pmatrix} = F(n-1,k) \begin{pmatrix} P_{UD} \\ P_D \end{pmatrix}$$

or as

$$F(n,k) = F(n-1,k-1) \begin{pmatrix} P_U & 0 \\ P_{DU} & 0 \end{pmatrix} + F(n-1,k) \begin{pmatrix} 0 & P_{UD} \\ 0 & P_D \end{pmatrix}.$$

The initial conditions are

$$F(n,0) = \begin{pmatrix} 0 & 0 \\ 0 & (P_D)^n \end{pmatrix}, \quad \text{for } n \geq 0.$$

Since $V_n \leq n+1$, we have, for all $k \geq n+1$,

$$F_{i,j}(n,k) = \mathbb{P}\{Z_n = j \mid Z_0 = i\} = (P^n)_{i,j}.$$

For $n \geq 0$ and $k \geq 0$, we introduce the matrix $f(n,k) = (f_{i,j}(n,k))$ defined by

$$f_{i,j}(n,k) = \mathbb{P}\{V_n = k, Z_n = j \mid Z_0 = i\},$$

which gives $f(n,k) = F(n,k) - F(n,k-1)$. As we did for matrices P and F, we decompose the matrix $f(n,k)$ with respect to the partition $\{U,D\}$ of the state space, S. For $k = n+1$, we obtain

$$
\begin{aligned}
f_{i,j}(n,n+1) &= \mathbb{P}\{V_n = n+1, Z_n = j \mid Z_0 = i\} \\
&= \mathbb{P}\{Z_n = j, Z_{n-1} \in U, \ldots, Z_1 \in U = j \mid Z_0 = i\} \\
&= \sum_{l \in U} \mathbb{P}\{Z_n = j, Z_{n-1} = l, Z_{n-2} \in U, \ldots, Z_1 \in U = j \mid Z_0 = i\} \\
&= \sum_{l \in U} \mathbb{P}\{Z_{n-1} = l, Z_{n-2} \in U, \ldots, Z_1 \in U = j \mid Z_0 = i\} P_{l,j} \\
&= \sum_{l \in U} f_{i,l}(n-1,n) P_{l,j}.
\end{aligned}
$$

We thus have, in matrix notation, $f_{UD}(n,n+1) = 0$, $f_{DU}(n,n+1) = 0$, $f_D(n,n+1) = 0$, and

$$
f_U(n,n+1) = f_U(n-1,n)P_U.
$$

Since $f_U(0,1) = I$, we finally obtain, for every $n \geq 0$,

$$
f_U(n,n+1) = (P_U)^n. \tag{6.1}
$$

6.2 Order statistics and Poisson process

We consider the order statistics of uniform random variables on $[0,t)$. Formally, let $U_1, \ldots, U_n$ be n i.i.d. uniform random variables on $[0,t)$. We thus have, for all $l = 1, \ldots, n$ and $x \in \mathbb{R}$,

$$
\mathbb{P}\{U_l \leq x\} = \begin{cases} 0 & \text{if } x \leq 0, \\ x/t & \text{if } x \in (0,t), \\ 1 & \text{if } x \geq t. \end{cases}
$$

If the random variables $U_1, \ldots, U_n$ are rearranged in ascending order of magnitude and written as

$$
U_{(1)} \leq U_{(2)} \leq \cdots \leq U_{(n)},
$$

we call $U_{(l)}$ the lth order statistic, $l = 1, \ldots, n$.

Let $F_l(x)$ be the distribution of $U_{(l)}$. We then have, for $x \in (0,t)$,

$$
\begin{aligned}
F_l(x) &= \mathbb{P}\{U_{(l)} \leq x\} \\
&= \mathbb{P}\{\text{at least } l \text{ of the } U_k \text{ are less than or equal to } x\} \\
&= \sum_{k=l}^{n} \binom{n}{k} \left(\frac{x}{t}\right)^k \left(1 - \frac{x}{t}\right)^{n-k}. \tag{6.2}
\end{aligned}
$$

The density function, $f_l(x)$, of $U_{(l)}$ is thus given, for $x \in (0,t)$, by

$$f_l(x) = \frac{n!}{(l-1)!(n-l)!} \frac{1}{t} \left(\frac{x}{t}\right)^l \left(1 - \frac{x}{t}\right)^{n-l}.$$

Consider now the spacings $Y_1 = U_{(1)}, Y_2 = U_{(2)} - U_{(1)}, \dots, Y_n = U_{(n)} - U_{(n-1)}$. As shown in [27], their joint density function $f(x_1,\dots,x_n)$ is given by

$$f(x_1, x_2, \dots, x_n) = \begin{cases} \dfrac{n!}{t^n} & \text{if } \displaystyle\sum_{i=1}^n x_i \leq t, \\ 0 & \text{otherwise.} \end{cases} \tag{6.3}$$

Let $\{N_t, t \in \mathbb{R}\}$ be a Poisson process with rate λ, and let $T_1, T_1 + T_2, \dots, T_1 + \cdots + T_n$, be the first n instants of jumps of $\{N_t\}$. The conditional distribution of the times $T_1, \dots, T_n$ under the condition $N_t = n$ is given by the following theorem.

THEOREM 6.2.1.   *For any nonnegative numbers x_i with $0 \leq x_1 + \cdots + x_n \leq t$,*

$$\mathbb{P}\{T_1 \leq x_1, \dots, T_n \leq x_n \mid N_t = n\} = \frac{n!}{t^n} x_1 \dots x_n.$$

Proof. We use a similar proof to that of [48, page 126]. The times, T_k, are independent and identically distributed following an exponential distribution with rate λ. We thus have, since $x_1 + \cdots + x_n \leq t$,

$$\mathbb{P}\{T_1 \leq x_1, \dots, T_n \leq x_n, N_t = n\}$$

$$= \mathbb{P}\{T_1 \leq x_1, \dots, T_n \leq x_n, T_1 + \cdots + T_{n+1} > t\}$$

$$= \int_0^{x_1} \cdots \int_0^{x_n} \int_{t-(t_1+\cdots+t_n)}^{\infty} \lambda^{n+1} e^{-\lambda(t_1+\cdots+t_{n+1})} dt_{n+1} \dots dt_1$$

$$= \lambda^{n+1} \int_0^{x_1} \cdots \int_0^{x_n} e^{-\lambda(t_1+\cdots+t_n)} \left[\frac{-e^{-\lambda t_{n+1}}}{\lambda}\right]_{t-(t_1+\cdots+t_n)}^{\infty} dt_n \dots dt_1$$

$$= \lambda^n e^{-\lambda t} x_1 \dots x_n.$$

But

$$\mathbb{P}\{N_t = n\} = e^{-\lambda t} \frac{(\lambda t)^n}{n!},$$

hence

$$\mathbb{P}\{T_1 \leq x_1, \dots, T_n \leq x_n \mid N_t = n\} = \frac{n!}{t^n} x_1 \dots x_n,$$

which completes the proof.   ∎

The conditional density of $T_1, \dots, T_n$, given that $N_t = n$, is thus equal to the joint density $f(x_1, \dots, x_n)$, given by (6.3), of the order statistics from the uniform distribution on $(0,t)$.

The symmetric role played by the variables $x_1, \ldots, x_n$ shows that the random variables, $T_1, \ldots, T_n$, are exchangeable. This remark leads to the following result. In the proof of Theorem 6.2.1, the variable T_{n+1} was the $(n+1)$th occurrence time of the Poisson process, $\{N_t\}$. Here, for fixed n and t, we redefine T_{n+1} as $T_{n+1} = t - (T_0 + \cdots + T_n)$.

THEOREM 6.2.2.   *For $1 \leq l \leq n$, we consider a subset of indices*

$$I = \{i_1, \cdots, i_l\} \subset \{1, \ldots, n, n+1\}$$

and a real number $s \in (0,t)$; then,

$$\mathbb{P}\{T_{i_1} + \cdots + T_{i_l} \leq s \mid N_t = n\} = \sum_{k=l}^{n} \binom{n}{k} \left(\frac{x}{t}\right)^k \left(1 - \frac{x}{t}\right)^{n-k}.$$

Proof.   If $n+1 \notin I$, we have

$$\mathbb{P}\{T_{i_1} + \cdots + T_{i_l} \leq s \mid N_t = n\} = \int \cdots \int_{E_s} f(x_1, \ldots, x_n) dx_1 \ldots dx_n$$

$$= \frac{n!}{t^n} \int \cdots \int_{E_s} dx_1 \ldots dx_n$$

$$= \mathbb{P}\{T_1 + \cdots + T_l \leq s \mid N_t = n\}$$

$$= \mathbb{P}\{Y_1 + \cdots + Y_l \leq s\}$$

$$= \mathbb{P}\{U_{(l)} \leq s\}$$

$$= \sum_{k=l}^{n} \binom{n}{k} \left(\frac{s}{t}\right)^k \left(1 - \frac{s}{t}\right)^{n-k} \quad \text{by (6.2),}$$

where

$$E_s = \{(x_1, \ldots, x_n) \mid 0 \leq x_1 + \cdots + x_l \leq s \quad \text{and} \quad 0 \leq x_1 + \cdots + x_n \leq t\}.$$

If $n+1 \in I$, we define $J = \{1, \ldots, n, n+1\} \setminus I$, so that $n+1 \notin J$, and we obtain, since $T_1 + \cdots + T_{n+1} = t$,

$$\mathbb{P}\{T_{i_1} + \cdots + T_{i_l} \leq s \mid N_t = n\}$$

$$= \mathbb{P}\left\{t - \sum_{j \in J} T_j \leq s \mid N_t = n\right\}$$

$$= \mathbb{P}\left\{ \sum_{j \in J} T_j > t - s \mid N_t = n \right\}$$

$$= 1 - \mathbb{P}\left\{ \sum_{j \in J} T_j \leq t - s \mid N_t = n \right\}$$

$$= 1 - \mathbb{P}\{T_1 + \cdots + T_{n-l+1} \leq t - s \mid N_t = n\}$$

$$= 1 - \mathbb{P}\{Y_1 + \cdots + Y_{n-l+1} \leq t - s\}$$

$$= 1 - \mathbb{P}\{U_{(n-l+1)} \leq t - s\}$$

$$= 1 - \sum_{k=n-l+1}^{n} \binom{n}{k} \left(\frac{t-s}{t}\right)^k \left(1 - \frac{t-s}{t}\right)^{n-k} \quad \text{by (6.2)}$$

$$= \sum_{k=l}^{n} \binom{n}{k} \left(\frac{s}{t}\right)^k \left(1 - \frac{s}{t}\right)^{n-k},$$

which completes the proof.   ∎

6.3   The continuous-time case

We consider the Markov chain $X = \{X_t, t \geq 0\}$, and the occupation time, W_t, of the subset U in $[0, t)$, that is

$$W_t = \int_0^t 1_{\{X_u \in U\}} \, du. \tag{6.4}$$

This random variable represents the total time spent by process X in the subset U during the interval $[0, t)$. It has attracted much attention as W_t/t is also known as the interval availability in reliability and dependability theory. In [28] an expression for the distribution of W_t was obtained using uniform order statistics on $[0, t)$. From a computational point of view that expression is very interesting and various methods have been developed to compute it even in the case of denumerable state spaces, see [28], [102], [95], [97] and the references therein. We recall here how the joint distribution of the pair (W_t, X_t) was obtained in [28] using the forward and backward equations associated with the uniformized Markov chain of process X.

To avoid large expressions, we use, for every $i \in S$, the notation $\mathbb{P}_i$ to denote the conditional probability given that Z_0 or X_0 is equal to i, that is $\mathbb{P}_i\{\cdot\} = \mathbb{P}\{\cdot \mid X_0 = i\} = \mathbb{P}\{\cdot \mid Z_0 = i\}$, since X_0 and Z_0 have the same distribution, α. The joint distribution of the pair (W_t, X_t) is given in the following theorem by using the distribution of V_n, the total number of states of U visited during the n first transitions of Z, that we obtained in the previous section.

THEOREM 6.3.1. *For every $i,j \in S$, for $t > 0$ and $s \in [0,t)$, we have*

$$\mathbb{P}_i\{W_t \leq s, X_t = j\}$$

$$= \sum_{n=0}^{\infty} e^{-\lambda t} \frac{(\lambda t)^n}{n!} \sum_{k=0}^{n} \binom{n}{k} \left(\frac{s}{t}\right)^k \left(1 - \frac{s}{t}\right)^{n-k} \mathbb{P}_i\{V_n \leq k, Z_n = j\}. \tag{6.5}$$

Proof. For $s < t$, since processes $\{X_t\}$ and $\{Z_{N_t}\}$ are equivalent, we have

$$\mathbb{P}\{W_t \leq s, X_t = j \mid X_0 = i\}$$

$$= \sum_{n=0}^{\infty} \mathbb{P}_i\{W_t \leq s, N_t = n, X_t = j\}$$

$$= \sum_{n=0}^{\infty} \mathbb{P}_i\{W_t \leq s, N_t = n, Z_n = j\}$$

$$= \sum_{n=0}^{\infty} \mathbb{P}_i\{N_t = n\}\mathbb{P}_i\{W_t \leq s, Z_n = j \mid N_t = n\}$$

$$= \sum_{n=0}^{\infty} e^{-\lambda t} \frac{(\lambda t)^n}{n!} \mathbb{P}_i\{W_t \leq s, Z_n = j \mid N_t = n\},$$

where this last equality follows from the independence of processes $\{N_t\}$ and $\{Z_n\}$. The second term in this sum is obtained by writing

$$\mathbb{P}_i\{W_t \leq s, Z_n = j \mid N_t = n\}$$

$$= \sum_{l=0}^{n+1} \mathbb{P}_i\{W_t \leq s, V_n = l, Z_n = j \mid N_t = n\}$$

$$= \sum_{l=0}^{n+1} \mathbb{P}_i\{V_n = l, Z_n = j \mid N_t = n\}\mathbb{P}_i\{W_t \leq s \mid V_n = l, Z_n = j, N_t = n\}$$

$$= \sum_{l=0}^{n+1} \mathbb{P}_i\{V_n = l, Z_n = j\}\mathbb{P}_i\{W_t \leq s \mid V_n = l, Z_n = j, N_t = n\}$$

$$= \sum_{l=0}^{n} \mathbb{P}_i\{V_n = l, Z_n = j\}\mathbb{P}_i\{W_t \leq s \mid V_n = l, Z_n = j, N_t = n\}.$$

The third equality follows from the independence of processes $\{Z_n\}$ and $\{N_t\}$. The last equality follows from the fact that if $l = n+1$, we trivially have

$$\{V_n = n+1, N_t = n\} \subset \{W_t = t\},$$

and so we obtain

$$\mathbb{P}\{W_t \leq s \mid V_n = n+1, Z_n = j, N_t = n\} = 0, \text{ since } s < t.$$

Let us consider now the expression $\mathbb{P}_i\{W_t \leq s \mid V_n = l, Z_n = j, N_t = n\}$. We denote by T_1, $T_1 + T_2, \ldots, T_1 + \cdots + T_n$ the n first instants of jumps of the Poisson process $\{N_t\}$ and we set $T_{n+1} = t - (T_1 + \cdots + T_n)$. Then,

$$\mathbb{P}_i\{W_t \leq s \mid V_n = l, Z_n = j, N_t = n\}$$
$$= \mathbb{P}_i\{T_{i_1} + \cdots + T_{i_l} \leq s \mid V_n = l, Z_n = j, N_t = n\}$$
$$= \mathbb{P}\{T_{i_1} + \cdots + T_{i_l} \leq s \mid N_t = n\},$$

where the distinct indices $i_1, \ldots, i_l \in \{1, \ldots, n, n+1\}$ correspond to the l states of U visited during the n first transitions of Z and the last equality is due to the independence of the processes $\{Z_n\}$ and $\{N_t\}$. Note that for $l = 0$, we obtain the correct result, which is equal to 1 using the convention $\sum_a^b(\ldots) = 0$ if $a > b$.

From Theorem 6.2.2 we have, for $l = 0, \ldots, n$,

$$\mathbb{P}\{T_{i_1} + \cdots + T_{i_l} \leq s \mid N_t = n\} = \sum_{k=l}^n \binom{n}{k} \left(\frac{s}{t}\right)^k \left(1 - \frac{s}{t}\right)^{n-k}.$$

Again, the convention $\sum_a^b(\ldots) = 0$ for $a > b$ allows us to cover the cases $l = 0$ and $l = n + 1$. Finally, we obtain

$$\mathbb{P}_i\{W_t \leq s \mid V_n = l, Z_n = j, N_t = n\} = \sum_{k=l}^n \binom{n}{k} \left(\frac{s}{t}\right)^k \left(1 - \frac{s}{t}\right)^{n-k}.$$

That is,

$$\mathbb{P}\{W_t \leq s, X_t = j \mid X_0 = i\}$$
$$= \sum_{n=0}^{\infty} e^{-\lambda t} \frac{(\lambda t)^n}{n!} \sum_{l=0}^n \sum_{k=l}^n \binom{n}{k} \left(\frac{s}{t}\right)^k \left(1 - \frac{s}{t}\right)^{n-k} \mathbb{P}\{V_n = l, Z_n = j \mid Z_0 = i\}$$
$$= \sum_{n=0}^{\infty} e^{-\lambda t} \frac{(\lambda t)^n}{n!} \sum_{k=0}^n \binom{n}{k} \left(\frac{s}{t}\right)^k \left(1 - \frac{s}{t}\right)^{n-k} \sum_{l=0}^k \mathbb{P}\{V_n = l, Z_n = j \mid Z_0 = i\}$$
$$= \sum_{n=0}^{\infty} e^{-\lambda t} \frac{(\lambda t)^n}{n!} \sum_{k=0}^n \binom{n}{k} \left(\frac{s}{t}\right)^k \left(1 - \frac{s}{t}\right)^{n-k} \mathbb{P}\{V_n \leq k, Z_n = j \mid Z_0 = i\},$$

which completes the proof.   ∎

We decompose the initial probability distribution α of X and the infinitesimal generator A of X with respect to partition $\{U, D\}$ of S as

$$\alpha = \begin{pmatrix} \alpha_U & \alpha_D \end{pmatrix} \quad \text{and} \quad A = \begin{pmatrix} A_U & A_{UD} \\ A_{DU} & A_D \end{pmatrix}.$$

From Relation (6.5), for $t > 0$, the distribution $\mathbb{P}\{W_t \leq s, X_t = j \mid X_0 = i\}$ is continuously differentiable with respect to the variables s and t, for $t > 0$ and $s \in (0, t)$. By taking the limit when s tends towards t, with $s < t$, in Relation (6.5), we obtain

$$\mathbb{P}\{W_t < t, X_t = j \mid X_0 = i\}$$

$$= \sum_{n=0}^{\infty} e^{-\lambda t} \frac{(\lambda t)^n}{n!} \mathbb{P}\{V_n \leq n, Z_n = j \mid Z_0 = i\}$$

$$= \sum_{n=0}^{\infty} e^{-\lambda t} \frac{(\lambda t)^n}{n!} ((P^n)_{i,j} - \mathbb{P}\{V_n = n+1, Z_n = j \mid Z_0 = i\})$$

$$= \sum_{n=0}^{\infty} e^{-\lambda t} \frac{(\lambda t)^n}{n!} ((P^n)_{i,j} - (P^n_U)_{i,j} \mathbb{1}_{\{i,j \in U\}})$$

$$= (e^{At})_{i,j} - (e^{A_U t})_{i,j} \mathbb{1}_{\{i,j \in U\}},$$

where the third equality is obtained using Relation (6.1). Since $W_t \leq t$, we obtain

$$\mathbb{P}\{W_t = t, X_t = j \mid X_0 = i\} = (e^{A_U t})_{i,j} \mathbb{1}_{\{i,j \in U\}}. \tag{6.6}$$

The jump at time t is thus

$$\mathbb{P}\{W_t = t\} = \alpha_U e^{A_U t} \mathbb{1}.$$

Note that Relation (6.6) can also be obtained from Relation (6.4) by observing that $W_t = t$ is equivalent to $X_u \in U$ for all $u \in [0,t)$, which is in turn equivalent to $S_{U,1} > t$. Thus, we obtain

$$\mathbb{P}\{W_t = t, X_t = j \mid X_0 = i\} = \mathbb{P}\{S_{U,1} > t, X_t = j \mid X_0 = i\} = (e^{A_U t})_{i,j} \mathbb{1}_{\{i,j \in U\}}.$$

COROLLARY 6.3.2. *For $t > 0$ and $s \in [0,t)$, we have*

$$\mathbb{P}\{W_t \leq s\} = \sum_{n=0}^{\infty} e^{-\lambda t} \frac{(\lambda t)^n}{n!} \sum_{k=0}^{n} \binom{n}{k} \left(\frac{s}{t}\right)^k \left(1 - \frac{s}{t}\right)^{n-k} \mathbb{P}\{V_n \leq k\}. \tag{6.7}$$

Proof. For $t > 0$ and $s \in [0,t)$, we have

$$\mathbb{P}\{W_t \leq s\} = \sum_{i \in S} \mathbb{P}\{X_0 = i\} \sum_{j \in S} \mathbb{P}\{W_t \leq s, X_t = j \mid X_0 = i\}$$

and, for $0 \leq k \leq n$,

$$\mathbb{P}\{V_n \leq k\} = \sum_{i \in S} \mathbb{P}\{Z_0 = i\} \sum_{j \in S} \mathbb{P}\{V_n \leq k, Z_n = j \mid Z_0 = i\}.$$

The result follows from Theorem 6.3.1 and because X_0 and Z_0 have the same distribution. ∎

The derivative of $\mathbb{P}\{W_t \leq s, X_t = j \mid X_0 = i\}$ is given in the following corollary.

COROLLARY 6.3.3.   *For $t > 0$ and $s \in (0,t)$, we have*

$$\frac{d\mathbb{P}\{W_t \leq s, X_t = j \mid X_0 = i\}}{ds}$$

$$= \lambda \sum_{n=0}^{\infty} e^{-\lambda t} \frac{(\lambda t)^n}{n!} \sum_{k=0}^{n} \binom{n}{k} \left(\frac{s}{t}\right)^k \left(1 - \frac{s}{t}\right)^{n-k} \mathbb{P}_i\{V_{n+1} = k+1, Z_{n+1} = j\}. \qquad (6.8)$$

Proof.   In order to simplify the writing, we define

$$G_{i,j}(t,s) = \mathbb{P}\{W_t \leq s, X_t = j \mid X_0 = i\},$$
$$F_{i,j}(n,k) = \mathbb{P}\{V_n \leq k, Z_n = j \mid Z_0 = i\},$$

and

$$f_{i,j}(n,k) = \mathbb{P}\{V_n = k, Z_n = j \mid Z_0 = i\}.$$

From Relation (6.5), for $t > 0$ and $s \in (0,t)$, we have

$$\frac{dG_{i,j}(t,s)}{ds} = \frac{1}{t} \sum_{n=1}^{\infty} e^{-\lambda t} \frac{(\lambda t)^n}{n!} \sum_{k=1}^{n} \frac{n!}{(k-1)!(n-k)!} \left(\frac{s}{t}\right)^{k-1} \left(1 - \frac{s}{t}\right)^{n-k} F_{i,j}(n,k)$$

$$- \frac{1}{t} \sum_{n=1}^{\infty} e^{-\lambda t} \frac{(\lambda t)^n}{n!} \sum_{k=0}^{n-1} \frac{n!}{k!(n-k-1)!} \left(\frac{s}{t}\right)^k \left(1 - \frac{s}{t}\right)^{n-k-1} F_{i,j}(n,k)$$

$$= \frac{1}{t} \sum_{n=1}^{\infty} e^{-\lambda t} \frac{(\lambda t)^n}{n!} \sum_{k=0}^{n-1} \frac{n!}{k!(n-k-1)!} \left(\frac{s}{t}\right)^k \left(1 - \frac{s}{t}\right)^{n-k-1} F_{i,j}(n,k+1)$$

$$- \frac{1}{t} \sum_{n=1}^{\infty} e^{-\lambda t} \frac{(\lambda t)^n}{n!} \sum_{k=0}^{n-1} \frac{n!}{k!(n-k-1)!} \left(\frac{s}{t}\right)^k \left(1 - \frac{s}{t}\right)^{n-k-1} F_{i,j}(n,k)$$

$$= \frac{1}{t} \sum_{n=1}^{\infty} e^{-\lambda t} \frac{(\lambda t)^n}{n!} \sum_{k=0}^{n-1} \frac{n!}{k!(n-k-1)!} \left(\frac{s}{t}\right)^k \left(1 - \frac{s}{t}\right)^{n-k-1} f_{i,j}(n,k+1)$$

$$= \lambda \sum_{n=0}^{\infty} e^{-\lambda t} \frac{(\lambda t)^n}{n!} \sum_{k=0}^{n} \binom{n}{k} \left(\frac{s}{t}\right)^k \left(1 - \frac{s}{t}\right)^{n-k} f_{i,j}(n+1,k+1),$$

which completes the proof.   ∎

Notes

The first algorithm for the computation of the interval availability distribution using the uniformization technique was proposed in [28]. The computational complexity of this method was improved in [95]. In [97], the authors extended the algorithm to the case of infinite state spaces. The interval availability distribution was analyzed using operational periods in [93]. For a two-state system with exponential failures and phase-type repairs, the distribution of interval availability may be expressed in closed-form as shown in [103].

7 Linear combination of occupation times – performability

As recognized in a large number of studies, the quantitative evaluation of fault-tolerant computer systems needs to deal simultaneously with aspects of both performance and dependability. For this purpose, Meyer [66] developed the concept of performability, which may be interpreted as the cumulative performance over a finite mission time. The increasing need to evaluate cumulative measures comes from the fact that in highly available systems, steady-state measures can be very poor, even if the mission time is not small. Considering, for instance, critical applications, it is crucial for the user to ensure that the probability that the system will achieve a given performance level is high enough.

Formally, the system fault-repair behavior is assumed to be modeled by a homogeneous Markov chain. Its state space is divided into disjoint subsets, which represent the different configurations of the system. A performance level or reward rate is associated with each of these configurations. This reward rate quantifies the ability of the system to perform in the corresponding configuration. Performability is then the accumulated reward over the mission time.

So, we consider a homogeneous Markov chain, $X = \{X_u, u \geq 0\}$, with finite state space, S. The process X is characterized by its infinitesimal generator, A, and its initial probability distribution, α. As usual, we denote by P the transition probability matrix of the Markov chain, $Z = \{Z_n, n \geq 0\}$, obtained by uniformization of Markov chain, X. The matrix, P, is given by $P = I + A/\lambda$, where I is the identity matrix and λ satisfies $\lambda \geq \max\{-A_{i,i}, i \in S\}$. The rate, λ, is the rate of the Poisson process $\{N_u, u \geq 0\}$, independent of Z, which counts the number of transitions of process $\{Z_{N_u}, u \geq 0\}$. It was shown in Chapter 3 that the processes $\{Z_{N_u}\}$ and X are stochastically equivalent.

A constant performance level or reward rate, $\rho(i)$, is associated with each state, i, of S. We consider the random variable, Y_t, defined by

$$Y_t = \int_0^t \rho(X_u)\,du.$$

We denote by $m + 1$ the number of distinct rewards and their values by

$$r_0 < r_1 < \cdots < r_{m-1} < r_m.$$

We then have $Y_t \in [r_0 t, r_m t]$ with probability 1. Without loss of generality, we may set $r_0 = 0$. This is done by considering the random variable $Y_t - r_0 t$ instead of Y_t and the reward rates $r_i - r_0$ instead of r_i. As in Chapter 4, the state space, S, is partitioned into subsets $B_0, \ldots, B_m$. The subset B_l contains the states with reward rate r_l, that is $B_l = \{i \in S \mid \rho(i) = r_l\}$. With this notation,

$$Y_t = \sum_{l=1}^{m} r_l \int_0^t 1_{\{X_u \in B_l\}} \, du = \sum_{l=1}^{m} r_l W_t^l, \tag{7.1}$$

where W_t^l represents the total time spent by the process X in the subset B_l during the interval $[0, t)$. The distribution of W_t^l was the subject of Chapter 6.

As the distribution of each W_t^l has at most two jumps at 0 and t, the distribution of Y_t has at most $m + 1$ jumps at the points $r_0 t = 0, r_1 t, \ldots, r_m t$. For $t > 0$, the jump at point $x = r_l t$ is equal to the probability that the process X, starting in subset B_l, stays in the subset B_l during all of $[0, t)$, that is

$$\mathbb{P}\{Y_t = r_l t\} = \alpha_{B_l} e^{A_{B_l B_l} t} 1_{B_l} \text{ for } t > 0,$$

where 1_{B_l} is the column vector of dimension $|B_l|$ with all its entries equal to 1 and $A_{B_l B_l}$ is the submatrix of A containing the $A_{i,j}$ for $i, j \in B_l$. For every $i, j \in S$ and $t > 0$, we define the functions $F_{i,j}(t, x)$ by

$$F_{i,j}(t, x) = \mathbb{P}\{Y_t > x, X_t = j \mid X_0 = i\}.$$

The joint distribution $\mathbb{P}\{W_t^0 \le x_0, \ldots, W_t^m \le x_m\}$ was obtained in [104] where it is shown that it is differentiable with respect to t and also with respect to $x_0, \ldots, x_m$, for $t > 0$ and $x_0, \ldots, x_m \in (0, t)$. Thus, the distribution $F_{i,j}(t, x)$ is differentiable with respect to x and t in the domain

$$E = \{(t, x) \mid t > 0 \quad \text{and} \quad x \in \bigcup_{l=1}^{m} (r_{l-1} t, r_l t)\}.$$

We introduce the matrix $F(t, x) = (F_{i,j}(t, x))_{i,j \in S}$ and, using the partition $B_m, B_{m-1}, \ldots, B_0$, the matrices A, P, and $F(t, x)$ can be written in block form as

$$A = \left(A_{B_u B_v} \right)_{0 \le u, v \le m},$$

$$P = \left(P_{B_u B_v} \right)_{0 \le u, v \le m},$$

and

$$F = \left(F_{B_u B_v}(t, x) \right)_{0 \le u, v \le m}.$$

Note that for $t > 0$ and $0 \le l \le m$, we have, from Relation (6.6),

$$\mathbb{P}\{Y_t = r_l t, X_t = j \mid X_0 = i\} = \mathbb{P}\{W_t^l = t, X_t = j \mid X_0 = i\}$$

$$= (e^{A_{B_l B_l} t})_{i,j} 1_{\{i, j \in B_l\}}, \tag{7.2}$$

i.e.

$$\mathbb{P}\{Y_t = r_l t, X_t = j \mid X_0 = i\} = \sum_{n=0}^{\infty} e^{-\lambda t} \frac{(\lambda t)^n}{n} ((P_{B_l B_l})^n)_{i,j} 1_{\{i, j \in B_l\}}. \tag{7.3}$$

The initial conditions are given, for $t > 0$, by

$$F_{i,j}(t,0) = \mathbb{P}\{X_t = j \mid X_0 = i\} - \mathbb{P}\{Y_t = 0, X_t = j \mid X_0 = i\},$$

that is, in matrix notation,

$$F_{B_u B_v}(t,0) = (e^{At})_{B_u B_v} - e^{A_{B_0 B_0} t} \mathbf{1}_{\{u=v=0\}},$$

which can also be written as

$$F_{B_u B_v}(t,0) = \sum_{n=0}^{\infty} e^{-\lambda t} \frac{(\lambda t)^n}{n} \left[(P^n)_{B_u B_v} - (P_{B_0 B_0})^n \mathbf{1}_{\{u=v=0\}} \right]. \tag{7.4}$$

7.1　　Backward and forward equations

We derive backward and forward equations satisfied by the distribution of the pair (Y_t, X_t). First, we recall some useful results in the following lemma. Remember that $\{N_t\}$ is a Poisson process with rate λ, independent of the Markov chain Z. We denote by $N_{t,t+s}$ the number of transitions during the interval $[t, t+s)$.

LEMMA 7.1.1.

$$\mathbb{P}\{N_{t,t+s} = 0 \mid X_t = j\} = e^{-\lambda s} \tag{7.5}$$

$$\mathbb{P}\{X_{t+s} = j, N_{t,t+s} = 1 \mid X_t = i\} = P_{i,j} \lambda s e^{-\lambda s} \tag{7.6}$$

$$\mathbb{P}\{N_{t,t+s} \geq 2 \mid X_0 = i\} = o(s). \tag{7.7}$$

Proof. Recall that the processes $\{X_t\}$ and $\{Z_{N_t}\}$ are equivalent and $Z_0 = X_0$. For Relation (7.5), by homogeneity, we have

$$\mathbb{P}\{N_{t,t+s} = 0 \mid X_t = j\} = \mathbb{P}\{N_s = 0 \mid X_0 = j\}$$
$$= \mathbb{P}\{N_s = 0 \mid Z_0 = j\}$$
$$= \mathbb{P}\{N_s = 0\}$$
$$= e^{-\lambda s}.$$

In the same way, for Relation (7.6), we have

$$\mathbb{P}\{X_{t+s} = j, N_{t,t+s} = 1 \mid X_t = i\} = \mathbb{P}\{X_s = j, N_s = 1 \mid X_0 = i\}$$
$$= \mathbb{P}\{Z_1 = j, N_s = 1 \mid Z_0 = i\}$$
$$= \mathbb{P}\{N_s = 1 \mid Z_0 = i, Z_1 = j\} P_{i,j}$$
$$= \mathbb{P}\{N_s = 1\} P_{i,j}$$
$$= P_{i,j} \lambda s e^{-\lambda s}.$$

For Relation (7.7), we have

$$\mathbb{P}\{N_{t,t+s} \geq 2 \mid X_t = i\} = \mathbb{P}\{N_s \geq 2 \mid X_0 = i\}$$
$$= \mathbb{P}\{N_s \geq 2\}$$
$$= 1 - e^{-\lambda s} - \lambda s e^{-\lambda s}$$
$$= o(s).$$

This completes the proof. ∎

The following theorem establishes the forward equation for the pair (Y_t, X_t).

THEOREM 7.1.2.   *For $t > 0$, $i, j \in S$, $1 \leq h \leq m$ and $x \in (r_{h-1}t, r_h t)$, we have*

$$\frac{\partial F_{i,j}(t,x)}{\partial t} = -\rho(j) \frac{\partial F_{i,j}(t,x)}{\partial x} + \sum_{k \in S} F_{i,k}(t,x) A_{k,j}. \qquad (7.8)$$

Proof.   By conditioning on the number of transitions in $[t, t+s)$, we have

$$\mathbb{P}_i\{Y_{t+s} > x, X_{t+s} = j\} = \mathbb{P}_i\{Y_{t+s} > x, X_{t+s} = j, N_{t,t+s} = 0\}$$
$$+ \mathbb{P}_i\{Y_{t+s} > x, X_{t+s} = j, N_{t,t+s} = 1\}$$
$$+ \mathbb{P}_i\{Y_{t+s} > x, X_{t+s} = j, N_{t,t+s} \geq 2\}.$$

We separately consider these three terms. For the first term, since $X_{t+s} = j$ and $N_{t,t+s} = 0$ is equivalent to $X_t = j$ and $N_{t,t+s} = 0$, we have

$$\mathbb{P}_i\{Y_{t+s} > x, X_{t+s} = j, N_{t,t+s} = 0\}$$
$$= \mathbb{P}_i\{Y_{t+s} > x, X_t = j, N_{t,t+s} = 0\}$$
$$= \mathbb{P}_i\{Y_{t+s} > x \mid X_t = j, N_{t,t+s} = 0\} \mathbb{P}_i\{X_t = j, N_{t,t+s} = 0\}$$
$$= \mathbb{P}_i\{Y_t > x - \rho(j)s \mid X_t = j, N_{t,t+s} = 0\} \mathbb{P}_i\{X_t = j, N_{t,t+s} = 0\}$$
$$= \mathbb{P}_i\{Y_t > x - \rho(j)s \mid X_t = j\} \mathbb{P}_i\{X_t = j, N_{t,t+s} = 0\}$$
$$= \mathbb{P}_i\{Y_t > x - \rho(j)s, X_t = j\} \mathbb{P}_i\{N_{t,t+s} = 0 \mid X_t = j\}$$
$$= \mathbb{P}\{N_{t,t+s} = 0 \mid X_t = j\} F_{i,j}(t, x - \rho(j)s)$$
$$= e^{-\lambda s} F_{i,j}(t, x - \rho(j)s)$$
$$= (1 - \lambda s) F_{i,j}(t, x - \rho(j)s) + o(s).$$

The third equality follows from the fact that, if $X_t = j$ and $N_{t,t+s} = 0$, we have $Y_{t+s} = Y_t + \rho(j)s$. The fourth and sixth follow from the Markov property, and the seventh from Relation (7.5). The last equality is due to the fact that $F_{i,j}(t, x - \rho(j)s) \in [0, 1]$. For the second term, denoted by $G(s)$, we define

$$G_k(s) = \mathbb{P}_i\{Y_{t+s} > x \mid X_t = k, X_{t+s} = j, N_{t,t+s} = 1\}.$$

We then have

$$G(s) = \mathbb{P}_i\{Y_{t+s} > x, X_{t+s} = j, N_{t,t+s} = 1\}$$
$$= \sum_{k \in S} G_k(s) \mathbb{P}_i\{X_t = k, X_{t+s} = j, N_{t,t+s} = 1\}.$$

Let us define $\rho_{\max} = \max_{i \in S}\{\rho(i)\}$. As $\min_{i \in S}\{\rho(i)\} = 0$, we have

$$Y_t \leq Y_{t+s} \leq Y_t + \rho_{\max}s,$$

and thus

$$\mathbb{P}_i\{Y_t > x \mid X_t = k, X_{t+s} = j, N_{t,t+s} = 1\} \leq G_k(s),$$

and

$$G_k(s) \leq \mathbb{P}_i\{Y_t > x - \rho_{\max}s \mid X_t = k, X_{t+s} = j, N_{t,t+s} = 1\}.$$

Using the Markov property,

$$\mathbb{P}_i\{Y_t > x \mid X_t = k\} \leq G_k(s) \leq \mathbb{P}_i\{Y_t > x - \rho_{\max}s \mid X_t = k\}.$$

We obtain

$$\sum_{k \in S} F_{i,k}(t,x) U_{k,j}(s) \leq G(s) \leq \sum_{k \in S} F_{i,k}(t,x - \rho_{\max}s) U_{k,j}(s),$$

where $U_{k,j}(s) = \mathbb{P}\{X_{t+s} = j, N_{t,t+s} = 1 \mid X_t = k\}$. From Relation (7.6),

$$\lim_{s \to 0} \frac{U_{k,j}(s)}{s} = \lambda P_{k,j},$$

so we obtain

$$\lim_{s \to 0} \frac{G(s)}{s} = \lambda \sum_{k \in S} F_{i,k}(t,x) P_{k,j}.$$

For the third term, we have, by Relation (7.7),

$$\mathbb{P}_i\{Y_{t+s} > x, X_{t+s} = j, N_{t,t+s} \geq 2\} \leq \mathbb{P}_i\{N_{t,t+s} \geq 2\} = o(s).$$

Combining the three terms, we obtain

$$\frac{F_{i,j}(t+s,x) - F_{i,j}(t,x)}{s}$$
$$= \frac{(1 - \lambda s)F_{i,j}(t, x - \rho(j)s) - F_{i,j}(t,x)}{s} + \frac{G(s)}{s} + \frac{o(s)}{s}$$
$$= \frac{F_{i,j}(t, x - \rho(j)s) - F_{i,j}(t,x)}{s} - \lambda F_{i,j}(t, x - \rho(j)s) + \frac{G(s)}{s} + \frac{o(s)}{s}.$$

If now s tends to 0, we obtain

$$\frac{\partial F_{i,j}(t,x)}{\partial t} = -\rho(j)\frac{\partial F_{i,j}(t,x)}{\partial x} - \lambda F_{i,j}(t,x) + \lambda \sum_{k \in S} F_{i,k}(t,x) P_{k,j}.$$

Since $P = I + A/\lambda$, we obtain

$$\frac{\partial F_{i,j}(t,x)}{\partial t} = -\rho(j)\frac{\partial F_{i,j}(t,x)}{\partial x} + \sum_{k \in S} F_{i,k}(t,x)A_{k,j},$$

which completes the proof. ∎

COROLLARY 7.1.3. *For $t > 0$, $i,j \in S$, $1 \le h \le m$ and $x \in (r_{h-1}t, r_h t)$, we have*

$$F_{i,j}(t,x) = \sum_{k \in S} \int_0^t F_{i,k}(t-u, x-\rho(j)u)\lambda e^{-\lambda u} du P_{k,j} + e^{-\lambda t}1_{\{x < \rho_j t\}}1_{\{i=j\}}. \tag{7.9}$$

Proof. Consider Equation (7.8) and the functions $\varphi_{i,j}$ defined by

$$\varphi_{i,j}(u) = F_{i,j}(t-u, x-\rho(j)u)e^{-\lambda u}.$$

Differentiating with respect to u yields

$$\varphi'_{i,j}(u) = e^{-\lambda u}\left[-\frac{\partial F_{i,j}}{\partial t} - \rho(j)\frac{\partial F_{i,j}}{\partial x}\right](t-u, x-\rho(j)u)$$
$$- F_{i,j}(t-u, x-\rho(j)u)\lambda e^{-\lambda u}$$

which gives, by Equation (7.8) and the relation $A = -\lambda(I - P)$,

$$\varphi'_{i,j}(u) = -\sum_{k \in S} F_{i,k}(t-u, x-\rho(j)u)A_{k,j}e^{-\lambda u} - F_{i,j}(t-u, x-\rho(j)u)\lambda e^{-\lambda u}$$
$$= -\sum_{k \in S} F_{i,k}(t-u, x-\rho(j)u)\lambda e^{-\lambda u}P_{k,j}.$$

Integrating that expression between 0 and t gives

$$\varphi_{i,j}(t) - \varphi_{i,j}(0) = -\sum_{k \in S}\int_0^t F_{i,k}(t-u, x-\rho(j)u)\lambda e^{-\lambda u} du P_{k,j}.$$

Finally, we have $\varphi_{i,j}(0) = F_{i,j}(t,x)$ and

$$\varphi_{i,j}(t) = F_{i,j}(0, x-\rho(j)t)e^{-\lambda t} = e^{-\lambda t}1_{\{x < \rho(j)t\}}1_{\{i=j\}},$$

which completes the proof. ∎

We next derive the backward equation for the evolution of the distribution of the pair (Y_t, X_t).

THEOREM 7.1.4. *For $t > 0$, $i,j \in S$, $1 \le h \le m$, and $x \in (r_{h-1}t, r_h t)$,*

$$F_{i,j}(t,x) = \sum_{k \in S} P_{i,k}\int_0^t F_{k,j}(t-u, x-\rho(i)u)\lambda e^{-\lambda u} du + e^{-\lambda t}1_{\{x < \rho_i t\}}1_{\{i=j\}}. \tag{7.10}$$

Proof. Let T_1 be the sojourn time in the initial state. We have

$$F_{i,j}(t,x) = \int_0^\infty \mathbb{P}\{Y_t > x, X_t = j \mid T_1 = u, X_0 = i\} \lambda e^{-\lambda u} du.$$

If $u \geq t$ and $X_0 = i$, we have $Y_t = \rho(i)t$ and $\mathbb{P}\{X_t = j \mid T_1 = u, X_0 = i\} = 1$, if $i = j$ and 0 otherwise. We thus obtain

$$F_{i,j}(t,x) = \int_0^t \mathbb{P}\{Y_t > x, X_t = j \mid T_1 = u, X_0 = i\} \lambda e^{-\lambda u} du + e^{-\lambda t} 1_{\{x < \rho_i t\}} 1_{\{i=j\}}.$$

Now,

$$\mathbb{P}\{Y_t > x, X_t = j \mid T_1 = u, X_0 = i\}$$
$$= \sum_{k \in S} \mathbb{P}\{Y_t > x, X_t = j \mid X_u = k, T_1 = u, X_0 = i\} \mathbb{P}\{X_u = k \mid T_1 = u, X_0 = i\}.$$

For the second factor in the sum, we have, from Theorem 3.1.6,

$$\mathbb{P}\{X_u = k \mid T_1 = u, X_0 = i\} = \mathbb{P}\{X_{T_1} = k \mid T_1 = u, X_0 = i\}$$
$$= \mathbb{P}\{Z_1 = k \mid T_1 = u, Z_0 = i\}$$
$$= P_{i,k}.$$

For the first factor, $T_1 = u$ and $X_0 = i$ imply that $Y_u = \rho(i)u$, so

$$\mathbb{P}\{Y_t > x, X_t = j \mid X_u = k, T_1 = u, X_0 = i\}$$
$$= \mathbb{P}\left\{ \int_u^t \rho(X_v) dv > x - \rho(i)u, X_t = j \mid X_u = k, T_1 = u, X_0 = i \right\}$$
$$= \mathbb{P}\left\{ \int_u^t \rho(X_v) dv > x - \rho(i)u, X_t = j \mid X_u = k \right\}$$
$$= \mathbb{P}\{Y_{t-u} > x - \rho(i)u, X_{t-u} = j \mid X_0 = k\}$$
$$= F_{k,j}(t - u, x - \rho(i)u),$$

where the second equality follows from the Markov property and the third by homogeneity. Combining these results, we obtain Relation (7.10). ∎

In the next theorem, we derive the backward equation satisfied by the distribution of the pair (Y_t, X_t).

THEOREM 7.1.5.   *For $t > 0$, $i, j \in S$, $1 \leq h \leq m$ and $x \in (r_{h-1}t, r_h t)$ we have*

$$\frac{\partial F_{i,j}(t,x)}{\partial t} = -\rho(i)\frac{\partial F_{i,j}(t,x)}{\partial x} + \sum_{k \in S} A_{i,k} F_{k,j}(t,x). \tag{7.11}$$

Proof. For every $0 \leq s \leq t$, we introduce the random variable

$$Y_{s,t} = \int_s^t \rho(X_u) \, du.$$

Then, we have $Y_t = Y_{0,t}$ and

$$Y_t = Y_s + Y_{s,t}.$$

By conditioning on the number of transitions in $[0, s)$, for $0 \leq s \leq t$, we have

$$\mathbb{P}_i\{Y_t > x, X_t = j\} = \mathbb{P}_i\{Y_t > x, X_t = j, N_s = 0\}$$
$$+ \mathbb{P}_i\{Y_t > x, X_t = j, N_s = 1\}$$
$$+ \mathbb{P}_i\{Y_t > x, X_t = j, N_s \geq 2\}.$$

We separately consider these three terms. For the first term, when $X_0 = i$, $N_s = 0$ implies that $X_s = i$, so

$$\mathbb{P}_i\{Y_t > x, X_t = j, N_s = 0\} = \mathbb{P}_i\{Y_t > x, X_t = j, X_s = i, N_s = 0\}$$
$$= \mathbb{P}_i\{Y_t > x, X_t = j | X_s = i, N_s = 0\} \mathbb{P}_i\{N_s = 0\}$$
$$= e^{-\lambda s} \mathbb{P}_i\{Y_s + Y_{s,t} > x, X_t = j | X_s = i, N_s = 0\}$$
$$= e^{-\lambda s} \mathbb{P}_i\{Y_{s,t} > x - \rho(i)s, X_t = j | X_s = i, N_s = 0\}$$
$$= e^{-\lambda s} \mathbb{P}\{Y_{s,t} > x - \rho(i)s, X_t = j | X_s = i\}$$
$$= e^{-\lambda s} \mathbb{P}\{Y_{0,t-s} > x - \rho(i)s, X_{t-s} = j | X_0 = i\}$$
$$= e^{-\lambda s} F_{i,j}(t - s, x - \rho(i)s)$$
$$= (1 - \lambda s) F_{i,j}(t - s, x - \rho(i)s) + o(s).$$

The third equality follows from the independence of processes $\{N_t\}$ and $\{Z_n\}$. The fourth equality follows from the fact that, if $X_s = i$ and $N_s = 0$, we have $Y_s = \rho(i)s$. The fifth follows from the Markov property and the sixth from the homogeneity of the Markov chain X.

For the second term, that we denote by $G(s)$, we have, using Relation (7.6),

$$G(s) = \mathbb{P}_i\{Y_t > x, X_t = j, N_s = 1\}$$
$$= \sum_{k \in S} \mathbb{P}_i\{Y_t > x, X_t = j, X_s = k, N_s = 1\}$$
$$= \sum_{k \in S} \mathbb{P}_i\{Y_t > x, X_t = j | X_s = k, N_s = 1\} \mathbb{P}_i\{X_s = k, N_s = 1\}$$
$$= \lambda s e^{-\lambda s} \sum_{k \in S} P_{i,k} \mathbb{P}_i\{Y_t > x, X_t = j | X_s = k, N_s = 1\}.$$

As we did for the proof of Theorem 7.1.2, we define $\rho_{\max} = \max_{i\in S}\{\rho(i)\}$ and we have $\min_{i\in S}\{\rho(i)\} = 0$. Since $0 \le Y_s \le \rho_{\max}s$, we have

$$Y_{s,t} \le Y_t \le \rho_{\max}s + Y_{s,t}.$$

By defining $G_k(s) = \mathbb{P}_i\{Y_t > x, X_t = j | X_s = k, N_s = 1\}$, we obtain

$$
\begin{aligned}
G_k(s) &\le \mathbb{P}_i\{Y_{s,t} > x - \rho_{\max}s, X_t = j | X_s = k, N_s = 1\} \\
&= \mathbb{P}\{Y_{s,t} > x - \rho_{\max}s, X_t = j | X_s = k\} \\
&= \mathbb{P}\{Y_{0,t-s} > x - \rho_{\max}s, X_{t-s} = j | X_0 = k\} \\
&= F_{k,j}(t - s, x - \rho_{\max}s),
\end{aligned}
$$

where the second and third equalities are due to the Markov property and the homogeneity of the Markov chain X, respectively. Using the same arguments, we obtain

$$G_k(s) \ge F_{k,j}(t - s, x),$$

which leads to

$$\lambda s e^{-\lambda s} \sum_{k\in S} P_{i,k} F_{k,j}(t - s, x) \le G(s) \le \lambda s e^{-\lambda s} \sum_{k\in S} P_{i,k} F_{k,j}(t - s, x - \rho_{\max}s).$$

Thus we have

$$\lim_{s\to 0} \frac{G(s)}{s} = \lambda \sum_{k\in S} P_{i,k} F_{k,j}(t, x).$$

For the third term, we have from Relation (7.7),

$$\mathbb{P}_i\{Y_t > x, X_t = j, N_s \ge 2\} \le \mathbb{P}_i\{N_s \ge 2\} = \mathbb{P}\{N_s \ge 2\} = o(s).$$

Combining the three terms, we obtain

$$
\begin{aligned}
\frac{F_{i,j}(t,x) - F_{i,j}(t - s, x)}{s} &= \frac{(1 - \lambda s)F_{i,j}(t - s, x - \rho(i)s) - F_{i,j}(t - s, x)}{s} \\
&\quad + \frac{G(s)}{s} + \frac{o(s)}{s} \\
&= \frac{F_{i,j}(t - s, x - \rho(i)s) - F_{i,j}(t - s, x)}{s} \\
&\quad - \lambda F_{i,j}(t - s, x - \rho(i)s) + \frac{G(s)}{s} + \frac{o(s)}{s}.
\end{aligned}
$$

If now s tends to 0, we obtain

$$\frac{\partial F_{i,j}(t,x)}{\partial t} = -\rho(i)\frac{\partial F_{i,j}(t,x)}{\partial x} - \lambda F_{i,j}(t,x) + \lambda \sum_{k\in S} P_{i,k} F_{k,j}(t,x).$$

Since $P = I + A/\lambda$, we obtain

$$\frac{\partial F_{i,j}(t,x)}{\partial t} = -\rho(i)\frac{\partial F_{i,j}(t,x)}{\partial x} + \sum_{k \in S} A_{i,k}F_{k,j}(t,x),$$

which completes the proof. ∎

Let D be the diagonal matrix with the reward rates $\rho(i)$ on the diagonal. In matrix notation, the forward and backward Equations (7.8) and (7.11) become

$$\frac{\partial F(t,x)}{\partial t} = -\frac{\partial F(t,x)}{\partial x}D + F(t,x)A \qquad (7.12)$$

and

$$\frac{\partial F(t,x)}{\partial t} = -D\frac{\partial F(t,x)}{\partial x} + AF(t,x). \qquad (7.13)$$

These are hyperbolic partial differential equations having a unique solution on the domain E with the initial conditions given by Relation (7.4) and the jumps given by Relation (7.3); see for instance [79].

7.2 Solution

The solution to Equation (7.12) is given by the following theorem.

THEOREM 7.2.1. *For every $t > 0$ and $x \in [r_{h-1}t, r_h t)$, for $1 \le h \le m$,*

$$F(t,x) = \sum_{n=0}^{\infty} e^{-\lambda t}\frac{(\lambda t)^n}{n!} \sum_{k=0}^{n} \binom{n}{k} x_h^k (1 - x_h)^{n-k} C^{(h)}(n,k), \qquad (7.14)$$

where $x_h = \dfrac{x - r_{h-1}t}{(r_h - r_{h-1})t}$ and the matrices $C^{(h)}(n,k) = \left(C_{B_u B_v}^{(h)}(n,k)\right)_{0 \le u,v \le m}$ are given by the following recurrence relations.

- *For $0 \le u \le m$ and $1 < h \le v \le m$:*
  - *for $n \ge 0$: $C_{B_u B_v}^{(1)}(n,0) = (P^n)_{B_u B_v}$, $C_{B_u B_v}^{(h)}(n,0) = C_{B_u B_v}^{(h-1)}(n,n)$,*
  - *for $1 \le k \le n$:*

$$C_{B_u B_v}^{(h)}(n,k) = \frac{r_v - r_h}{r_v - r_{h-1}} C_{B_u B_v}^{(h)}(n,k-1)$$

$$+ \frac{r_h - r_{h-1}}{r_v - r_{h-1}} \sum_{w=0}^{m} C_{B_u B_w}^{(h)}(n-1,k-1)P_{B_w B_v}. \qquad (7.15)$$

- *For $0 \le u \le m$ and $0 \le v \le h-1 < m-1$:*
  - *for $n \ge 0$: $C_{B_u B_v}^{(m)}(n,n) = 0_{B_u B_v}$, $C_{B_u B_v}^{(h)}(n,n) = C_{B_u B_v}^{(h+1)}(n,0)$,*

– *for* $0 \le k \le n-1$:

$$C_{B_u B_v}^{(h)}(n,k) = \frac{r_{h-1} - r_v}{r_h - r_v} C_{B_u B_v}^{(h)}(n,k+1)$$

$$+ \frac{r_h - r_{h-1}}{r_h - r_v} \sum_{w=0}^{m} C_{B_u B_w}^{(h)}(n-1,k) P_{B_w B_v}. \tag{7.16}$$

Proof. For $t > 0$ and $x \in [r_{h-1}t, r_h t)$, for $1 \le h \le m$, we write the solution to Equation (7.12) as

$$F(t,x) = \sum_{n=0}^{\infty} e^{-\lambda t} \frac{(\lambda t)^n}{n!} \sum_{k=0}^{n} \binom{n}{k} x_h^k (1 - x_k)^{n-k} C^{(h)}(n,k),$$

and we establish the relations that the matrices $C^{(h)}(n,k)$ must satisfy. So,

$$\frac{\partial F(t,x)}{\partial t} = -\lambda F(t,x) + \frac{\lambda}{r_h - r_{h-1}} \sum_{n=0}^{\infty} e^{-\lambda t} \frac{(\lambda t)^n}{n!} \sum_{k=0}^{n} \binom{n}{k} x_h^k (1 - x_h)^{n-k}$$

$$\times \left[r_h C^{(h)}(n+1,k) - r_{h-1} C^{(h)}(n+1,k+1) \right], \tag{7.17}$$

and

$$\frac{\partial F(t,x)}{\partial x} = \frac{\lambda}{r_h - r_{h-1}} \sum_{n=0}^{\infty} e^{-\lambda t} \frac{(\lambda t)^n}{n!} \sum_{k=0}^{n} \binom{n}{k} x_h^k (1 - x_h)^{n-k}$$

$$\times \left[C^{(h)}(n+1,k+1) - C^{(h)}(n+1,k) \right].$$

Since $A = -\lambda(I - P)$, we obtain $F(t,x)A = -\lambda F(t,x) + \lambda F(t,x)P$, that is,

$$F(t,x)A = -\lambda F(t,x) + \lambda \sum_{n=0}^{\infty} e^{-\lambda t} \frac{(\lambda t)^n}{n!} \sum_{k=0}^{n} \binom{n}{k} x_h^k (1 - x_h)^{n-k} C^{(h)}(n,k) P.$$

It follows that if matrices $C^{(h)}(n,k)$ satisfy

$$C^{(h)}(n+1,k+1)[D - r_{h-1} I]$$

$$= C^{(h)}(n+1,k)[D - r_h I] + (r_h - r_{h-1}) C^{(h)}(n,k) P, \tag{7.18}$$

then Equation (7.12) is satisfied. For every $1 \le h \le m$ and $0 \le u \le m$, the recurrence Relation (7.18) can also be written as follows.

If $h \le v \le m$, then

$$C_{B_u B_v}^{(h)}(n,k) = \frac{r_v - r_h}{r_v - r_{h-1}} C_{B_u B_v}^{(h)}(n,k-1)$$

$$+ \frac{r_h - r_{h-1}}{r_v - r_{h-1}} \sum_{w=0}^{m} C_{B_u B_w}^{(h)}(n-1,k-1) P_{B_w B_v},$$

and if $0 \leq v \leq h-1$, then

$$C_{B_u B_v}^{(h)}(n,k) = \frac{r_{h-1} - r_v}{r_h - r_v} C_{B_u B_v}^{(h)}(n,k+1)$$

$$+ \frac{r_h - r_{h-1}}{r_h - r_v} \sum_{w=0}^{m} C_{B_u B_w}^{(h)}(n-1,k) P_{B_w B_v}.$$

To obtain the initial conditions for the $C^{(h)}(n,k)$, we consider the jumps of $F(t,x)$. We first consider the jump at $x = r_0 t = 0$. For $t > 0$, at $x = 0$, that is, for $h = 1$, Relation (7.14) yields

$$F(t,0) = \sum_{n=0}^{\infty} e^{-\lambda t} \frac{(\lambda t)^n}{n!} C^{(1)}(n,0).$$

It follows from Relation (7.4) that for $0 \leq u,v \leq m$,

$$C_{B_u B_v}^{(1)}(n,0) = (P^n)_{B_u B_v} - (P_{B_0 B_0})^n 1_{\{u=v=0\}}. \tag{7.19}$$

In particular, this implies that for every $0 \leq u \leq m$,

$$C_{B_u B_v}^{(1)}(n,0) = (P^n)_{B_u B_v}, \quad \text{for } 1 \leq v \leq m.$$

Next, we consider the jumps at $x = r_h t$, $1 \leq h \leq m-1$. For $t > 0$, $1 \leq h \leq m-1$ and $i,j \in S$, we have

$$F_{i,j}(t, r_h t) = \lim_{x \to r_h t, x < r_h t} F_{i,j}(t,x) - \mathbb{P}\{Y_t = r_h t, X_t = j \mid X_0 = i\}.$$

From Relations (7.14) and (7.3), we obtain

$$C_{B_u B_v}^{(h+1)}(n,0) = C_{B_u B_v}^{(h)}(n,n) - (P_{B_h B_h})^n 1_{\{u=v=h\}}. \tag{7.20}$$

In particular, this implies that for every $0 \leq u \leq m$,

$$C_{B_u B_v}^{(h)}(n,0) = C_{B_u B_v}^{(h-1)}(n,n), \quad \text{for } 1 < h \leq v \leq m,$$

and

$$C_{B_u B_v}^{(h)}(n,n) = C_{B_u B_v}^{(h+1)}(n,0), \quad \text{for } 0 \leq v \leq h-1 < m-1.$$

Finally, we consider the jump at $x = r_m t$, that is, for $h = m$. For $t > 0$,

$$0 = F_{i,j}(t, r_m t) = \lim_{x \to r_m t, x < r_m t} F_{i,j}(t,x) - \mathbb{P}\{Y_t = r_m t, X_t = j \mid X_0 = i\},$$

which, as in the preceding case, leads to

$$C_{B_u B_v}^{(m)}(n,n) = (P_{B_m B_m})^n 1_{\{u=v=m\}}. \tag{7.21}$$

This implies that for every $0 \leq u \leq m$,

$$C_{B_u B_v}^{(m)}(n,n) = 0, \quad \text{for } 0 \leq v \leq m-1,$$

which completes the proof. ∎

Note that, by the way, we obtained Relation (7.17) which gives the density function of Y_t, for $t > 0$ and $x \neq r_l t$, $l = 0, \ldots, m$.

COROLLARY 7.2.2. *For $1 \leq h \leq m$, $n \geq 0$ and $0 \leq k \leq n$, the matrices $C^{(h)}(n,k) = \left(C_{B_u B_v}^{(h)}(n,k) \right)_{0 \leq u,v \leq m}$ satisfy the following recurrence relations.*

- *For $1 \leq h \leq u \leq m$ and $0 \leq v \leq m$:*
  - *for $n \geq 0$: $C_{B_u B_v}^{(1)}(n,0) = (P^n)_{B_u B_v}$, $C_{B_u B_v}^{(h)}(n,0) = C_{B_u B_v}^{(h-1)}(n,n)$,*
  - *for $1 \leq k \leq n$:*

$$C_{B_u B_v}^{(h)}(n,k) = \frac{r_u - r_h}{r_u - r_{h-1}} C_{B_u B_v}^{(h)}(n,k-1)$$

$$+ \frac{r_h - r_{h-1}}{r_u - r_{h-1}} \sum_{w=0}^{m} P_{B_u B_w} C_{B_w B_v}^{(h)}(n-1,k-1).$$

- *For $0 \leq u \leq h-1 < m-1$ and $0 \leq v \leq m$:*
  - *for $n \geq 0$: $C_{B_u B_v}^{(m)}(n,n) = 0_{B_u B_v}$, $C_{B_u B_v}^{(h)}(n,n) = C_{B_u B_v}^{(h+1)}(n,0)$,*
  - *for $0 \leq k \leq n-1$:*

$$C_{B_u B_v}^{(h)}(n,k) = \frac{r_{h-1} - r_u}{r_h - r_u} C_{B_u B_v}^{(h)}(n,k+1)$$

$$+ \frac{r_h - r_{h-1}}{r_h - r_u} \sum_{w=0}^{m} P_{B_u B_w} C_{B_w B_v}^{(h)}(n-1,k).$$

Proof. The proof is the same as that of Theorem 7.2.1 using (7.13) and (7.14). We thus find that the matrices $C^{(h)}(n,k)$ satisfy the relation

$$[D - r_{h-1}I]C^{(h)}(n+1,k+1)$$

$$= [D - r_h I]C^{(h)}(n+1,k) + (r_h - r_{h-1})PC^{(h)}(n,k). \tag{7.22}$$

For every $1 \leq h \leq m$ and $0 \leq v \leq m$, Relation (7.22) may also be written as follows.
If $h \leq u \leq m$, then

$$C_{B_u B_v}^{(h)}(n,k) = \frac{r_u - r_h}{r_u - r_{h-1}} C_{B_u B_v}^{(h)}(n,k-1)$$

$$+ \frac{r_h - r_{h-1}}{r_u - r_{h-1}} \sum_{w=0}^{m} P_{B_u B_w} C_{B_w B_v}^{(h)}(n-1,k-1),$$

and if $0 \leq u \leq h-1$, then

$$C_{B_u B_v}^{(h)}(n,k) = \frac{r_{h-1} - r_u}{r_h - r_u} C_{B_u B_v}^{(h)}(n,k+1)$$

$$+ \frac{r_h - r_{h-1}}{r_h - r_u} \sum_{w=0}^{m} P_{B_u B_w} C_{B_w B_v}^{(h)}(n-1,k).$$

As in the proof of Theorem 7.2.1, we consider the jumps of $F(t,x)$. Relation (7.19) implies that, for every $0 \leq v \leq m$,

$$C_{B_u B_v}^{(1)}(n,0) = (P^n)_{B_u B_v}, \quad \text{for } 1 \leq u \leq m.$$

Relation (7.20) implies that, for every $0 \leq v \leq m$,

$$C_{B_u B_v}^{(h)}(n,0) = C_{B_u B_v}^{(h-1)}(n,n), \quad \text{for } 1 < h \leq u \leq m,$$

and

$$C_{B_u B_v}^{(h)}(n,n) = C_{B_u B_v}^{(h+1)}(n,0), \quad \text{for } 0 \leq u \leq h-1 < m-1.$$

Finally, (7.21) implies that, for every $0 \leq v \leq m$,

$$C_{B_u B_v}^{(m)}(n,n) = 0, \quad \text{for } 0 \leq u \leq m-1,$$

which completes the proof. ∎

The following corollary gives an upper bound for the matrices $C^{(h)}(n,k)$. If M and K are square matrices of the same dimension, the notation $M \leq K$ means element-wise inequality.

COROLLARY 7.2.3. *For every $n \geq 0$, $0 \leq k \leq n$ and $1 \leq h \leq m$,*

$$0 \leq C^{(h)}(n,k) \leq P^n.$$

Proof. The proof is made using a two-stage induction; first over n, then, for fixed n, over k, using the recurrence relation in Theorem 7.2.1, or equivalently in Corollary 7.2.2. The result clearly holds for $n = 0$. Note that in Relation (7.15), that is, for $h \leq v$, we have

$$0 \leq \frac{r_v - r_h}{r_v - r_{h-1}} = 1 - \frac{r_h - r_{h-1}}{r_v - r_{h-1}} \leq 1,$$

and in Relation (7.16), that is, for $v \leq h-1$, we have

$$0 \leq \frac{r_{h-1} - r_v}{r_h - r_v} = 1 - \frac{r_h - r_{h-1}}{r_h - r_v} \leq 1.$$

Consider first the case $v \leq h-1$. The result holds for the pair (n,n), since $C_{B_u B_v}^{(m)}(n,n) = 0$. Suppose that the result holds for $n-1$ and for the pair $(n,k+1)$, then from

Relation (7.16), we obtain $C^{(h)}_{B_uB_v}(n,k) \geq 0$ and

$$C^{(h)}_{B_uB_v}(n,k) = \frac{r_{h-1}-r_v}{r_h-r_v} C^{(h)}_{B_uB_v}(n,k+1)$$

$$+ \frac{r_h - r_{h-1}}{r_h - r_v} \sum_{w=0}^{m} C^{(h)}_{B_uB_w}(n-1,k) P_{B_wB_v}$$

$$\leq \frac{r_{h-1}-r_v}{r_h-r_v}(P^n)_{B_uB_v} + \frac{r_h-r_{h-1}}{r_h-r_v} \sum_{w=0}^{m}(P^{n-1})_{B_uB_w}P_{B_wB_v}$$

$$= \frac{r_{h-1}-r_v}{r_h-r_v}(P^n)_{B_uB_v} + \frac{r_h-r_{h-1}}{r_h-r_v}(P^n)_{B_uB_v}$$

$$= (P^n)_{B_uB_v}.$$

The same argument is used in the case $h \leq v$ from Relation (7.15). Moreover, the relations

$$C^{(h)}_{B_uB_v}(n,0) = C^{(h-1)}_{B_uB_v}(n,n), \quad \text{for } 1 < h \leq v \leq m,$$

and

$$C^{(h)}_{B_uB_v}(n,n) = C^{(h+1)}_{B_uB_v}(n,0), \quad \text{for } 0 \leq v \leq h-1 < m-1,$$

are used to account for both cases $v \leq h-1$ and $h \leq v$. ∎

In numerical procedures, that result is particularly useful in avoiding overflow problems, since all entries of matrices $C^{(h)}(n,k)$ are in the interval $(0,1)$. Before giving a sketch of the algorithm to compute the distribution of Y_t, we propose an analytic expression of this distribution on a small example.

7.3 Examples

7.3.1 A two-state example

We consider a single-component system which fails after a time exponentially distributed with rate λ. After the failure, the system is considered down forever. Let X_t be the number of properly functioning components at time t. The state space is $\{1,0\}$, where 0 is the absorbing state. The infinitesimal generator, A, of the continuous-time Markov chain $X = \{X_t, t \geq 0\}$ is given by

$$A = \begin{pmatrix} -\lambda & \lambda \\ 0 & 0 \end{pmatrix}.$$

The graph of the Markov chain is depicted in Figure 7.1.

Matrix P is the transition probability matrix of the Markov chain $Z = \{Z_n, n \geq 0\}$ obtained after uniformization of Markov chain X. It is given by $P = I + A/\beta$, where I is the identity matrix and β satisfies $\beta \geq \max\{-A_{i,i}, i \in S\} = \lambda$.

Figure 7.1   A single-component system

We choose $\beta = \lambda$. Matrix P is thus given by

$$P = \begin{pmatrix} 0 & 1 \\ 0 & 1 \end{pmatrix}.$$

A reward rate, $\rho(j)$, is associated with state j and we consider the performability of the system, i.e. the random variable Y_t defined by

$$Y_t = \int_0^t \rho(X_s)ds.$$

For our example, we assume that $\rho(1) = 1$ and $\rho(0) = 0$. In fact, we have in this case $Y_t = W_t$ which has been studied in Chapter 6. This example being simple, we obtain the explicit joint distribution of Y_t and X_t using the integral Equations (7.9) and (7.10).

Recall that, for $t \geq 0$ and $x \geq 0$, $F_{i,j}(t,x)$ is defined as

$$F_{i,j}(t,x) = \mathbb{P}\{Y_t > x, X_t = j \mid X_0 = i\}.$$

Equations (7.9) and (7.10) can be written as

$$F_{i,j}(t,x) = \sum_{k=0}^{1} \int_0^t F_{i,k}(t-u, x-\rho(j)u)\lambda e^{-\lambda u} du P_{k,j} + e^{-\lambda t}1_{\{x<\rho_j t\}}1_{\{i=j\}}, \tag{7.23}$$

$$F_{i,j}(t,x) = \sum_{k=0}^{1} P_{i,k} \int_0^t F_{k,j}(t-u, x-\rho(i)u)\lambda e^{-\lambda u} du + e^{-\lambda t}1_{\{x<\rho_i t\}}1_{\{i=j\}}. \tag{7.24}$$

For $t > 0$, as stated in Relation (7.2), only functions $F_{i,i}(t,x)$ have jumps. These jumps are given by

$$\mathbb{P}\{Y_t = 0, X_t = 0 \mid X_0 = 0\} = 1$$
$$\mathbb{P}\{Y_t = t, X_t = 1 \mid X_0 = 1\} = e^{-\lambda t}.$$

Since state 1 cannot be reached from state 0, we have

$$F_{0,1}(t,x) = 0.$$

Applying either (7.23) or (7.24) when $i = j = 1$, we obtain

$$F_{1,1}(t,x) = e^{-\lambda t}1_{\{x<t\}}. \tag{7.25}$$

If $i = j = 0$ we have $Y_t = 0$ and thus

$$F_{0,0}(t,x) = 1_{\{x<0\}}. \tag{7.26}$$

Applying (7.24) when $i = 1$ and $j = 0$, we obtain

$$F_{1,0}(t,x) = \int_0^t F_{0,0}(t-u,x-u)\lambda e^{-\lambda u}du$$

$$= \int_0^t e^{-\lambda u}1_{\{x<u\}}du$$

$$= \int_0^t e^{-\lambda u}du 1_{\{x<0\}} + \int_x^t e^{-\lambda u}du 1_{\{0\le x<t\}}$$

$$= (1-e^{-\lambda t})1_{\{x<0\}} + (e^{-\lambda x}-e^{-\lambda t})1_{\{0\le x<t\}}.$$

Suppose now that the initial state is state 1 with probability 1, we have

$$\mathbb{P}\{Y_t > x\} = F_{1,1}(t,x) + F_{1,0}(t,x) = 1_{\{x<0\}} + e^{-\lambda x}1_{\{0\le x<t\}}.$$

Note the jump at point $x = t$ in the distribution of Y_t. Indeed, since $Y_t \le t$, we have

$$\mathbb{P}\{Y_t = t\} = \mathbb{P}\{Y_t \ge t\} - \mathbb{P}\{Y_t > t\} = \lim_{x \to t, x<t} \mathbb{P}\{Y_t > x\} = e^{-\lambda t}.$$

This example is so simple that it could have been solved more simply using the fact that when $X_0 = 1$ we have $Y_t = \min\{T_1, t\}$, where T_1 is the first and the only jump of Markov chain X. Also, when $X_0 = 0$ we have $Y_t = 0$. That is why we consider in the next subsection another simple example which leads to a non-trivial distribution of Y_t.

If we define $F_{i,j}(x)$ as

$$F_{i,j}(x) = \lim_{t \to \infty} F_{i,j}(t,x),$$

we obtain

$$F_{0,1}(x) = F_{1,1}(x) = 0, F_{0,0}(x) = 1_{\{x<0\}} \quad \text{and} \quad F_{1,0}(x) = 1_{\{x<0\}} + e^{-\lambda x}1_{\{x\ge 0\}},$$

and since the initial state is state 1,

$$\mathbb{P}\{Y_\infty > x\} = F_{1,1}(x) + F_{1,0}(x) = 1_{\{x<0\}} + e^{-\lambda x}1_{\{x\ge 0\}}.$$

As expected, this coincides with the distribution of Y_∞ given by

$$\mathbb{P}\{Y_\infty > x\} = \alpha_U e^{R_U^{-1}A_U x}1_U,$$

where $U = \{1\}$ and

$$\alpha_U = 1, \; R_U = 1, \; A_U = -\lambda, \quad \text{and} \quad 1_U = 1.$$

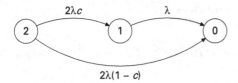

Figure 7.2   A two-component system

7.3.2   A three-state example

This example was already used in Chapter 1, Section 1.2.3. We consider a two-component parallel redundant system so that the times to failure of each component are independent and exponentially distributed with rate λ. We assume that not all faults are recoverable and that c is the coverage factor denoting the conditional probability that the system recovers, given that a fault has occurred. When both components have failed, the system is considered to have failed and no recovery is possible. Let X_t be the number of properly functioning components at time t. The state space is $\{2,1,0\}$ where 0 is the absorbing state. The infinitesimal generator, A, of the continuous-time Markov chain $X = \{X_t, \ t \geq 0\}$ is given by

$$A = \begin{pmatrix} -2\lambda & 2\lambda c & 2\lambda(1-c) \\ 0 & -\lambda & \lambda \\ 0 & 0 & 0 \end{pmatrix}.$$

The graph of the Markov chain is depicted in Figure 7.2.

Matrix P is the transition probability matrix of the Markov chain $Z = \{Z_n, \ n \geq 0\}$ obtained after uniformization of Markov chain X. It is given by $P = I + A/\beta$, where I is the identity matrix and β satisfies $\beta \geq \max\{-A_{i,i}, \ i \in S\} = 2\lambda$. We choose $\beta = 2\lambda$. Matrix P is thus given by

$$P = \begin{pmatrix} 0 & c & 1-c \\ 0 & 1/2 & 1/2 \\ 0 & 0 & 1 \end{pmatrix}.$$

As before, a reward rate $\rho(j)$ is associated with state j and we consider the performability of the system, i.e. the random variable Y_t defined by

$$Y_t = \int_0^t \rho(X_s)\,ds.$$

For this example, we assume that $\rho(2) = r > 1$, $\rho(1) = 1$ and $\rho(0) = 0$. This example being relatively simple, we obtain the explicit joint distribution of Y_t and X_t using the integral Equations (7.9) and (7.10).

Recall that, for $t \geq 0$ and $x \geq 0$, $F_{i,j}(t,x)$ is defined by

$$F_{i,j}(t,x) = \mathbb{P}\{Y_t > x, X_t = j \mid X_0 = i\}.$$

Equations (7.9) and (7.10) can be written as

$$F_{i,j}(t,x) = \sum_{k=0}^{2} \int_0^t F_{i,k}(t-u,x-\rho(j)u)2\lambda e^{-2\lambda u} du P_{k,j} + e^{-2\lambda t} 1_{\{x<\rho_j t\}} 1_{\{i=j\}}, \qquad (7.27)$$

$$F_{i,j}(t,x) = \sum_{k=0}^{2} P_{i,k} \int_0^t F_{k,j}(t-u,x-\rho(i)u)2\lambda e^{-2\lambda u} du + e^{-2\lambda t} 1_{\{x<\rho_i t\}} 1_{\{i=j\}}. \qquad (7.28)$$

For $t > 0$, as stated in Relation (7.2), only functions $F_{i,i}(t,x)$ have jumps. These jumps are given by

$$\mathbb{P}\{Y_t = 0, X_t = 0 \mid X_0 = 0\} = 1$$
$$\mathbb{P}\{Y_t = t, X_t = 1 \mid X_0 = 1\} = e^{-\lambda t}$$
$$\mathbb{P}\{Y_t = rt, X_t = 2 \mid X_0 = 2\} = e^{-2\lambda t}.$$

Since state 2 cannot be reached from states 0 and 1, and since state 1 cannot be reached from state 0, we have

$$F_{1,2}(t,x) = F_{0,2}(t,x) = F_{0,1}(t,x) = 0.$$

Applying either (7.27) or (7.28) when $i = j = 2$, we obtain

$$F_{2,2}(t,x) = e^{-2\lambda t} 1_{\{x<rt\}}. \qquad (7.29)$$

Applying (7.28) when $i = j = 1$, we obtain

$$F_{1,1}(t,x) = \int_0^t F_{1,1}(t-u,x-u)\lambda e^{-2\lambda u} du + e^{-2\lambda t} 1_{\{x<t\}}.$$

But, the model being acyclic, $F_{1,1}(t,x)$ is as in previous example, which contained two states,

$$F_{1,1}(t,x) = e^{-\lambda t} 1_{\{x<t\}}. \qquad (7.30)$$

It is easily checked that this is the solution to the integral equation above. If $i = j = 0$, we have $Y_t = 0$ and thus

$$F_{0,0}(t,x) = 1_{\{x<0\}}. \qquad (7.31)$$

Applying (7.28) again when $i = 2$ and $j = 1$, we obtain

$$F_{2,1}(t,x) = c \int_0^t F_{1,1}(t-u,x-ru)2\lambda e^{-2\lambda u} du.$$

From Relation (7.30), we have

$$F_{1,1}(t-u,x-ru) = e^{-\lambda(t-u)} 1_{\{x-ru<t-u\}}.$$

Now, we have, since $r > 1$,

$$x - ru < t - u \iff u > \frac{x-t}{r-1}.$$

So, in the above integral equation, u varies from 0 to t if $x \leq t$, since in that case we have $(x-t)/(r-1) < 0$ and u varies from $(x-t)/(r-1)$ to t if $t < x < rt$ since in that case we have $0 < (x-t)/(r-1) < t$. This leads to

$$F_{2,1}(t,x) = c \int_0^t e^{-\lambda(t-u)} 2\lambda e^{-2\lambda u} du 1_{\{x \leq t\}}$$

$$+ c \int_{(x-t)/(r-1)}^t e^{-\lambda(t-u)} 2\lambda e^{-2\lambda u} du 1_{\{t < x < rt\}},$$

which gives

$$F_{2,1}(t,x) = 2ce^{-\lambda t}(1 - e^{-\lambda t})1_{\{x \leq t\}} + 2ce^{-\lambda t}(e^{-\lambda(x-t)/(r-1)} - e^{-\lambda t})1_{\{t < x < rt\}}. \qquad (7.32)$$

Applying (7.28) when $i = 1$ and $j = 0$, we obtain

$$F_{1,0}(t,x) = \frac{1}{2} \int_0^t F_{1,0}(t-u, x-u) 2\lambda e^{-2\lambda u} du$$

$$+ \frac{1}{2} \int_0^t F_{0,0}(t-u, x-u) 2\lambda e^{-2\lambda u} du$$

$$= \frac{1}{2} \int_0^t F_{1,0}(t-u, x-u) 2\lambda e^{-2\lambda u} du + \int_0^t \lambda e^{-2\lambda u} 1_{\{u > x\}} du.$$

But, as for $F_{1,1}(t,x)$, the model being acyclic, $F_{1,0}(t,x)$ is as in previous example, which contained two states,

$$F_{1,0}(t,x) = (1 - e^{-\lambda t})1_{\{x < 0\}} + (e^{-\lambda x} - e^{-\lambda t})1_{\{0 \leq x < t\}}. \qquad (7.33)$$

It is also easily checked that this is the solution to the integral equation above.

It remains to evaluate the function $F_{2,0}(t,x)$. Applying (7.28) again when $i = 2$ and $j = 0$, we obtain

$$F_{2,0}(t,x) = c \int_0^t F_{1,0}(t-u, x-ru) 2\lambda e^{-2\lambda u} du$$

$$+ (1-c) \int_0^t F_{0,0}(t-u, x-ru) 2\lambda e^{-2\lambda u} du.$$

We saw above that

$$F_{1,0}(t-u, x-ru) = (1 - e^{-\lambda(t-u)})1_{\{u > x/r\}}$$

$$+ (e^{-\lambda(x-ru)} - e^{-\lambda(t-u)})1_{\{(x-t)/(r-1) < u \leq x/r\}}$$

and

$$F_{0,0}(t-u, x-ru) = 1_{\{x-ru < 0\}} = 1_{\{u > x/r\}}.$$

So we have to distinguish the cases $x < 0$, $0 \leq x \leq t$, and $t < x < rt$.
If $x < 0$, we have

$$F_{2,0}(t,x) = c \int_0^t (1 - e^{-\lambda(t-u)})2\lambda e^{-2\lambda u}\,du + (1-c) \int_0^t 2\lambda e^{-2\lambda u}\,du$$

$$= 1 - 2ce^{-\lambda t} + (2c-1)e^{-2\lambda t}.$$

If $0 \leq x \leq t$, we have

$$F_{2,0}(t,x)$$

$$= c \int_0^{x/r} (e^{-\lambda(x-ru)} - e^{-\lambda(t-u)})2\lambda e^{-2\lambda u}\,du$$

$$+ c \int_{x/r}^t (1 - e^{-\lambda(t-u)})2\lambda e^{-2\lambda u}\,du + (1-c)\int_{x/r}^t 2\lambda e^{-2\lambda u}\,du$$

$$= \begin{cases} \dfrac{2c}{2-r}e^{-\lambda x} + \left(1 - \dfrac{2c}{2-r}\right)e^{-2\lambda x/r} - 2ce^{-\lambda t} + (2c-1)e^{-2\lambda t} & \text{if } r \neq 2, \\[3mm] (\lambda cx + 1)e^{-\lambda x} - 2ce^{-\lambda t} + (2c-1)e^{-2\lambda t} & \text{if } r = 2. \end{cases}$$

If $t < x < rt$, we have

$$F_{2,0}(t,x)$$

$$= c \int_{(x-t)/(r-1)}^{x/r} (e^{-\lambda(x-ru)} - e^{-\lambda(t-u)})2\lambda e^{-2\lambda u}\,du$$

$$+ c \int_{x/r}^t (1 - e^{-\lambda(t-u)})2\lambda e^{-2\lambda u}\,du + (1-c)\int_{x/r}^t 2\lambda e^{-2\lambda u}\,du$$

$$= \begin{cases} \dfrac{2c(r-1)}{2-r}e^{-\lambda(x+(r-2)t)/(r-1)} + \left(1 - \dfrac{2c}{2-r}\right)e^{-2\lambda x/r} + (2c-1)e^{-2\lambda t} & \text{if } r \neq 2, \\[3mm] (\lambda c(2t-x) - 2c + 1)e^{-\lambda x} + (2c-1)e^{-2\lambda t} & \text{if } r = 2. \end{cases}$$

If the initial state is state 2 with probability 1, we have

$$\mathbb{P}\{Y_t > x\} = F_{2,2}(t,x) + F_{2,1}(t,x) + F_{2,0}(t,x).$$

Here again, note the jump at point $x = rt$ in the distribution of Y_t. Indeed, since $Y_t \leq rt$, we have

$$\mathbb{P}\{Y_t = rt\} = \mathbb{P}\{Y_t \geq rt\} - \mathbb{P}\{Y_t > rt\} = \lim_{x \to t, x < t} \mathbb{P}\{Y_t > x\} = e^{-2\lambda t}.$$

Defining again $F_{i,j}(x)$ as

$$F_{i,j}(x) = \lim_{t \to \infty} F(t,x),$$

we obtain

$$F_{2,2}(x) = F_{1,1}(t,x) = 0, \quad F_{0,0}(x) = 1_{\{x<0\}},$$
$$F_{1,2}(x) = F_{0,2}(t,x) = F_{0,1}(t,x) = F_{2,1}(x) = 0,$$
$$F_{1,0}(x) = 1_{\{x<0\}} + e^{-\lambda x} 1_{\{x\geq0\}}$$

and

$$F_{2,0}(x) = \begin{cases} 1_{\{x<0\}} + \dfrac{2c}{2-r}e^{-\lambda x} + \left(1 - \dfrac{2c}{2-r}\right)e^{-2\lambda x/r}1_{\{x\geq0\}} & \text{if } r\neq2, \\[4mm] 1_{\{x<0\}} + (\lambda cx+1)e^{-\lambda x}1_{\{x\geq0\}} & \text{if } r=2. \end{cases}$$

Since the initial state is state 2, we obtain

$$\mathbb{P}\{Y_\infty > x\} = F_{2,2}(x) + F_{2,1}(x) + F_{2,0}(x).$$

As expected, this coincides with the distribution of Y_∞ given by

$$\mathbb{P}\{Y_\infty > x\} = \alpha_U e^{R_U^{-1}A_U x}1_U,$$

where $U = \{2,1\}$ and

$$\alpha_U = (1,0), \ R_U = \begin{pmatrix} r & 0 \\ 0 & 1 \end{pmatrix}, A_U = \begin{pmatrix} -2\lambda & 2\lambda c \\ 0 & -\lambda \end{pmatrix} \quad \text{and} \quad 1_U = \begin{pmatrix} 1 \\ 1 \end{pmatrix}.$$

Indeed, we obtain

$$\mathbb{P}\{Y_\infty > x\} = \begin{cases} 1_{\{x<0\}} + \dfrac{2c}{2-r}e^{-\lambda x} + \left(1 - \dfrac{2c}{2-r}\right)e^{-2\lambda x/r}1_{\{x\geq0\}} & \text{if } r\neq2, \\[4mm] 1_{\{x<0\}} + (\lambda cx+1)e^{-\lambda x}1_{\{x\geq0\}} & \text{if } r=2. \end{cases}$$

7.4   Algorithmic aspects

In this section, we consider the computation of matrix $F(t,x)$ from Theorem 7.2.1.

Let $\varepsilon > 0$ be the desired precision for the computation of each entry of matrix $F(t,x)$. We define the integer N by

$$N = \min\left\{n \geq 0 \;\middle|\; \sum_{i=0}^{n} e^{-\lambda t}\frac{(\lambda t)^i}{i!} \geq 1-\varepsilon\right\}. \tag{7.34}$$

We thus have, from Relation (7.14),

$$F(t,x) = \sum_{n=0}^{N} e^{-\lambda t} \frac{(\lambda t)^n}{n!} \sum_{k=0}^{n} \binom{n}{k} x_h^k (1-x_h)^{n-k} C^{(h)}(n,k) + e(N).$$

From the first inequality of Corollary 7.2.3, we obtain that the matrix remainder $e(N) = \left(e_{i,j}(N)\right)_{i,j \in S}$ of the series satisfies, for every $i,j \in S$,

$$e_{i,j}(N) = \sum_{n=N+1}^{\infty} e^{-\lambda t} \frac{(\lambda t)^n}{n!} \sum_{k=0}^{n} \binom{n}{k} x_h^k (1-x_h)^{n-k} \left[C^{(h)}(n,k)\right]_{i,j}$$

$$\leq \sum_{n=N+1}^{\infty} e^{-\lambda t} \frac{(\lambda t)^n}{n!} (P^n)_{i,j} \sum_{k=0}^{n} \binom{n}{k} x_h^k (1-x_h)^{n-k}$$

$$= \sum_{n=N+1}^{\infty} e^{-\lambda t} \frac{(\lambda t)^n}{n!} (P^n)_{i,j}.$$

For $h = 1,\dots,m$ and $x \in [r_{h-1}t, r_h t)$, we introduce the matrix $F^\varepsilon(t,x)$ defined by

$$F^\varepsilon(t,x) = \sum_{n=0}^{N} e^{-\lambda t} \frac{(\lambda t)^n}{n!} \sum_{k=0}^{n} \binom{n}{k} x_h^k (1-x_h)^{n-k} C^{(h)}(n,k).$$

Since matrix P^n is stochastic, we have shown that for every $i \in S$, we have

$$\sum_{j \in S} (F_{i,j}(t,x) - F^\varepsilon_{i,j}(t,x)) \leq \varepsilon.$$

The main effort goes into the computation of the matrices $C^{(h)}(n,k)$. With regards to storage requirements, since the values of the $C^{(h)}(n,k)$ at step n depend only on their values at step $n-1$, we need to store only two arrays of $(N+1)m$ matrices. At step n, we need to compute $n+1$ matrices for each $h = 1,\dots,m$. That can easily be seen from the algorithmic description in Table 7.2. The procedure **Accumulate**(n) is used to compute the approximate matrix $F^\varepsilon(t,x)$ for $h = 1,\dots,m$ and $x \in [r_{h-1}t, r_h t)$. This procedure, described in Table 7.1, involves a fixed value $t > 0$ and M distinct values of x, denoted by $x(i)$, $1 \leq i \leq M$. We initialize $F^\varepsilon(t,x(i)) = 0$, we denote by h_i the index such that $x(i) \in [r_{h_i-1}t, r_{h_i}t)$, and we define

$$x_{h_i} = \frac{x(i) - r_{h_i-1}t}{(r_{h_i} - r_{h_i-1})t}.$$

This computational method avoids numerical problems since all the computed quantities are between 0 and 1 and require only additions and multiplications of nonnegative quantities. This leads to a stable algorithm whose precision, ε, can be specified in advance.

Table 7.1. The procedure **Accumulate**(n)

$$
\text{for } i = 1 \text{ to } M \text{ do}
$$

$$
F^{\varepsilon}(t,x(i),n) = e^{-\lambda t} \frac{(\lambda t)^n}{n!} \sum_{k=0}^{n} \binom{n}{k} x_{h_i}^k (1 - x_{h_i})^{n-k} C^{(h_i)}(n,k)
$$

$$
F^{\varepsilon}(t,x(i)) = F^{\varepsilon}(t,x(i)) + F^{\varepsilon}(t,x(i),n)
$$

$$
\text{endfor}
$$

Table 7.2. Computation of the matrices $C^{(h)}(n,k)$ and $F^{\varepsilon}(t,x)$

for $h = 1$ to m do $\forall\, u,v = 0,\ldots,m$, $C^{(h)}_{B_u B_v}(0,0) = 0_{B_u B_v}$ endfor
for $h = 1$ to m do $\forall\, v = h,\ldots,m$, $C^{(h)}_{B_v B_v}(0,0) = I_{B_v B_v}$ endfor
Accumulate(0)
for $n = 1$ to N do
$\quad \forall\, u = 0,\ldots,m$, $\forall\, v = 1,\ldots,m$, $C^{(1)}_{B_u B_v}(n,0) = (P^n)_{B_u B_v}$
$\quad$ for $h = 1$ to m do
$\quad\quad$ for $k = 1$ to n do
$\quad\quad\quad \forall\, u = 0,\ldots,m$, $\forall\, v = h,\ldots,m$, compute Relation (7.15)
$\quad\quad$ endfor
$\quad\quad \forall\, u = 0,\ldots,m$, $\forall\, v = h+1,\ldots,m$, $C^{(h+1)}_{B_u B_v}(n,0) = C^{(h)}_{B_u B_v}(n,n)$
$\quad$ endfor
$\quad \forall\, u = 0,\ldots,m$, $\forall\, v = 0,\ldots,m-1$, $C^{(m)}_{B_u B_v}(n,n) = 0_{B_u B_v}$
$\quad$ for $h = m$ **downto** 1 do
$\quad\quad$ for $k = n-1$ **downto** 0 do
$\quad\quad\quad \forall\, u = 0,\ldots,m$, $\forall\, v = 0,\ldots,h-1$, compute Relation (7.16)
$\quad\quad$ endfor
$\quad\quad \forall\, u = 0,\ldots,m$, $\forall\, v = 0,\ldots,h-2$, $C^{(h-1)}_{B_u B_v}(n,n) = C^{(h)}_{B_u B_v}(n,0)$
$\quad$ endfor
$\quad$ **Accumulate**(n)
endfor

Note that, for a fixed ε, the integer N, defined in (7.34), is an increasing function of t, say $N(t)$. So, if the matrix $F(t,x)$ is to be computed at L different t-values, say $t_1 < \cdots < t_L$, we only need to evaluate the matrices $C^{(h)}(n,k)$ for $n = 0, 1, \ldots, N(t_L)$, as these matrices do not depend on the values of $t_1, \ldots, t_L$. The main effort required for the computation of matrix $F(t,x)$ is in the computation of matrices $C^{(h)}(n,k)$. For matrix P we use a compact storage. If d denotes the connectivity degree of matrix P, that is the maximum number of non-zero entries in each row, then the computational cost of one matrix $C^{(h)}(n,k)$ is $O(d|S|^2)$. The number of such matrices that have to be computed (see Table 7.2) is equal to $m(N+1)(N+2)/2$. The total computational effort required is thus $O(md|S|^2N^2/2)$. For the storage requirements, it is easy to see, from Table 7.2, that we need to store two arrays of $m(N+1)$ matrices for the recursive computation of matrices $C^{(h)}(n,k)$. Thus the storage complexity is $O(m|S|^2N)$.

Note also that if we only want to compute the distribution $\mathbb{P}\{Y_t > x\}$, there is no need to evaluate the matrices $C^{(h)}(n,k)$. Since $\mathbb{P}\{Y_t > x\} = \alpha F(t,x)\mathbb{1}$, where α is the initial probability distribution of the Markov chain X and $\mathbb{1}$ is the column vector of dimension $|S|$ with all components equal to 1, it is sufficient to evaluate the row vectors $b^{(h)}(n,k) = \alpha C^{(h)}(n,k)$. The algorithm thereby becomes more efficient, as the matrix-matrix products are replaced by vector-matrix products. Thus the end product of the algorithm is the vector $g(t,x) = \alpha F(t,x)$ and the complexity is reduced by a factor m. The jth entry, $g_j(t,x)$, of row vector $g(t,x)$ represents the joint probability distribution of Y_t and X_t, i.e.

$$g_j(t,x) = \mathbb{P}\{Y_t > x, X_t = j\}.$$

In the same way, using Corollary 7.2.2 instead of Theorem 7.2.1, we obtain the backward version of the previous algorithm. This algorithm can be used to compute the column vector $h(t,x) = F(t,x)\mathbb{1}$, the ith entry of which represents the conditional distribution of Y_t given that the initial state of X is state i, i.e.

$$h_i(t,x) = \mathbb{P}\{Y_t > x \mid X_0 = i\}.$$

To do this it is sufficient to evaluate the column vectors $c^{(h)}(n,k) = C^{(h)}(n,k)\mathbb{1}$ in the backward version of our algorithm.

7.4.1   Numerical examples

Consider the Markov chain, X, with state space, $S = \{1,2,3\}$, infinitesimal generator, A, and the reward matrix, D, given by

$$A = \begin{pmatrix} -1 & 1 & 0 \\ 0.5 & -1 & 0.5 \\ 0 & 1 & -1 \end{pmatrix} \quad \text{and} \quad D = \begin{pmatrix} 2 & 0 & 0 \\ 0 & 1 & 0 \\ 0 & 0 & 0 \end{pmatrix}.$$

We take as the uniformization rate $\lambda = \max\{-A_{i,i}, i \in S\} = 1$. We also have $m = 2$ and $0 \le Y_t \le 2t$ with probability 1. We consider the distribution function $G_{i,j}(t,x)$ defined by

$$G_{i,j}(t,x) = \mathbb{P}\{Y_t \le x, X_t = j \mid X_0 = i\}.$$

We then have

$$G_{i,j}(t,x) = \mathbb{P}\{X_t = j \mid X_0 = i\} - \mathbb{P}\{Y_t > x, X_t = j \mid X_0 = i\}$$
$$= (e^{At})_{i,j} - F_{i,j}(t,x).$$

Matrix $G(t,x) = (G_{i,j}(t,x))_{i,j \in S}$ is thus given by

$$G(t,x) = e^{At} - F(t,x).$$

For the error tolerance $\varepsilon = 10^{-10}$, we obtain the following results.
For $t = 1$, we have

$$G(1,0) = \begin{pmatrix} 0.0000000000 & 0.0000000000 & 0.0000000000 \\ 0.0000000000 & 0.0000000000 & 0.0000000000 \\ 0.0000000000 & 0.0000000000 & 0.3678794412 \end{pmatrix}.$$

Note the high precision of the algorithm: the entry $G_{3,3}(1,0)$ is the jump corresponding to the Markov chain staying in the state 3 up to time $t = 1$. This jump is equal to $\exp(-1) \approx 0.36787944117$.

$$G(1,0.5) = \begin{pmatrix} 0.0000621571 & 0.0020075639 & 0.0122319275 \\ 0.0010037819 & 0.0244638551 & 0.1008688160 \\ 0.0122319275 & 0.2017376321 & 0.4402204650 \end{pmatrix},$$

$$G(1,1-10^{-12}) = \begin{pmatrix} 0.0010069669 & 0.0163117922 & 0.0499470501 \\ 0.0081558961 & 0.0998941002 & 0.2080102831 \\ 0.0499470501 & 0.4160205662 & 0.4667665745 \end{pmatrix},$$

$$G(1,1) = \begin{pmatrix} 0.0010069669 & 0.0163117922 & 0.0499470501 \\ 0.0081558961 & 0.4677735414 & 0.2080102831 \\ 0.0499470501 & 0.4160205662 & 0.4667665745 \end{pmatrix}.$$

Again, note the high precision of the algorithm: except for entry $(2,2)$, all the entries of $G(1,1-10^{-12})$ and $G(1,1)$ are equal. The difference between the two entries $(2,2)$ is equal to 0.3678794412; it corresponds to the jump at $x = 1$, which is the probability that the Markov chain stays in state 2 beyond time $t = 1$, that is $\exp(-1) \approx 0.36787944117$.

$$G(1,1.5) = \begin{pmatrix} 0.0275530764 & 0.2305947262 & 0.0876621727 \\ 0.1152973631 & 0.5432037865 & 0.2151623972 \\ 0.0876621727 & 0.4303247945 & 0.4677113843 \end{pmatrix},$$

$$G(1,2^-) = \begin{pmatrix} 0.0998941002 & 0.4323323583 & 0.0998941002 \\ 0.2161661792 & 0.5676676416 & 0.2161661792 \\ 0.0998941002 & 0.4323323583 & 0.4677735414 \end{pmatrix}.$$

The entries of the matrix $G(1,2^-)$ are

$$G_{i,j}(1,2^-) = P\{Y_t < 2, X_1 = j \mid X_0 = i\}.$$

Since $G(1,2) = e^A$, we easily obtain the jump

$$P\{Y_t = 2 \mid X_0 = 1\} = \sum_{j=1}^{3} G_{1,j}(1,2) - \sum_{j=1}^{3} G_{1,j}(1,2^-)$$

$$= 1 - \sum_{j=1}^{3} G_{1,j}(1,2^-) = 0.3678794412,$$

whose value is $\exp(-1) \approx 0.36787944117$.

For $t = 100$, the algorithm gives $G_{i,j}(100,0) = 0$ for every $1 \le i,j \le 3$ and

$$G(100,50) = \begin{pmatrix} 0.0000000072 & 0.0000000261 & 0.0000000234 \\ 0.0000000130 & 0.0000000467 & 0.0000000414 \\ 0.0000000234 & 0.0000000827 & 0.0000000724 \end{pmatrix},$$

$$G(100,100) = \begin{pmatrix} 0.1050278103 & 0.2299261526 & 0.1250000000 \\ 0.1149630763 & 0.2500000000 & 0.1350369237 \\ 0.1250000000 & 0.2700738474 & 0.1449721896 \end{pmatrix},$$

$$G(100,150) = \begin{pmatrix} 0.2499999276 & 0.4999999172 & 0.2499999766 \\ 0.2499999586 & 0.4999999533 & 0.2499999869 \\ 0.2499999766 & 0.4999999739 & 0.2499999928 \end{pmatrix},$$

$$G(100,200^-) = \begin{pmatrix} 0.2500000000 & 0.5000000000 & 0.2500000000 \\ 0.2500000000 & 0.5000000000 & 0.2500000000 \\ 0.2500000000 & 0.5000000000 & 0.2500000000 \end{pmatrix}.$$

Note that in this case the jumps are numerically invisible, since $\exp(-100) \approx 0.372 \times 10^{-43}$.

Notes

The distribution of the random variable Y_t has been studied in several papers. Some of these papers ([67], [36], [42], [33], [19], [43], [75]) are restricted to the case of acyclic Markov chains, which are used to model non-repairable systems.

In [46], Iyer *et al.* proposed an algorithm to compute recursively the moments of the accumulated reward over the mission time, with a polynomial computational complexity in the number of states. In [108], the distribution of this random variable was derived using Laplace transforms and numerical inversion procedures to obtain the result in the time domain. De Souza e Silva and Gail [29] proposed a method based on the uniformization technique, however their method exhibits an exponential computational complexity in the number of reward rates.

Using the same technique, Donatiello and Grassi [32] obtained an algorithm with a polynomial computational complexity. However, this algorithm seems to be numerically unstable since the coefficients computed in their recursion can have positive and negative signs and are unbounded, which can lead to severe numerical errors and overflow problems. More recently, De Souza e Silva and Gail [30] also obtained an algorithm with a polynomial computational complexity which is linear in a parameter that is smaller than the number of rewards, but their algorithm seems to have the same instability problem due to the use of both positive and negative coefficients. In [30], they extended their method to the case where the model has not only a reward rate associated with each state but also an impulse rate associated with each transition. Such models are also discussed in [6] and in [59]. An analysis of the moments of the accumulated reward is available in [17]. Pattipati *et al.* [78] obtained the distribution of the accumulated reward for non-homogeneous Markov chains as the solution of a system of linear hyperbolic partial differential equations which is numerically solved using a discretization approach.

Two different architectures have been compared using the expected accumulated reward in [47]. When the Markov chain is absorbing, the distribution of the accumulated reward over $[0, \infty)$ has been obtained in [18] and is given in Chapter 5. Other performability measures using the successive sojourn times in operational states have been studied in [88], [87], and [96].

The results of this chapter are mainly based on the work published in papers [74] and [104]. The algorithm proposed in this chapter has been improved in [74] and [16].

8   Stationarity detection

In this chapter we first consider the point availability and the expected interval availability of a system modeled by an irreducible continuous-time homogeneous Markov chain $X = \{X_u, u \geq 0\}$ over a finite state space denoted by S. In a second step, we deal with the corresponding point and expected performability measures.

As we did in the previous chapters, we denote by A the infinitesimal generator and by α the initial probability distribution. We denote by $Z = \{Z_n, n \geq 0\}$ the Markov chain obtained after uniformization of X, as described in Chapter 3. Its initial probability distribution is α and its transition probability matrix, P, is related to matrix A by the relation $P = I + A/\lambda$, where I is the identity matrix and λ satisfies $\lambda \geq \max\{-A_{i,i}, i \in S\}$. The rate λ is the rate of the Poisson process $\{N_u, u \geq 0\}$, independent of Z, which counts the number of transitions of process $\{Z_{N_u}, u \geq 0\}$. As shown in Chapter 3, processes $\{Z_{N_u}\}$ and X are stochastically equivalent.

8.1   Point availability

Let U be a proper subset of S representing the operational states of a system modeled by X. The point availability is defined by the function

$$\mathrm{PAV}(t) = \mathbb{P}\{X_t \in U\}.$$

As shown in Chapter 3, we have

$$\mathrm{PAV}(t) = \alpha e^{At}\mathbb{1}_U = \sum_{n=0}^{+\infty} e^{-\lambda t}\frac{(\lambda t)^n}{n!}\alpha P^n \mathbb{1}_U, \tag{8.1}$$

where $\mathbb{1}_U$ is a column vector whose ith entry is 1 if $i \in U$ and 0 otherwise. We denote by V_n the column vector defined by $V_n = P^n \mathbb{1}_U$. It follows that for every $n \geq 0$, we have $V_n = PV_{n-1}$ and $V_0 = \mathbb{1}_U$. In the following, we define for every $n \geq 0$, $v_n = \alpha P^n \mathbb{1}_U = \alpha V_n$.

8.1.1 The classical uniformization method

The classical way to compute the function $\mathrm{PAV}(t)$ is based on Relation (8.1). Let ε be a given specified error tolerance and N be defined as

$$N = \min\left\{n \in \mathbb{N} \;\middle|\; \sum_{j=0}^{n} e^{-\lambda t}\frac{(\lambda t)^j}{j!} \geq 1 - \varepsilon\right\}. \tag{8.2}$$

Then we obtain

$$\mathrm{PAV}(t) = \sum_{n=0}^{N} e^{-\lambda t}\frac{(\lambda t)^n}{n!}v_n + e(N),$$

where the rest of the series $e(N)$ satisfies

$$e(N) = \sum_{n=N+1}^{\infty} e^{-\lambda t}\frac{(\lambda t)^n}{n!}v_n \leq \sum_{n=N+1}^{\infty} e^{-\lambda t}\frac{(\lambda t)^n}{n!} = 1 - \sum_{n=0}^{N} e^{-\lambda t}\frac{(\lambda t)^n}{n!} \leq \varepsilon.$$

The computation of integer N can be made without any numerical problems even for large values of λt using, for instance, the method described in [9] and [105].

The truncation level N is in fact a function of t, say $N(t)$. For a fixed value of ε, $N(t)$ is an increasing function of t. It follows that if we want to compute $\mathrm{PAV}(t)$ for J distinct values of t, denoted by $t_1 < \cdots < t_J$, we only need to compute v_n for $n = 1, \ldots, N(t_J)$ since the values of v_n are independent of the parameter t.

The pseudo-code of the classical uniformization method can then be written as shown in Table 8.1.

Table 8.1 Classical algorithm for the computation of $\mathrm{PAV}(t)$

> **input :** ε, $t_1 < \cdots < t_J$
> **output :** $\mathrm{PAV}(t_1), \ldots, \mathrm{PAV}(t_J)$
> Compute N from Relation (8.2) with $t = t_J$
> $V_0 = \mathbb{1}_U$; $v_0 = \alpha V_0$
> **for** $n = 1$ **to** N **do**
> $V_n = PV_{n-1}$; $v_n = \alpha V_n$
> **endfor**
> **for** $j = 1$ **to** J **do**
> $$\mathrm{PAV}(t_j) = \sum_{n=0}^{N} e^{-\lambda t_j}\frac{(\lambda t_j)^n}{n!}v_n$$
> **endfor**

8.1.2 Stationarity detection

An approach to detect the stationarity of Markov chains has been proposed in [63], [73], and [109]. This approach is based on the uniformization method. The state probability

vectors of the uniformized Markov chain are successively computed and the iterates
that are spaced m iterations apart are compared. When the difference between two such
iterates is small enough, the computation is stopped. The main problem with this method
is that, unlike the standard uniformization, there is no ability to specify a priori an
error bound that is easily computable. Nevertheless, the experiments performed by the
authors of these papers on a variety of models show that this approach is very efficient.

The stationarity detection that we consider is based on the control of the sequence of
vectors, $V_n = P^n \mathbb{1}_U$. Let the row vector π denote the stationary probability distribution
of the Markov chain X. This vector satisfies $\pi A = 0$ and $\pi P = \pi$. The steady-state
availability is given by $\text{PAV}(\infty) = \pi \mathbb{1}_U$. To ensure the convergence of the sequence
of vectors, V_n, we require that the uniformization rate λ satisfies $\lambda > \max\{-A_{i,i}, i \in S\}$,
since this guarantees that the transition probability matrix, P, is aperiodic. We then have,
for every $i \in S$,

$$\lim_{n \to \infty} V_n(i) = \pi \mathbb{1}_U.$$

We describe now the test used to detect that, for a given value of n, the entries of vector
V_n are close to $\pi \mathbb{1}_U$. For every $n \geq 0$, we define

$$m_n = \min_{i \in S} V_n(i) \quad \text{and} \quad M_n = \max_{i \in S} V_n(i).$$

Note that, since $V_0 = \mathbb{1}_U$, we have $M_0 = 1$ and $m_0 = 0$. The following result gives bounds
of the steady-state availability, $\text{PAV}(\infty) = \pi \mathbb{1}_U$.

LEMMA 8.1.1.   *The sequences m_n and M_n are non-decreasing and non-increasing,
respectively, and, for every $n \geq 0$, we have*

$$\left| v_n - \frac{M_n + m_n}{2} \right| \leq \frac{M_n - m_n}{2} \quad \text{and} \quad \left| \pi \mathbb{1}_U - \frac{M_n + m_n}{2} \right| \leq \frac{M_n - m_n}{2}.$$

Moreover, both sequences m_n and M_n converge to $\pi \mathbb{1}_U$.

Proof.   For every $i \in S$, we have

$$V_{n+1}(i) = \sum_{j \in S} P(i,j) V_n(j).$$

It follows that $m_n \leq V_{n+1}(i) \leq M_n$ and so we have $m_n \leq m_{n+1}$ and $M_{n+1} \leq M_n$,
which shows that the sequences m_n and M_n are non-decreasing and non-increasing,
respectively.

Since $v_n = \sum_{j \in S} \alpha(j) V_n(j)$, we have $m_n \leq v_n \leq M_n$, which is equivalent to

$$\left| v_n - \frac{M_n + m_n}{2} \right| \leq \frac{M_n - m_n}{2}.$$

Writing $\pi \mathbb{1}_U = \pi P^n \mathbb{1}_U = \pi V_n = \sum_{j \in S} \pi(j) V_n(j)$, we obtain in the same way $m_n \leq \pi \mathbb{1}_U \leq M_n$, which is equivalent to

$$\left| \pi \mathbb{1}_U - \frac{M_n + m_n}{2} \right| \leq \frac{M_n - m_n}{2}.$$

The state space S being finite and the fact that for every $i \in S$, $V_n(i)$ converges to $\pi \mathbb{1}_U$ show that both sequences m_n and M_n converge to $\pi \mathbb{1}_U$. ∎

Remark. We have assumed that the Markov chain X is irreducible. If X is not irreducible but contains transient states and an absorbing state denoted by a with $a \in D$, where $D = S \setminus U$, then we have $\pi \mathbb{1}_U = 0$ and, for every $i \in S$, we have $V_n(i) \longrightarrow 0$ when $n \longrightarrow \infty$. Now, since a is absorbing, we have $V_n(a) = 0$ for every $n \geq 0$ and so we also have $m_n = 0$ for every $n \geq 0$. Thus, in this case, it is sufficient to consider the sequence M_n which is non-increasing and converges to 0. ∎

This lemma shows that the difference $M_n - m_n$ converges to 0, that is, for a fixed error tolerance, $\varepsilon > 0$, there exists an integer, k, such that for $n \geq k$ we have $M_n - m_n \leq \varepsilon$. Since $m_n \leq M_n$, we have $m_n \leq m_{n+1} \leq M_{n+1} \leq M_n$, and so the sequence $(M_n - m_n)$ is non-increasing. We can then define the following integer

$$K = \inf\{n \geq 0 \mid M_n - m_n \leq \varepsilon/2\}.$$

Using integer K, Relation (8.1) can be written as

$$\text{PAV}(t) = \sum_{n=0}^{K} e^{-\lambda t} \frac{(\lambda t)^n}{n!} v_n + \frac{M_K + m_K}{2} \left(1 - \sum_{n=0}^{K} e^{-\lambda t} \frac{(\lambda t)^n}{n!} \right) + e_1(K), \qquad (8.3)$$

where

$$e_1(K) = \sum_{n=K+1}^{\infty} e^{-\lambda t} \frac{(\lambda t)^n}{n!} v_n - \frac{M_K + m_K}{2} \sum_{n=K+1}^{\infty} e^{-\lambda t} \frac{(\lambda t)^n}{n!}.$$

Using Lemma 8.1.1, the remainder $e_1(K)$ satisfies

$$|e_1(K)| \leq \sum_{n=K+1}^{\infty} e^{-\lambda t} \frac{(\lambda t)^n}{n!} \left| v_n - \frac{M_K + m_K}{2} \right| \leq \varepsilon/4. \qquad (8.4)$$

This last inequality follows from the fact that, for $n \geq K$, we have from Lemma 8.1.1 $m_K \leq m_n \leq v_n \leq M_n \leq M_K$ and so

$$\left| v_n - \frac{M_K + m_K}{2} \right| \leq \frac{M_K - m_K}{2} \leq \varepsilon/4.$$

The time, K, can be interpreted as the discrete time to stationarity with respect to the subset U.

For every $t \geq 0$ and for every integer $l \geq 0$, we denote by $F_l(t)$ the function defined by

$$F_l(t) = \sum_{n=0}^{l} e^{-\lambda t} \frac{(\lambda t)^n}{n!} (M_n - m_n).$$

It is easy to check that for a fixed value of l, the function $F_l(t)$ decreases, from 1 to 0 over the interval $[0, \infty)$. We can then define for every integer $l \geq 0$ and for every $\varepsilon > 0$, the time T_l as

$$T_l = \inf\{t \geq 0 \mid F_l(t) \leq \varepsilon/4\}.$$

We then have the following theorem.

THEOREM 8.1.2.   *For every $\varepsilon > 0$ and for every $t \geq T_K$, we have*

$$|\text{PAV}(t) - \pi \mathbb{1}_U| \leq 3\varepsilon/4 \tag{8.5}$$

$$\left| \pi \mathbb{1}_U - \frac{M_K + m_K}{2} \right| \leq \varepsilon/4 \tag{8.6}$$

$$\left| \text{PAV}(t) - \frac{M_K + m_K}{2} \right| \leq \varepsilon. \tag{8.7}$$

Proof.   First note that, from Lemma 8.1.1, we have $m_n \leq v_n \leq M_n$ and $m_n \leq \pi \mathbb{1}_U \leq M_n$, for every $n \geq 0$. It follows that $|v_n - \pi \mathbb{1}_U| \leq M_n - m_n$ for every $n \geq 0$. We then have

$$|\text{PAV}(t) - \pi \mathbb{1}_U| = \left| \sum_{n=0}^{\infty} e^{-\lambda t} \frac{(\lambda t)^n}{n!} v_n - \pi \mathbb{1}_U \right|$$

$$\leq \sum_{n=0}^{\infty} e^{-\lambda t} \frac{(\lambda t)^n}{n!} |v_n - \pi \mathbb{1}_U|$$

$$\leq \sum_{n=0}^{\infty} e^{-\lambda t} \frac{(\lambda t)^n}{n!} (M_n - m_n)$$

$$= F_K(t) + \sum_{n=K+1}^{\infty} e^{-\lambda t} \frac{(\lambda t)^n}{n!} (M_n - m_n).$$

Since $t \geq T_K$, we have $F_K(t) \leq \varepsilon/4$. In the second term, since $n \geq K$, we have $M_n - m_n \leq M_K - m_K \leq \varepsilon/2$ and so we obtain Relation (8.5).

Relation (8.6) is immediate from Lemma 8.1.1. Finally, combining Relations (8.5) and (8.6), we obtain Relation (8.7).   ∎

The time T_K can be interpreted as the continuous time to stationarity with respect to the subset U.

8.1.3 The new algorithm

Using these results, we obtain the algorithm given in Table 8.2. To simplify the writing of this algorithm, we define

$$G_l(t) = \sum_{n=0}^{l} e^{-\lambda t} \frac{(\lambda t)^n}{n!} v_n, \quad H_l(t) = 1 - \sum_{n=0}^{l} e^{-\lambda t} \frac{(\lambda t)^n}{n!}, \quad S_l = \frac{M_l + m_l}{2}.$$

Note that it is not necessary to compute the continuous time to stationarity, T_K, with a high precision. It is sufficient to obtain an upper bound of T_K such as for instance $\lceil T_K \rceil$, which is the smallest integer greater than or equal to T_K.

It must also be noted that, in this algorithm, the truncation step, N, is a function of the time, t_J, as in the classical uniformization algorithm but the times to stationarity, K, and T_K, are independent of the time parameter, when the discrete time, K, is reached.

The computational time complexity of both algorithms is essentially due to the computation of vectors, V_n. To compute these vectors, the classical algorithm requires N matrix-vector products and our new algorithm requires only $\min(K,N)$ matrix-vector products.

8.2 Expected interval availability analysis

In this section we show how the new algorithm proposed above for the point availability computation can be adapted to compute the expected interval availability using stationarity detection.

The expected interval availability represents the mean percentage of time during which the system is in operation over a finite observation period $(0,t)$. The interval availability over $(0,t)$ was denoted in Chapter 1 by $IA(t)$ (see Section 1.2.2) and written in Chapter 6 as W_t/t, where

$$W_t = \int_0^t 1_{\{X_u \in U\}} du.$$

The expected interval availability, denoted here by $EIAV(t)$, over $(0,t)$ is then defined by $EIAV(t) = \mathbb{E}\{W_t/t\}$. We thus have

$$EIAV(t) = \frac{1}{t} \int_0^t PAV(s) ds.$$

Using Relation (8.1) and by integration over $(0,t)$, we obtain

$$EIAV(t) = \sum_{n=0}^{+\infty} e^{-\lambda t} \frac{(\lambda t)^n}{n!} \frac{1}{n+1} \sum_{k=0}^{n} \alpha P^k 1_U.$$

We denote by V'_n the column vector defined by

$$V'_n = \frac{1}{n+1} \sum_{k=0}^{n} P^k 1_U,$$

and we define $v'_n = \alpha V'_n$. Using the definition of V_n and v_n from the previous section, we obtain, for every $n \geq 0$,

$$V'_n = \frac{1}{n+1} \sum_{k=0}^{n} V_k \quad \text{and} \quad v'_n = \frac{1}{n+1} \sum_{k=0}^{n} v_k.$$

It follows that V'_n and v'_n are recursively given, for $n \geq 1$, by

$$V'_n = \frac{n}{n+1} V'_{n-1} + \frac{1}{n+1} V_n,$$

and

$$v'_n = \frac{n}{n+1} v'_{n-1} + \frac{1}{n+1} v_n, \tag{8.8}$$

with $V'_0 = V_0 = \mathbb{1}_U$ and thus $v'_0 = v_0$. For every $n \geq 0$, we have $0 \leq v'_n \leq 1$. It follows that, using the truncation step N defined in Relation (8.2), we obtain the classical algorithm to compute the expected interval availability, by writing

$$\text{EIAV}(t) = \sum_{n=0}^{N} e^{-\lambda t} \frac{(\lambda t)^n}{n!} v'_n + e'(N),$$

where

$$e'(N) = \sum_{n=N+1}^{\infty} e^{-\lambda t} \frac{(\lambda t)^n}{n!} v'_n \leq \sum_{n=N+1}^{\infty} e^{-\lambda t} \frac{(\lambda t)^n}{n!} = 1 - \sum_{n=0}^{N} e^{-\lambda t} \frac{(\lambda t)^n}{n!} \leq \varepsilon.$$

This algorithm is basically the some as the one depicted in Table 8.1. More precisely, the computation of v_n in Table 8.1 must be followed by the Relation (8.8), with $v'_0 = v_0$, and in the last loop over j, v_n must be replaced by v'_n in order to obtain $\text{EIAV}(t_j)$ instead of $\text{PAV}(t_j)$.

8.2.1 Stationarity detection for the expected interval availability

Using the results obtained for the point availability, we can derive a new method to obtain the expected interval availability using stationarity detection. This method is based on the two following theorems. Both theorems will be used in the case where the discrete time to stationarity, K, is such that $K \leq N$. The first theorem states that to compute the expected interval availability, $\text{EIAV}(t)$, we only need the values of v'_n for $n \leq K$. The second theorem states that to compute the expected interval availability, $\text{EIAV}(t)$, for $t \geq T_K$, we only need the value $\text{EIAV}(t')$ at a time t' such that $t \geq t' \geq T_K$.

We denote by $G'_K(t)$ the function

$$G'_K(t) = \sum_{n=0}^{K} e^{-\lambda t} \frac{(\lambda t)^n}{n!} v'_n$$

and we recall that

$$H_K(t) = 1 - \sum_{n=0}^{K} e^{-\lambda t} \frac{(\lambda t)^n}{n!} \quad \text{and} \quad S_K = \frac{M_K + m_K}{2}.$$

THEOREM 8.2.1.   *For every $t \geq 0$, we have*

$$\left| \text{EIAV}(t) - \left[G'_K(t) + \frac{K+1}{\lambda t} (v'_K - S_K) H_{K+1}(t) + S_K H_K(t) \right] \right| \leq \varepsilon/4. \qquad (8.9)$$

Proof.   For every $t \geq 0$, we have

$$\text{EIAV}(t) = G'_K(t) + \phi(t),$$

where

$$\phi(t) = \sum_{n=K+1}^{\infty} e^{-\lambda t} \frac{(\lambda t)^n}{n!} v'_n.$$

For $n \geq K+1$, we have

$$v'_n = \frac{1}{n+1} \sum_{k=0}^{n} v_k$$

$$= \frac{1}{n+1} \left[\sum_{k=0}^{K} v_k + \sum_{k=K+1}^{n} v_k \right]$$

$$= \frac{K+1}{n+1} v'_K + \frac{1}{n+1} \sum_{k=K+1}^{n} v_k$$

$$= \frac{K+1}{n+1} v'_K + \frac{1}{n+1} \sum_{k=K+1}^{n} (v_k - S_K) + \frac{(n-K)}{n+1} S_K$$

$$= \frac{K+1}{n+1} v'_K + \frac{(n-K)}{n+1} S_K + x_n,$$

where

$$|x_n| = \left| \frac{1}{n+1} \sum_{k=K+1}^{n} (v_k - S_K) \right| \leq \frac{1}{n+1} \sum_{k=K+1}^{n} |v_k - S_K| \leq \frac{(n-K)}{n+1} \frac{\varepsilon}{4} \leq \varepsilon/4.$$

The inequality $|v_k - S_K| \leq \varepsilon/4$, for $k \geq K$, follows from Lemma 8.1.1; it has already been used to bound the error $e_1(K)$ in Relation (8.4). If $\psi(t)$ is the function defined by

$$\psi(t) = \sum_{n=K+1}^{\infty} e^{-\lambda t} \frac{(\lambda t)^n}{n!} x_n,$$

we obtain $|\psi(t)| \le \varepsilon/4$. We then have

$$\phi(t) = \sum_{n=K+1}^{\infty} e^{-\lambda t} \frac{(\lambda t)^n}{n!} \left(\frac{K+1}{n+1} v_K' + \frac{(n-K)}{n+1} S_K \right) + \psi(t).$$

By writing $n - K = n + 1 - (K + 1)$ in this last expression, we obtain

$$\phi(t) = \frac{K+1}{\lambda t}(v_K' - S_K)H_{K+1}(t) + S_K H_K(t) + \psi(t).$$

We then obtain

$$\left| \text{EIAV}(t) - \left[G_K'(t) + \frac{K+1}{\lambda t}(v_K' - S_K)H_{K+1}(t) + S_K H_K(t) \right] \right| = |\psi(t)|,$$

which completes the proof since $|\psi(t)| \le \varepsilon/4$.   ■

THEOREM 8.2.2.   *For every $\varepsilon > 0$, for every t and t' such that $t \ge t' \ge T_K$, we have*

$$\left| \text{EIAV}(t) - \left[\frac{t'}{t}\text{EIAV}(t') + \left(1 - \frac{t'}{t} \right) S_K \right] \right| \le \varepsilon. \tag{8.10}$$

Proof.   For every t and t' such that $t \ge t' \ge T_K$, we have

$$\text{EIAV}(t) = \frac{1}{t} \int_0^t \text{PAV}(s)\,ds$$

$$= \frac{1}{t} \left[\int_0^{t'} \text{PAV}(s)\,ds + \int_{t'}^{t} \text{PAV}(s)\,ds \right]$$

$$= \frac{1}{t} \left[\int_0^{t'} \text{PAV}(s)\,ds + (t - t')S_K + \int_{t'}^{t} [\text{PAV}(s) - S_K]\,ds \right]$$

$$= \frac{t'}{t}\text{EIAV}(t') + \left(1 - \frac{t'}{t} \right) S_K + \frac{1}{t} \int_{t'}^{t} [\text{PAV}(s) - S_K]\,ds.$$

Using Relation (8.7), we have, since $t' \ge T_K$,

$$\left| \frac{1}{t} \int_{t'}^{t} [\text{PAV}(s) - S_K]\,ds \right| \le \frac{1}{t} \int_{t'}^{t} |\text{PAV}(s) - S_K|\,ds \le \left(1 - \frac{t'}{t} \right) \varepsilon \le \varepsilon,$$

which completes the proof.   ■

Note that Theorem 8.2.2 is still valid if we replace T_K by $\lceil T_K \rceil$. So, as for the point availability, we can use $\lceil T_K \rceil$ instead of T_K to make the computation of the expected interval availability easier.

Using these two theorems, we obtain a new algorithm to compute the expected interval availability, which is similar to the one described in Table 8.2 for the point

Table 8.2 Algorithm for the computation of $PAV(t)$ using stationarity detection

---

input : $\varepsilon, t_1 < \cdots < t_J$
output : $PAV(t_1), \ldots, PAV(t_J)$
Compute N from Relation (8.2) with $t = t_J$
$V_0 = \mathbb{1}_U; v_0 = \alpha V_0$
$M_0 = 1; m_0 = 0; K = N+1$
for $n = 1$ **to** N **do**
    $V_n = PV_{n-1}; v_n = \alpha V_n$
    Compute M_n, m_n and S_n
    if $M_n - m_n \leq \varepsilon/2$
        $K = n$; **break**
    endif
endfor
if $K = N+1$
    for $j = 1$ **to** J **do** $PAV(t_j) = G_N(t_j)$ **endfor**
    $m_N \leq PAV(\infty) \leq M_N$
endif
if $K \leq N$
    Compute $T_K = \inf\{t \geq 0 \mid F_K(t) \leq \varepsilon/4\}$
    for $j = 1$ **to** J **do**
        if $t_j \leq T_K$ **then** $PAV(t_j) = G_K(t_j) + S_K H_K(t_j)$
        if $t_j > T_K$ **then** $PAV(t_j) = PAV(\infty) = S_K$
    endfor
endif

---

availability. It is sufficient to perform the following changes in the algorithm given in Table 8.2. The computation of v_n' given by Relation (8.8) must be added just after the computation of v_n, with $v_0' = v_0$. The relation $PAV(t_j) = G_K(t_j)$ must be replaced by $EIAV(t_j) = G_K'(t_j)$ and the computations of $PAV(t_j)$ in the case where $K \leq N$ must be replaced by those of $EIAV(t_j)$ given in Relation (8.9) for $t_j \leq T_K$ and in Relation (8.10) for $t_j > T_K$. To use Relation (8.10), we need $EIAV(t)$ for one value of t, such that $t \geq T_K$. Such a value can be obtained by using Relation (8.9) one more time for the smallest value of t_j, such that $t_j \geq T_K$. Note that we have the well-known stationary relation $PAV(\infty) = EIAV(\infty)$.

8.3   Numerical example

We consider a fault-tolerant multiprocessor system with finite buffer stages. This system was first considered in [67] for two processors without repair and was extended in [46] to include repairs for the computation of the moments of the performability. It was

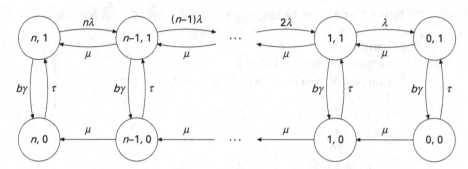

Figure 8.1   State-transition diagram for an n-processor system

also used in [51] to obtain the distribution of the performability. We use here the same model for the computation of the point availability with our new method. It consists of n identical processors and b buffer stages. Processors fail independently at rate λ and are repaired singly with rate μ. Buffer stages fail independently at rate γ and are repaired with rate τ. Processor failures cause a graceful degradation of the system and the number of operational processors is decreased by one. The system is in a failed state when all the processors have failed or any of the buffer stages has failed. No additional processor failures are assumed to occur when the system is in a failed state. The model is represented by a Markov chain with the state-transition diagram shown in Figure 8.1. The state space of the system is $S = \{(i,j) \mid 0 \le i \le n, j = 0, 1\}$. The component i of a state (i,j) means that there are i operational processors and the component j is zero if any of the buffer stages is failed, otherwise it is one. It follows that the set U of operational states is $U = \{(i,1) \mid 1 \le i \le n\}$.

We evaluate the point availability given that the system started in state $(n,1)$. The number of processors is fixed to 16, each with a failure rate $\lambda = 0.01$ per week and a repair rate $\mu = 0.1666$ per hour. The individual buffer stage failure rate is $\gamma = 0.22$ per week and its repair rate is $\tau = 0.1666$ per hour. The error tolerance is $\varepsilon = 0.00001$.

In Figure 8.2, we plot the point availability, PAV(t), as a function of t for different values of the number of buffer stages b. The largest value of t, that is the value of t_J in the algorithm, has been chosen to be 10 000 hours.

For the largest value of t we show in Table 8.3 the truncation step, $N = N(10000)$; the discrete time to stationarity, K; and the continuous time to stationarity, T_K, (in fact we give $\lceil T_K \rceil$) for different values of the number of buffer stages, b. This table shows for example that when $b = 16$ the classical algorithm needs 3581 matrix-vector products and our new algorithm needs only 18 matrix-vector products, the continuous time to stationarity being equal to 78. When $b = 1024$ the classical algorithm needs 15 616 matrix-vector products and our new algorithm needs only 86 matrix-vector products, the continuous time to stationarity being equal to 62. Moreover, our algorithm also computes the steady-state availability with a precision equal to $\varepsilon/4$. Table 8.3 also shows that both situations, $K < T_K$ and $K > T_K$, are possible.

In Table 8.4 we consider smaller values of t_J. The number of buffer stages is fixed to $b = 8$. For $t_J \le 10$ we have $N(10) \le 14$ and the discrete time to stationarity, K, is not

Table 8.3. Stationarity detection for different numbers of buffer stages

b	2	4	8	16	32	64	128	256	512	1024
N	3581	3581	3581	3581	3581	3581	3602	5334	8776	15616
K	19	19	18	18	18	18	18	28	48	86
$\lceil T_K \rceil$	81	81	80	78	77	75	77	70	66	62

Table 8.4. Stationarity detection for small values of time

t_J	10	20	30	40	50	60	70	80	90	100
$N(t_J)$	14	20	26	32	37	42	47	51	56	61

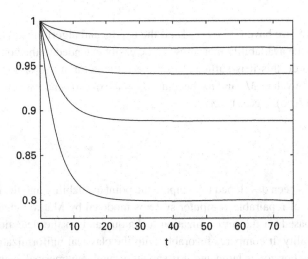

Figure 8.2   From top to bottom: PAV(t) for $b = 2, 4, 8, 16, 32$

reached. This means that $K > 14$. For $t_J \geq 20$ we have $N(t_J) \geq 20$ and the discrete time to stationarity is reached. Its value is $K = 18$ and the continuous time to stationarity is $\lceil T_K \rceil = 80$. Table 8.4 shows that even for small values of t_J ($t_J < T_K$), our algorithm can reduce the computation time compared with the classical technique. For instance when $t_J = 60$, the classical algorithm needs 42 matrix-vector products and our new algorithm needs only 18 matrix-vector products.

8.4   Extension to performability analysis

The method proposed for the computation of the point availability and the expected interval availability using steady-state availability detection can be extended to more general measures such as the point performability and the expected performability.

In performability modeling (see, for instance, [67], [46], [51], [18], [108], [74], [29], [66] and the references therein) reward rates are associated with states of the model to quantify the ability of the system to perform in the corresponding states. We denote by $\rho(i)$ the reward rate associated with state $i \in S$. The reward rates, $\rho(i)$, are assumed to be nonnegative real numbers. The point performability at time t, denoted by PP(t), and the expected performability, denoted by EP(t), are defined by

$$PP(t) = \sum_{i \in S} \rho(i)\mathbb{P}\{X_t = i\} \quad \text{and} \quad EP(t) = \frac{1}{t}\int_0^t PP(s)\,ds.$$

We define $\rho = \max_{i \in S} \rho(i)$ and $r(i) = \rho(i)/\rho$ and we denote by r the column vector whose ith entry is equal to $r(i)$. We then have PP(t) $= \rho f(t)$ and EP(t) $= \rho g(t)$, where

$$f(t) = \alpha e^{At}r \quad \text{and} \quad g(t) = \frac{1}{t}\int_0^t f(s)\,ds.$$

Since for every $i \in S$, we have $0 \leq r(i) \leq 1$, all the results and algorithms obtained for the computation of the availability measures can be easily extended to the computation of $f(t)$ and $g(t)$. To do this it is sufficient to replace the column vector $\mathbb{1}_U$ with the column vector r. The values M_0 and m_0 become $M_0 = \max_{i \in S} r(i)$ and $m_0 = \min_{i \in S} r(i)$. Moreover, we have $f(\infty) = g(\infty) = \pi r$.

8.5   Conclusions

A new algorithm has been developed to compute the point availability and the expected interval availability of repairable computer systems modeled by Markov chains. This new algorithm is based on the uniformization technique and on the detection of the steady-state availability. It compares favorably with the classical uniformization algorithm when the time horizon is large and it is shown through a numerical example that computational savings can be obtained even when the time horizon is small. Moreover, our algorithm gives the steady-state availability if the stationarity is reached and bounds of the steady-state availability otherwise. Finally, this method can be easily extended to the computation of more general measures such as the point performability and the expected performability.

9 Simulation techniques

Simulating a Markov model is the method of choice when analytic or numerical approaches are not possible. Basically, the former happens when the model is not regular enough, meaning that it does not have enough structure, the latter when its size precludes the use of numerical procedures. We also use the term Monte Carlo to refer to these simulation methods, using the term in its most common meaning, i.e. the use of randomness to solve deterministic problems. There are many important subfields in this wide area that have been the object of concerted research efforts, for instance the MCMC [83], [31] or the perfect simulation [81] approaches, and many procedures designed for specific application areas, such as queuing [52], [38], [109], financial problems [37], physics [40], and many more [45], [35], [4]. Monte Carlo techniques are extremely powerful, and they can deal, in principle, with any kind of model. They are not limited by the lack of structure in the model, nor by the model's size. However, they have their own difficulties, the main one being the case of rare events. If you want to analyze a specific metric quantifying some aspect of this rarity, the fact that the focused event appears with very small probability means that it will also be hard to observe in the simulation. If the system failure during a typical mission, say an eight-hour flight of some aircraft under analysis, occurs with a probability of 10^{-9}, then we need to simulate on average a billion (10^9) flights to observe a failure. As a rule of thumb, we need between one hundred billion and one thousand billion samples of that simulated flight in order to have some accuracy on, for instance, the estimation of the failure probability. We call this approach the standard one in simulation (also sometimes called "crude" simulation). For many systems, this is simply useless, because we will need unacceptable computing times to build this many samples. If only standard straightforward procedures are available, and they are not able to provide results because of the consequences of dealing with very small probabilities, other very specialized techniques exist, and they can, in many cases, solve these evaluation problems with enough accuracy and in a reasonable time. This chapter deals precisely with this type of situation.

Following the general lines of the book, we focus on two contributions of the authors in this domain, both on dependability problems and both concerning the problem of rare events. The first one deals with the analysis of what is probably the most basic dependability metric, the Mean Time To Failure (MTTF) of a complex multicomponent system (see Subsection 1.2.1). The second one concerns another basic metric of such a system, its reliability at time t, $R(t)$, that is, a transient measure of system behavior with respect to failures and repairs. These two fundamental metrics are concerned with the following context: the system can be either operational or failed. At time 0 it is operational, and the first system failure (that is, the first transition into a failed system state) happens after a random time, T. The MTTF is the expectation of T and the reliability at t is the probability that $T > t$, that is

$$\text{MTTF} = \mathbb{E}\{T\},$$

$$R(t) = \mathbb{P}\{T > t\}.$$

As said before, when evaluating a dependability metric using simulation, the main technical problem that must be addressed is *rarity*. Many systems in several areas (computer science, telecommunications, transporting, energy production, etc.) are highly dependable. This means that system failures are rare events, that is, events with small probabilities (how small depends on the specific technological area). This translates into the fact that the MTTF is usually a large number, and that $R(t)$ is usually close to 1, especially when t is small. For these systems, a standard simulation approach is at least problematic, and often simply impossible to apply.

The chapter starts with some generalities about the simulation of Markov models in order to estimate the MTTF and the reliability function following the standard approach (Subsections 9.1.1 and 9.1.2). We also refresh the basic concepts needed to describe the specialized techniques in the rest of the chapter. Section 9.2 is the main part of the chapter. We present some basic techniques (belonging to the Importance Sampling family) to estimate the MTTF. They also allow us to illustrate some important issues related to the rare event problem. Then, in Section 9.3, we present a method for estimating the reliability at time t, that is, a transient metric, a much less explored problem. The goal is to illustrate the techniques that can be followed to deal with rarity for metrics defined on finite time intervals. Conclusions are presented in Subsections 9.2.6 for the problems related to the estimation of the MTTF, and in 9.3.3 for the estimation of the reliability at time t.

9.1   The standard Monte Carlo method

This section briefly describes the standard (also called "crude", or "naive") Monte Carlo approach for the estimation of the MTTF and of the reliability at time t, $R(t)$.

9.1.1   Standard estimation of the MTTF

The first idea for estimating the MTTF consists of building M possible evolutions (trajectories) of the chain, stopping each when the global system fails. More formally, if U is the subset of operational (or up) states of the CTMC X, and D is the set of failed (or down) states, with the initial state $0 \in U$, the system is simulated starting from state 0 and the trajectory is stopped once D is entered. We can thus build M independent copies of $T = \inf\{t \geq 0 \mid X_t \in D\}$, denoted by $T_1, T_2, \ldots, T_M$, and estimate the MTTF by means of its unbiased estimator

$$\widetilde{\text{MTTF}} = \frac{1}{M} \sum_{m=1}^{M} T_m.$$

The copies being independent, we can use the Central Limit Theorem in the standard way to derive a confidence interval for the MTTF. If G is the c.d.f. of $N(0,1)$, the standard normal distribution, that is, if

$$\Phi(x) = \frac{1}{\sqrt{2\pi}} \int_{-\infty}^{x} \exp(-u^2/2) \, du,$$

if $\xi = \Phi^{-1}(1 - \alpha/2)$, and

$$\hat{\sigma} = \sqrt{\frac{1}{M-1} \sum_{m=1}^{M} (T_i - \widetilde{\text{MTTF}})^2},$$

then an α-level confidence interval for the MTTF is

$$\left(\widetilde{\text{MTTF}} - \xi \frac{\hat{\sigma}}{\sqrt{M}}, \ \widetilde{\text{MTTF}} + \xi \frac{\hat{\sigma}}{\sqrt{M}} \right).$$

For instance, if $\alpha = 0.05$, then $\xi \approx 1.96$. The number ξ is the $(1 - \alpha/2)$-quantile of the standard normal distribution.

The main difficulty here is that building each of these copies can be too costly, or even impossible to do, if the MTTF is too large. Another idea, that is very useful in designing more efficient techniques than the standard one, is to use the regenerative properties of Markov chains. Let us denote by τ_0 the return time to state 0, and by $\tau_D = T$ the first time at which the chain visits D. The regenerative approach to evaluate the MTTF consists of using the following well-known expression

$$\text{MTTF} = \frac{\mathbb{E}\{\min(\tau_0, \tau_D)\}}{\gamma}, \tag{9.1}$$

where $\gamma = \mathbb{P}\{\tau_D < \tau_0\}$. To see why this is true, denote $\tau_{\min} = \min(\tau_0, \tau_D)$, recall that $T = \tau_D$, and write

$$\mathbb{E}\{T\} = \mathbb{E}\{T - \tau_{\min} + \tau_{\min}\}$$
$$= \mathbb{E}\{\tau_{\min}\} + \mathbb{E}\{T - \tau_{\min}\}$$
$$= \mathbb{E}\{\tau_{\min}\} + \mathbb{E}\{T - \tau_{\min} \mid \tau_D < \tau_0\}\gamma + \mathbb{E}\{T - \tau_{\min} \mid \tau_D \geq \tau_0\}(1 - \gamma)$$
$$= \mathbb{E}\{\tau_{\min}\} + \mathbb{E}\{\tau_D - \tau_D \mid \tau_D < \tau_0\}\gamma + \mathbb{E}\{T - \tau_0 \mid \tau_D \geq \tau_0\}(1 - \gamma)$$
$$= \mathbb{E}\{\tau_{\min}\} + \mathbb{E}\{T\}(1 - \gamma),$$

from which we obtain $\mathbb{E}\{T\} = \mathbb{E}\{\tau_{\min}\}/\gamma$. The fact that $\mathbb{E}\{T - \tau_0 \mid \tau_D \geq \tau_0\} = \mathbb{E}\{T\}$ follows from the strong Markov property (see [41] for details).

We must then estimate two numbers, $\mathbb{E}\{\min(\tau_D, \tau_0)\}$ and γ. We can generate M independent cycles $C_1, C_2, \ldots, C_M$, that is, M sequences of adjacent states starting and ending with state 0, and not containing it in any other position, and estimate both $\mathbb{E}\{\min(\tau_D, \tau_0)\}$ and γ.

Observe first that the numerator and the denominator in the right-hand side of Equation (9.1) can be computed directly from the embedded discrete-time chain, Y, having transition probability matrix P. To formalize this, let us denote by $\mathcal{C}$ the set of all the cycles containing the initial state 0, and by $\mathcal{D}$ the subset of $\mathcal{C}$ composed of the cycles passing through D. The probability of a cycle $c \in \mathcal{C}$ is

$$q(c) = \prod_{(x,y) \in c} P_{x,y}. \tag{9.2}$$

An estimator of the MTTF is then

$$\widehat{\text{MTTF}} = \frac{\displaystyle\sum_{i=1}^{M} G(C_i)}{\displaystyle\sum_{i=1}^{M} H(C_i)} \tag{9.3}$$

where for any cycle c, $G(c)$ is the sum of the expectations of the sojourn times in all the states of c until reaching D or returning to 0, and $H(c)$ is equal to 1 if $c \in \mathcal{D}$, and to 0 otherwise. Observe that, to estimate the denominator in the expression of the MTTF when a cycle reaches D, the path construction can be stopped, since we already know that $\tau_D < \tau_0$.

Using the Central Limit Theorem, we have the convergence in distribution

$$\frac{\sqrt{M}(\widehat{\text{MTTF}} - \text{MTTF})}{\sigma / \bar{H}_M} \to N(0,1) \quad \text{as } M \to \infty, \tag{9.4}$$

with $\bar{H}_M = \dfrac{1}{M} \displaystyle\sum_{i=1}^{M} H(C_i)$, and

$$\sigma^2 = \sigma_q^2(G) - 2\,\text{MTTF}\,\text{Cov}_q(G,H) + \text{MTTF}^2 \sigma_q^2(H), \tag{9.5}$$

where $G = G(C)$ is the length in time of the generic random cycle C, $H = H(C)$ is the r.v. $1_{\{C \in \mathcal{D}\}}$, and $\sigma_q^2(.)$ and $\text{Cov}_q(.,.)$ denote variances and covariances under the probability measure q. A confidence interval can thus be obtained from this result using the standard procedure, as in the previous case.

The estimation of the numerator in Equation (9.1) presents no problem even in the rare event context since, in that case, $\mathbb{E}\{\min(\tau_D, \tau_0)\} \approx \mathbb{E}\{\tau_0\}$. The estimation of γ, however, is difficult or even impossible using the standard Monte Carlo scheme in the rare event case, because the expectation of the first time where we observe "$\tau_D < \tau_0$" in a cycle is large for highly reliable systems. In other words, for the main difficulty here, the case of rare events, we have just transformed the problem into a new one, now in discrete time, but also suffering from rarity. The point is that the formulation where the target is γ is very useful in designing more efficient procedures.

So, the main problem remains the estimation of γ. This is the object of Section 9.2. Before that, let us recall the basics of estimating the reliability at t.

9.1.2   Standard estimation of the reliability at t, $R(t)$

Recall that the reliability at t, $R(t)$, is the probability that the system works during $[0, t]$. Let us write it as

$$R(t) = \mathbb{E}\{W_t\},$$

where $W_t = 1_{\{X_s \in U, \; \forall s \in [0,t]\}}$.

The unbiased standard Monte Carlo estimator of the reliability parameter $R(t)$ is the sample mean $\widehat{R}(t)$ defined by

$$\widehat{R}(t) = \frac{1}{M} \sum_{i=1}^{M} W_t^{(i)},$$

where M is a fixed sample size and $W_t^{(1)}, \ldots, W_t^{(M)}$ are independent and identically distributed random variables with the same distribution as W_t.

If M is large enough, the Central Limit Theorem can be applied to obtain a confidence interval. More precisely, an α-level confidence interval is given by

$$\left(\widehat{R}(t) - \xi \sqrt{\operatorname{Var}\left\{\widehat{R}(t)\right\}}, \; \widehat{R}(t) + \xi \sqrt{\operatorname{Var}\left\{\widehat{R}(t)\right\}} \right),$$

i.e.,

$$\left(\widehat{R}(t) - \xi \sqrt{\frac{\operatorname{Var}\{W_t\}}{M}}, \; \widehat{R}(t) + \xi \sqrt{\frac{\operatorname{Var}\{W_t\}}{M}} \right) \tag{9.6}$$

where $\xi = \Phi^{-1}(1 - \alpha/2)$, and, since $W_t = W_t^2$,

$$\operatorname{Var}\left\{\widehat{R}(t)\right\} = \frac{\operatorname{Var}\{W_t\}}{M} = \frac{(1 - R(t))R(t)}{M}.$$

To obtain a practical confidence interval, since the value of $\operatorname{Var}\{W_t\}$ is unknown, we replace it in (9.6) by its standard unbiased estimator which we denote here by $\widehat{V}$:

$$\widehat{V} = \frac{1}{M-1} \sum_{i=1}^{M} \left(W_t^{(i)} - \widehat{R}(t) \right)^2.$$

The simulation algorithm consists of independently repeating the following experiment M times. The simulation starts at time 0 with the system in its initial state, and repeatedly finds the time of the next event (component failure or repair), updating the system state and advancing in time, until the chain reaches a state belonging to D or when the accumulated sojourn time in the different states is greater than t. In the first case the value of the sample is 0, in the second case it is 1. The estimation of $R(t)$ is the sum of these values divided by M. The standard simulation is easy to implement. Unfortunately, it becomes extremely wasteful in execution time when the system is highly reliable, as the number of experiments must be very large to obtain some failed experiments and a reasonable relative error.

In the case of highly reliable systems, the problem is fundamentally caused by the fact that the sojourn time in the operational states of the system is usually much longer than the considered interval, $[0,t]$. Then, the chain X stays in the operational states during all the interval $[0,t]$ for most of the samples; this means that a very large sample size is needed to obtain an accurate estimate of $R(t)$.

9.2   Estimating the MTTF of a multicomponent repairable system

Recall that our system is represented by a continuous-time Markov chain, $X = \{X_t, t \geq 0\}$, with values on a finite state space, S. Let us just observe here that this last assumption is not critical for simulation, and most techniques can handle models with denumerable state spaces.

After specifying the family of models that cover most of the applications of Markov chains in dependability analysis, we address the issue of formalizing the rarity phenomenon, in order to better understand the behavior of these techniques, one of the main goals of the section. We then review specific Monte Carlo schemes for the estimation of γ (see Subsection 9.1.1) in different and important cases.

The system is composed of a number of components, which can be either operational or failed. There are K classes of components. At time 0 the process is in a specific state denoted by 0 where the N_k components of class k are all operational. An operational class k component can fail, and this happens with failure rate $\lambda_k(x)$ if the chain is in state x.

The number of operational class k components when the process is in state x is $v_k(x) \leq N_k$. We have already stated that $v_k(0) = N_k$. The set of transitions in the Markovian model is partitioned into two disjoint sets: $\mathcal{F}$, the set of *failures* and $\mathcal{R}$, the set of *repairs*. Formally, any transition (x,y) in the model satisfies that either for all k, $v_k(y) \leq v_k(x)$ and for some k, $v_k(y) < v_k(x)$, in which case (x,y) is a failure, or for all k, $v_k(y) \geq v_k(x)$ and for some k, $v_k(y) > v_k(x)$, in which case (x,y) is a repair. To improve legibility, we denote $\lambda_{x,y} = Q_{x,y}$ when $(x,y) \in \mathcal{F}$ and $\mu_{x,y} = Q_{x,y}$ when $(x,y) \in \mathcal{R}$. We also denote by F_x the set of states that can be reached from x after a failure, and by R_x the set of states that can be reached from x after a repair, that is,

$$F_x = \{y \in S \mid (x,y) \in \mathcal{F}\}, \quad R_x = \{y \in S \mid (x,y) \in \mathcal{R}\}. \tag{9.7}$$

The states of X are also classified as operational or failed. We denote by U the set of operational (or "up") states and by D the set of failed (or "down") states; state 0 belongs to set U. We also have $R_0 = \emptyset$ (that is, no repairs from 0 since everything is assumed to work in that state).

To finish the description of the model, let us specify how transitions occur. After the failure of some operational class k component when the system's state is x, the system jumps to state y with probability $p(x, k; y)$. This allows us to take into account the case of *failure propagation*, that is, the situation where the failure of some component induces, with some probability, that a subset of components is shut down (for instance, the failure of the power supply can make some other components unoperational). The probabilities $p(x, k; y)$ are assumed to be defined for all x, y, k (in most cases, we have $p(x, k; y) = 0$).

Observe that

$$\forall (x, y) \in \mathcal{F}, \quad \lambda_{x,y} = \sum_{k=1}^{K} v_k(x) \lambda_k(x) p(x, k; y). \tag{9.8}$$

For the repairs, we use a simplifying assumption saying that from every state different from the initial one, there is at least one repair transition, that is,

$$\forall x \neq 0, \quad R_x \neq \emptyset. \tag{9.9}$$

This excludes the case of *delayed repairs*, corresponding to systems where the repair facilities are activated only when there are "enough" failed units.

9.2.1   The rarity parameter [106, 76]

Let us formalize the fact that failures are rare (or slow), and that repairs are frequent (or fast). Following Shahabuddin [106], we introduce a *rarity parameter*, ε. We assume that the failure rates of class k components have the following form:

$$\lambda_k(x) = a_k(x) \varepsilon^{i_k(x)} \tag{9.10}$$

where either the real $a_k(x)$ is strictly positive and the integer $i_k(x)$ is greater than or equal to 1, or $a_k(x) = i_k(x) = 0$. We naturally set $a_k(x) = 0$ if $v_k(x) = 0$. No particular assumption is necessary about the $p(x, k; y)$, so we write

$$p(x, k; y) = b_k(x, y) \varepsilon^{j_k(x,y)} \tag{9.11}$$

with real $b_k(x, y) \geq 0$, integer $j_k(x, y) \geq 0$, and $j_k(x, y) = 0$ when $b_k(x, y) = 0$. For the repair rates, we simply set

$$\mu_{x,y} = \Theta(1), \tag{9.12}$$

where $f(\varepsilon) = \Theta(\varepsilon^d)$ means that there exist two constants, $k_1, k_2 > 0$, such that $k_1 \varepsilon^d \leq |f(\varepsilon)| \leq k_2 \varepsilon^d$ (recall that for every state $x \neq 0$, there exists at least one state y such that $\mu_{x,y} > 0$).

The form of the failure rates of the components has the following consequence on the failure transitions in X: for all $(x, y) \in \mathcal{F}$,

$$\lambda_{x,y} = \Theta\left(\varepsilon^{m(x,y)}\right) \tag{9.13}$$

where
$$m(x,y) = \min_{k:a_k(x)b_k(x,y)>0} \{i_k(x) + j_k(x,y)\} \qquad (9.14)$$

(observe that if $F_x \neq \emptyset$, then for all $y \in F_x$ we necessarily have $m(x,y) \geq 1$).

Let us look now at the transition probabilities of the canonically embedded discrete-time Markov chain, Y. For any $x \neq 0$, since we assume that $R_x \neq \emptyset$, we have

$$(x,y) \in F \Longrightarrow P_{x,y} = \Theta\left(\varepsilon^{m(x,y)}\right), \quad m(x,y) \geq 1, \qquad (9.15)$$

and

$$(x,y) \in \mathcal{R} \Longrightarrow P_{x,y} = \Theta\left(1\right). \qquad (9.16)$$

For the initial state, we have that for all $y \in F_0$,

$$P_{0,y} = \Theta\left(\varepsilon^{m(0,y) - \min_{z \in F_0} m(0,z)}\right). \qquad (9.17)$$

Observe here that if $\operatorname{argmin}_{z \in F_0} m(0,z) = w \in D$, then we have $P_{0,w} = \Theta\left(1\right)$ and there is no rare event problem. This happens in particular if $F_0 \cap U = \emptyset$. So, the interesting case (the rare event situation) is the case where $P_{0,w} = o(1)$ for all $w \in F_0 \cap D$. In other words, the case of interest is when (i) $F_0 \cap U \neq \emptyset$ and (ii) $\exists y \in F_0 \cap U$ s.t. $\forall \omega \in F_0 \cap D$, $m(0,y) < m(0,\omega)$.

A simple consequence of the previous assumptions is that for any cycle c, its probability, $q(c)$, is $q(c) = \Theta\left(\varepsilon^h\right)$ where the integer h satisfies $h \geq 1$. If we define

$$\mathcal{C}_h = \{c \in \mathcal{C} \mid q(c) = \Theta\left(\varepsilon^h\right)\}, \qquad (9.18)$$

we then obtain

$$\gamma = \sum_{c \in D} q(c) = \Theta\left(\varepsilon^r\right) \qquad (9.19)$$

where $r = \operatorname{argmin}\{h \mid \mathcal{C}_h \neq \emptyset\} \geq 1$. This shows that γ decreases as $\varepsilon \to 0$, that is, the event of interest becomes rarer as $\varepsilon \to 0$. This parameter ε allows us to control how rare the event of interest is, and then to study the properties of the estimators as rarity increases, that is, as $\varepsilon \to 0$.

9.2.2   Importance sampling schemes

Recall that we denote by $Q_{x,y}$ the transition rate from state x to state y, by Y the discrete-time Markov chain canonically embedded in X at its jump times, and that $P_{x,y}$ denotes the transition probability associated with states x and y in Y. Formally,

$$P_{x,y} = \frac{Q_{x,y}}{\sum_{z \in S, z \neq x} Q_{x,z}}.$$

For the estimation of γ, we can follow an Importance Sampling scheme, based on the regenerative approach. The idea is to change the underlying measure such that *all* the cycles in the set D of interest receive a higher weight. This is not possible in general, and what we in fact do is to change the transition probability matrix P into a new P' such that we expect that the probability of *most of the cycles of interest* will increase. There is an element of heuristic reasoning in this, as in most Importance Sampling ideas.

Estimating γ following a standard Monte Carlo approach consists of building paths starting from state 0 and stopping when they return to 0 or when they reach subset D. If the cycles are denoted by $C_1, \ldots, C_M$, the estimator is

$$\hat{\gamma} = \frac{1}{M} \sum_{i=1}^{M} H(C_i),$$

where, for a cycle c, we denote by $H(c)$ (see page 204) the indicator of the event $\{c \in D\}$. If the dynamics of the chain is changed from matrix P to a new matrix P' such that whenever $P_{x,y} > 0$ we also have $P'_{x,y} > 0$, γ can also be estimated by the Importance Sampling estimator $\tilde{\gamma}$ defined by

$$\tilde{\gamma} = \frac{1}{M} \sum_{i=1}^{M} L(C_i) H(C_i),$$

where now the cycles are generated under the dynamics given by the new transition probabilities in P' and where the *likelihood ratio L* is defined by

$$L(c) = \prod_{(x,y) \in c} \frac{P_{x,y}}{P'_{x,y}}.$$

We now present several procedures designed to improve Monte Carlo efficiency with respect to the standard method. To simplify the description of the different schemes, let us introduce the following notation. For every state x, we denote by $f_x(y)$ the transition probability $P_{x,y}$, for each $y \in F_x$. In the same way, for all states x, let us denote $r_x(y) = P_{x,y}$ for each $y \in R_x$. As stated above, using an Importance Sampling scheme means that instead of P we use a different matrix P', leading to new $f'_x()$s and $r'_x()$s. The transition probabilities associated with the states of D are not concerned with the estimation of γ, since when a cycle reaches D, it is "stopped" as we explained before.

Failure Biasing (FB) [60, 21]

This is the most straightforward method: to (hopefully) increase the probability of most regenerative cycles including system failures, we increase the probability of the failure transitions. We must choose a parameter $\alpha \in (0, 1)$, which is equal to $f'_x(F_x) = \sum_{z \in F_x} f'_x(z)$ for all $x \neq 0$ (typically, $0.5 \leq \alpha \leq 0.9$). The transition probabilities are then changed as follows.

- $\forall x \in U, x \neq 0, \forall y \in F_x:$ $\quad f'_x(y) = \alpha \dfrac{f_x(y)}{f_x(F_x)},$
- $\forall x \in U, x \neq 0, \forall z \in R_x:$ $\quad r'_x(z) = (1 - \alpha) \dfrac{r_x(z)}{r_x(R_x)},$

where $f_x(F_x) = \sum_{z \in F_x} f_x(z)$ and $r_x(R_x) = \sum_{z \in R_x} r_x(z)$.

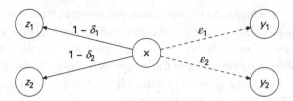

Figure 9.1   Failure biasing, before the change of measure; notation suggests that $\varepsilon_1, \varepsilon_2 \ll 1$

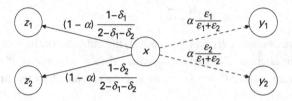

Figure 9.2   Failure biasing, after the change of measure

The $f_0()$ are not modified (since we already have $f_0(F_0) = \sum_{z \in F_0} f_0(z) = \Theta(1)$). Observe that the total probability of failure from x is now equal to α (that is, for any $x \in U - \{0\}, f'_x(F_x) = \alpha$).

Figure 9.1 symbolically represents the transition probabilities before the change of measure, and Figure 9.2 does the same with the chain after the change of measure.

Selective Failure Biasing (SFB) [41]

The idea here is to separate the failure transitions from x ($x \in U$) into two disjoint sets: those consisting of the *first* failure of a component of some class k (called *initial failures*), and the remaining ones (called *non-initial failures*). Following this, the set of states, F_x, is partitioned into two disjoint sets, IF_x and NIF_x, where

$$IF_x = \{y \mid (x, y) \text{ is an } initial \text{ failure}\},$$

$$NIF_x = \{y \mid (x, y) \text{ is a } non\text{-}initial \text{ failure}\}.$$

The idea is then to increase the probability of a non-initial failure, that is, to make the failure of some class k components more probable than in the original model, if there is at least one component of that class that has already failed.

To implement this, we must choose two parameters, $\alpha, \beta \in (0, 1)$ (typically, $0.5 \leq \alpha, \beta \leq 0.9$), and change the transition probabilities in the following way:

- $\forall x \in U, x \neq 0, \forall y \in IF_x,\ f'_x(y) = \alpha(1 - \beta) \dfrac{f_x(y)}{f_x(IF_x)},$

  and $\forall y \in NIF_x,\ f'_x(y) = \alpha\beta \dfrac{f_x(y)}{f_x(NIF_x)}.$

  For $x = 0$, we use the same formulas with $\alpha = 1$; in the same way, if $IF_x = \emptyset$, we set $\beta = 1$ and if $NIF_x = \emptyset$, we set $\beta = 0$.

- $\forall x \in U, x \neq 0, \forall y \in R_x,\ r'_x(y) = (1 - \alpha) \dfrac{r_x(y)}{r_x(R_x)}.$

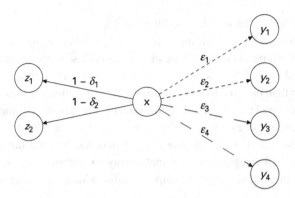

Figure 9.3   Selective failure biasing, before the change of measure; transitions (x, y_1) and (x, y_2) are initial failures; transitions (x, y_3) and (x, y_4) are non-initial failures

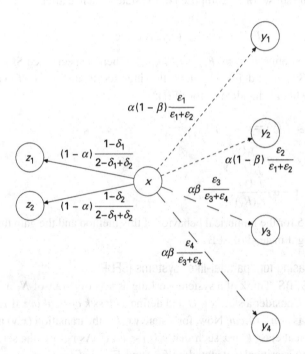

Figure 9.4   Selective failure biasing, after the change of measure; transitions (x, y_1) and (x, y_2) are initial failures; transitions (x, y_3) and (x, y_4) are non-initial failures

In this scheme, as in the FB method, the total failure probability becomes $f_x'(F_x) = \alpha$, but now we have a further refinement, leading to $f_x'(NIF_x) = \alpha\beta$ and $f_x'(IF_x) = \alpha(1-\beta)$. Figures 9.3 and 9.4 symbolically illustrate the transition probabilities before and after the change of measure.

Selective Failure Biasing for "series-like" systems (SFBS)

The implicit assumption in SFB is that the criteria used to define an operational state (which characterizes the type of system we are modeling) is close to the situation where the system is up if and only if, for each class k component, the number of operational components is greater than or equal to some threshold, l_k, and neither the initial number of components, N_k, nor the level, l_k, are "very dependent" on k. Now, assume that this last part of the assumptions does not hold, that is, assume that from the dependability point of view, the system is a series of l_k-out-of-N_k modules, but that the N_k and l_k are strongly dependent on integer k. A reasonable way to improve SFB is to make more probable the failures of the class k components for which $v_k(x)$ is closer to the threshold l_k.

Consider a state $x \in U$ and define a class k *critical in* x if $v_k(x) - l_k = \min_{k'=1,\dots,K}(v_{k'}(x) - l_{k'})$; otherwise, the class is *non-critical*. Now, for a state $y \in F_x$, the transition (x,y) is *critical* if there is some critical class k in x such that $v_k(y) < v_k(x)$. We denote by $F_{x,c}$ the subset of F_x composed of the states reached after a critical failure, that is,

$$F_{x,c} = \{y \in F_x \mid (x,y) \text{ is critical}\}.$$

In the same way, we define $F_{x,nc}$ by $F_{x,nc} = F_x \setminus F_{x,c}$. Then, a specialized SFB method, which we call SFBS, can be defined by the following modification of the $f_x()$ (we omit the frontiers' case which is handled as for SFB):

- $\forall x \in U, \forall y \in F_{x,nc}, \; f'_x(y) = \alpha(1 - \beta)\dfrac{f_x(y)}{f_x(F_{x,nc})},$

  and $\forall y \in F_{x,c}, \; f'_x(y) = \alpha\beta\dfrac{f_x(y)}{f_x(F_{x,c})}.$

- $\forall y \in R_x, \; r'_x(y) = (1 - \alpha)\dfrac{r_x(y)}{r_x(R_x)}.$

See Subsection 9.2.5 for the numerical behavior of this method and the gain that can be obtained when using it instead of SFB.

Selective Failure Biasing for "parallel-like" systems (SFBP)

This is the dual of SFBS. Think of a system working as sets of l_k-out-of-N_k modules in parallel, $1 \le k \le K$. Consider a state $x \in U$ and define a class k *critical in* x if $v_k(x) \ge l_k$; otherwise, the class is *non-critical*. Now, for a state $y \in F_x$, the transition (x,y) is *critical* if there is some critical class k in x such that $v_k(y) < v_k(x)$. As before, the set of states $y \in F_x$ such that (x,y) is critical is denoted by $F_{x,c}$, and $F_{x,nc} = F_x - F_{x,c}$.

A first idea is to follow the analogous scheme as for the SFBS case: using in the same way two parameters α and β, the principle is to accelerate the critical transitions first, then the non-critical ones, by means of the respective weights $\alpha\beta$ and $\alpha(1 - \beta)$. This leads to the same following rules as in the SFBS method:

- $\forall x \in U, \forall y \in F_{x,nc}, \; f'_x(y) = \alpha(1 - \beta)\dfrac{f_x(y)}{f_x(F_{x,nc})},$

  and $\forall y \in F_{x,c}, \; f'_x(y) = \alpha\beta\dfrac{f_x(y)}{f_x(F_{x,c})}.$

- $\forall y \in R_x, \; r'_x(y) = (1-\alpha)\dfrac{r_x(y)}{r_x(R_x)}.$

As we will see in Subsection 9.2.4, there is no need for the β parameter in practice, and the method we call SFBP is instead defined by the following rules:

- $\forall x \in U, \; \forall y \in F_{x,c}, \; f'_x(y) = \alpha\dfrac{f_x(y)}{f_x(F_{x,c})},$

  and $\forall y \in F_{x,nc}, \; f'_x(y) = (1-\alpha)\dfrac{f_x(y)}{r_x(R_x)+f_x(F_{x,nc})}.$

- $\forall y \in R_x, \; r'_x(y) = (1-\alpha)\dfrac{r_x(y)}{r_x(R_x)+f_x(F_{x,nc})}.$

As we see, we only accelerate the critical transitions, the non-critical ones are handled in the same way as the repairs.

Distance-based Selected Failure Biasing (DSFB) [14]

We assume that there may be some propagation of failures in the system. For all $x \in U$, its distance $d(x)$ to D is defined as the minimal number of components whose failure put the model in a down state, that is,

$$d(x) = \min_{y \in D} \sum_k (v_k(x) - v_k(y)).$$

Obviously, for any $y \in F_x$ we have $d(y) \le d(x)$. A failure (x,y) is said to be *dominant* if and only if $d(x) > d(y)$ and it is *non-dominant* if and only if $d(x) = d(y)$. The *criticality* of $(x,y) \in \mathcal{F}$ is defined by

$$c(x,y) = d(x) - d(y) \ge 0.$$

The idea of this algorithm is to take into account the different criticalities to control more deeply the failure transitions in the Importance Sampling scheme. It is assumed, of course, that the user can compute the distances, $d(x)$, for any operational state x with low cost. This is often possible, though not always.

Define recursively the following partition of F_x:

$$F_{x,0} = \{y \in F_x \mid c(x,y) = 0\},$$

and $F_{x,l}$ is the set of states $y \in F_x$ such that $c(x,y)$ is the smallest criticality value greater than $c(x,w)$ for any $w \in F_{x,l-1}$. More formally, if we define, for all $l \ge 1$,

$$G_{x,l} = F_x - F_{x,0} - F_{x,1} - \cdots - F_{x,l-1},$$

then we have

$$F_{x,l} = \{y \in G_{x,l} \mid y = \mathrm{argmin}\{c(x,z), \; z \in G_{x,l}\}\}.$$

Let us denote by V_x the number of criticality values that are greater than 0, corresponding to failures starting from x, that is,

$$V_x = \mathrm{argmax}\{l \ge 1 \mid F_{x,l} \ne \emptyset\}.$$

The method proposed by Carrasco [14] has three parameters, α, β, and $\beta_c \in (0, 1)$. The new transition probabilities are

- $\forall x \in U, \forall y \in F_{x,0}, f_x'(y) = \alpha(1-\beta)\dfrac{f_x(y)}{f_x(F_{x,0})};$

  $\forall l$ such that $1 \leq l < V_x, \forall y \in F_{x,l}, f_x'(y) = \alpha\beta(1-\beta_c)\beta_c^{l-1}\dfrac{f_x(y)}{f_x(F_{x,l})};$

  $\forall y \in F_{x,V_x}, f_x'(y) = \alpha\beta\beta_c^{V_x-1}\dfrac{f_x(y)}{f_x(F_{x,V_x})};$

- $\forall x \neq 0, \forall y \in R_x, r_x'(y) = (1-\alpha)\dfrac{r_x(y)}{r_x(R_x)}.$

As before, we must define what happens at the "frontiers" of the transformation. If $F_{x,0} = \emptyset$, then we set $\beta = 1$. If $x = 0$, then we set $\alpha = 1$.

It seems intuitively clear that we must, in general, give a higher weight to the failures with higher criticalities. This is not the case of the approach originally proposed by Carrasco [14].

Just by exchanging the order of the weights of the failures arriving at the set $F_{x,l}$, $l \geq 1$, we obtain a new version which gives higher probabilities to failure transitions with higher criticalities. The Distance-based Selective Failure Biasing (DSFB) which we define here corresponds to the following change of dynamics:

- $\forall x \in U, \forall y \in F_{x,0}, f_x'(y) = \alpha(1-\beta)\dfrac{f_x(y)}{f_x(F_{x,0})};$

  $\forall y \in F_{x,1}, f_x'(y) = \alpha\beta\beta_c^{V_x-1}\dfrac{f_x(y)}{f_x(F_{x,1})};$

  $\forall l$ such that $2 \leq l \leq V_x, \forall y \in F_{x,l}, f_x'(y) = \alpha\beta(1-\beta_c)\beta_c^{V_x-l}\dfrac{f_x(y)}{f_x(F_{x,l})};$

- $\forall x \neq 0, \forall y \in R_x, r_x'(y) = (1-\alpha)\dfrac{r_x(y)}{r_x(R_x)}.$

9.2.3 Balanced Importance Sampling methods

All the previous methods classify the transitions from a fixed state into a number of disjoint sets, and assign modified global probabilities to each of these sets; but they do not modify the relative weights of the transitions belonging to the same set. An alternative is to assign uniform probabilities to all transitions from x leading to the same subset of F_x. This can be done independently of the number and the definition of those sets, so that we can find *balanced versions* of all the previously mentioned methods. The reason for this is the following. In [106], the concept of bounded relative error is defined as follows.

DEFINITION 9.1. *Let σ^2 denote the variance of an estimator of γ and let ξ be the $(1-\alpha/2)$-quantile of the standard normal distribution. Then the relative error for a sample size M, at level α, is defined by*

$$RE = \xi\frac{\sigma}{\gamma\sqrt{M}}. \tag{9.20}$$

We say that the estimator has bounded relative error (BRE) if the RE remains bounded as $\varepsilon \to 0$ (that is, if $\sigma/\sqrt{M}$ remains bounded as $\varepsilon \to 0$).

This is a desired property, guaranteeing that the quality of the estimators doesn't degrade when the event of interest becomes increasingly rare. If the estimator enjoys this property, only a fixed number of iterations is required to obtain a confidence interval having a given error, no matter how rarely failures occur.

From Shahabuddin's work [106], or using Theorem 2 in [76], it can be shown that any of the balanced versions of the previously described methods has BRE (the proof needs Assumption (9.9) stating that $R_x \neq \emptyset$ for any state $x \neq 0$).

In the following, we give explicitly the rules for specifying balanced versions of FB, SFB, SFBS, SFBP, and DSFB. In spite of the similarity in the descriptions, we provide this list to help the reader who wants to implement one of these techniques. Before looking at the balanced versions in detail, let us observe that sometimes the systems are already balanced, that is, there are no significant differences between the magnitudes of the transition probabilities. In these cases, the unbalanced and balanced versions of the same method basically behave in the same manner.

Balanced FB

Analyzing the FB method, it was proved (first by Shahabuddin in [106]) that balancing it improves its behavior when there are transition probabilities from the same state x which differ by orders of magnitude. The Balanced FB method is then defined by

- $\forall x \neq 0, \forall y \in F_x, \; f'_x(y) = \alpha \dfrac{1}{|F_x|}$;

- $\forall x \neq 0, \forall y \in R_x, \; r'_x(y) = (1 - \alpha) \dfrac{r_x(y)}{r_x(R_x)}$.

If $x = 0$, then we set $\alpha = 1$ in the change of measure.

Balanced SFB

The Balanced SFB scheme consists of the following rules:

- $\forall x \neq 0, \forall y \in IF_x, \; f'_x(y) = \alpha(1 - \beta) \dfrac{1}{|IF_x|}$,

  and $\forall y \in NIF_x, \; f'_x(y) = \alpha\beta \dfrac{1}{|NIF_x|}$;

  for $x = 0$, we use the same formulas with $\alpha = 1$; in the same way, if $IF_x = \emptyset$, we set $\beta = 1$ and if $NIF_x = \emptyset$, we set $\beta = 0$.

- $\forall x \neq 0, \forall y \in R_x, \; r'_x(y) = (1 - \alpha) \dfrac{r_x(y)}{r_x(R_x)}$.

Balanced SFBS

We describe now the transformations associated with the Balanced SFBS scheme, except for the repairs and the frontier cases, which are as in the Balanced SFB method:

- $\forall x, \forall y \in F_{x,nc}, \; f'_x(y) = \alpha(1 - \beta) \dfrac{1}{|F_{x,nc}|}$,

  and $\forall y \in F_{x,c}, \; f'_x(y) = \alpha\beta \dfrac{1}{|F_{x,c}|}$.

Balanced SFBP

The Balanced SFBP method is defined by the following rules:

- $\forall x \in U, \ \forall y \in F_{x,c}, \ f'_x(y) = \alpha \dfrac{1}{|F_{x,c}|},$

  and $\forall y \in F_{x,nc}, \ f'_x(y) = (1-\alpha) \dfrac{1}{|R_x| + |F_{x,nc}|}.$

- $\forall y \in R_x, \ r'_x(y) = (1-\alpha) \dfrac{1}{|R_x| + |F_{x,nc}|}.$

It can be observed that, for the Balanced SFBP scheme, we do not take the repair probabilities proportionally to the original ones. Indeed, we have grouped repairs and non-initial failures, so taking the new transition probabilities proportional to the original ones would give rare events for the non-initial failures. Thus this small change, that is, a uniform distribution over $F_{x,nc} \cup R_x$, balances all the transitions.

Balanced DSFB

The Balanced DSFB scheme is

- $\forall x \in U, \quad \forall y \in F_{x,0}, \quad f'_x(y) = \alpha(1-\beta) \dfrac{1}{|F_{x,0}|};$

  $\forall y \in F_{x,1}, \quad f'_x(y) = \alpha\beta\beta_c^{V_x-1} \dfrac{1}{|F_{x,1}|};$

  $\forall l$ such that $2 \leq l \leq V_x, \quad \forall y \in F_{x,l}, \quad f'_x(y) = \alpha\beta(1-\beta_c)\beta_c^{V_x-l} \dfrac{1}{|F_{x,l}|};$

- $\forall x \neq 0, \quad \forall y \in R_x, \quad r'_x(y) = (1-\alpha) \dfrac{r_x(y)}{r_x(R_x)}.$

9.2.4   On the selection of the appropriate technique to use

Given a specified system, we can wonder which scheme, among the several described in Subsection 9.2.2, is the most appropriate. We provide here some hints on the selection of the most appropriate scheme, with two specific goals. First, we explain why we do not use a β parameter in the SFBP scheme, as we do in the SFB and SFBS cases. Second, we make some asymptotic comparisons of the discussed techniques. We consider only balanced schemes because they are the only ones, among the methods described in Subsection 9.2.2, to verify in general the BRE property (as well as other properties not discussed here; see the bibliographic remarks at the end of the chapter).

The asymptotic efficiency (as $\varepsilon \to 0$) is controlled by two quantities: the asymptotic variance of the estimator and the mean number of transitions needed by the embedded chain Y to hit D when it does it before returning to 0.

On the SFBP choice

- We want to compare the variance of the two considered choices in SFBP (with or without a β parameter), in the case of a system structured as a set of l_k-out-of-N_k modules in parallel, $k = 1, \ldots, K$, that is the case of interest. To do this, let us denote by $f'_{x,\beta}(y)$ the transition probability associated with an SFBP scheme using a β parameter. Let s be the integer such that $\sigma^2 = \Theta(\varepsilon^s)$. We can observe that the most important paths for the variance estimation, that is, the paths $c \in \mathcal{D}$ verifying $q^2(c)/q'(c) = \Theta(\varepsilon^s)$, are typically composed of critical transitions (x, y) for which the failure SFBP probability $f'_x(y)$ (without using β) satisfies

$$f'_x(y) = f'_{x,\beta}(y)/\beta, \tag{9.21}$$

that is, the transitions are now driving the model closer to the failure states. So, if we denote σ^2 (resp. σ_β^2) the variance of the estimator without (resp. with) the β parameter, then $\sigma^2/\sigma_\beta^2 < 1$ for ε small enough.

- Let us denote by $|c|$ the number of transitions in cycle $c \in \mathcal{D}$ until hitting $\mathcal{D}$. The expected number of transitions necessary to hit $\mathcal{D}$ under the modified measure q' is

$$\mathbb{E}\{T\} = \sum_{c \in \mathcal{D}} |c| q'(c). \tag{9.22}$$

From Equation (9.21), we see that $\mathbb{E}\{T\}$ is smaller if we do not use the β parameter.

Both arguments support the idea of using only the α parameter in the SFBP scheme.

Comparison of balanced schemes

Using the balanced schemes, all the variances are of the same order (i.e. $O(\varepsilon^{2r})$, where r is given in Equation (9.19)) because each path is in $\Theta(1)$ (see Shahabuddin [106] or Nakayama [76] for a proof). Then, we can point out the following facts:

- The variances are of the same order with all the balanced schemes. Nevertheless the constant of the $\Theta(\cdot)$ may be quite different. The analysis of this constant is much more difficult in this general case than for the SFBP schemes previously presented and appears to depend too much on the specific model parameters to allow any kind of general claim about it.
- The preceding point suggests that the choice between the different methods should be based mainly on the mean hitting time to $\mathcal{D}$ given in Equation (9.22). To obtain the shortest computational time, our experience is the following:
  - if there are many propagation faults in the system, we suggest the use of the balanced DSFB scheme;
  - if there are no (or very few) propagation faults and if the system is working as a series of l_k-out-of-N_k modules, $k = 1, \ldots, K$, the balanced SFBS scheme seems the appropriate one;
  - if there are no (or very few) propagation faults and if the system is working as a set of l_k-out-of-N_k modules in parallel, $k = 1, \ldots, K$, we suggest the use of the balanced SFBP method;
  - for the case of a poorly structured system, or one where it is not clear if the structure function is of the series type, or of the parallel one, the general balanced FB scheme can also lead to a useful variance reduction.

9.2.5   Numerical illustrations

We are not going to compare all the methods discussed before in their balanced versions. Our aim is to get a feeling for what can be obtained in practice, and to give some general guidelines to choose among the different methods. In all cases, the estimated measure is $\gamma = \mathbb{P}\{\tau_D < \tau_0\}$.

First, let us consider methods FB, SFB, and SFBS. When the modeled systems have a structure close to a series of l_k-out-of-N_k modules, it seems clear that both SFB and SFBS are better than FB. If the values $N_k - l_k$ (that is, the number of redundant components for class k) do not (or do slightly) depend on k, SFB and SFBS should have more or less the same behavior; but when some components have significant differences in these values, SFBS should outperform SFB. To look at how these rules of thumb work out in a particular case, we study two versions of a Tandem computer, described in [49] (we follow here a later description made in [61]). This computer is composed of a multiprocessor, p; a dual disk controller, dc; two RAID disks, dd; two fans, f; two power supplies, ps; and one dual interprocessor bus, b. In addition to a CPU, each processor contains its own memory. When a component in a dual fails, the subsystem is reconfigured into a simplex. This Tandem computer requires all subsystems, one fan, and one power supply for it to be operational. The failure rates are equal to 5ε, 2ε, 4ε, 0.1ε, 3ε, and 0.3ε in failures per hour (fph) for the processors, the disk controller, the disks, the fans, the power supplies, and the bus, respectively, with $\varepsilon = 10^{-5}$ fph. There is only one repairman and the repair rates are equal to 30 in repairs per hour (rph), for all the components, except for the bus, which has a repair rate of 15 rph.

We first consider a version of this computer where both the multiprocessor and the disks have two units, and only one is needed for the system to be operational. In this case, $N_k = 2$ and $l_k = 1$ for all k (here, $K = 6$). Table 9.1 presents the relative variances and computing times for the SFB and the SFBS methods, with respect to the basic FB one, observed when estimating γ with a sample size $M = 10^5$, and parameters $\alpha = 0.7$, $\beta = 0.8$. As expected, we can observe that for this situation, algorithms SFB and SFBS are basically equivalent (both in precision and in execution time); their variance is an order of magnitude smaller than the variance of the FB algorithm, which is also slower. The slight difference in the execution time between SFB and SFBS comes from the fact that in the latter there are a few supplementary computations to do, with basically the same cycle structure.

Let us now consider this same architecture, but with a four-unit multiprocessor (only one of the four processors is required to have an operational system), and with each RAID being composed of five disks, only three of which are required for the RAID to be operational. In this case, N_k and l_k vary with k. Table 9.2 presents the relative variances and computing times for the SFB and the SFBS methods, observed when estimating γ with a sample size $M = 10^5$, and parameters $\alpha = 0.7$, $\beta = 0.8$. As in the previous case, the FB algorithm is the least performant; but now we observe how SFBS obtains a better precision and a lower computational cost than SFB.

Consider now a model of a replicated database; there are four sites, and each site has a whole copy of the database on a RAID disk cluster. We take all clusters identical, with

Table 9.1. Methods SFB and SFBS for a series of l_k-out-of-N_k modules without any dependence on k. We plot the ratio (in %) between the variance of the SFB and SFBS techniques and the variance of the FB one, and the same with the execution times.

Method	RelVar	RelTime
SFB	11.5%	53.1%
SFBS	11.5%	56.7%

Table 9.2. Methods SFB and SFBS for a series of l_k-out-of-N_k modules without any dependence on k. We plot the ratio (in %) between the variance of the SFB and SFBS algorithms and the variance of the FB one, and the same with the execution times.

Method	RelVar	RelTime
SFB	15.6%	53.5%
SFBS	10.5%	45.9%

Table 9.3. Methods SFB and SFBP for an l_k-out-of-N_k modules in parallel without any dependence on k. We plot the ratio (in %) between the variance of the SFB and SFBP methods and the variance of the FB one, and the same with the execution times.

Method	RelVar	RelTime
SFB	4028%	113.1%
SFBP	41.0%	67.1%

the same redundancies 7-out-of-9, and with a failure rate for each disk of $\varepsilon = 10^{-2}$. There is one repairman per class, and the repair rate is 1. We consider that the system is up if there is at least one copy of the database in working order. The structure function of this system is then a set of l_k-out-of-N_k blocks in parallel. In Table 9.3 we compare the behavior of the SFB and SFBP algorithms for this system, where all component classes k have the same redundancy; the SFBP method performs much better than both the FB and SFB methods.

Next, we analyze a model of a system exhibiting failure propagations, the fault-tolerant database system presented in [69]. The components of this system are: a front-end, a database, and two processing subsystems formed by a switch, a memory, and two processors. These components may fail with rates 1/2400, 1/2400, 1/2400,

Table 9.4. Methods SFB and DSFB on a system with failure propagations. We plot the ratio (in %) between the variance of the SFB and DSFB procedures and the variance of the FB one, and the same with the execution times.

Method	RelVar	RelTime
SFB	100.2%	103.4%
DSFB	2.72%	40.5%

1/2400, and 1/120, respectively (say in fph). There is a single repairman who gives priority to the front-end and the database, followed by the switches and memory units, followed by the processors; all with a repair rate of 1 rph. If a processor fails it contaminates the database with probability 0.001. The system is operational if the front-end, the database, and a processing subsystem are up; a processing subsystem is up if the switch, the memory, and a processor are up. In Table 9.4 we illustrate the results obtained with the FB, SFB, and DSFB techniques using $\alpha = 0.7$, $\beta = 0.8$, $\beta_c = 0.2$, for a sample size $M = 10^5$. The DSFB technique performs much better in this context, both in precision (a two-order reduction in the variance) and in computational effort. Its reduced execution time is due to the fact that, as it goes much faster to the states where the system is down than with the other methods, the cycle lengths are much shorter. Our last example illustrates the use of simulation techniques to evaluate a model with a very large state space. The system is similar to one presented by Nicola *et al.* in [77], but has more components and as a result the underlying Markov chain has a larger state space. The system is composed of two sets of four processors each, four sets of two dual-ported controllers, and eight sets of disk arrays composed of four units. Each controller cluster is in charge of two disk arrays; each processor has access to all the controller clusters. The system is up if there are at least one processor (of either class), one controller of each cluster, and three disks of each array, in operational order.

We omit the time unit again (or say we measure time in hours, thus failures and repair rates in fph and rph, respectively). The failure rates for the processors and the controllers are 1/2000; for the disk arrays we consider four different failure rates (each corresponds to two arrays), namely 1/4000, 1/5000, 1/8000, and 1/10000. We consider a single case of failure propagation: when a processor of a cluster fails, there is a probability of 0.1 that a processor of the other cluster is affected. Each failure has two modes; the repair rates depend on the mode, and take the value 1 for the first mode and 0.5 for the second.

The system has more than 7.4×10^{14} states. This precludes even the generation of the state space, and makes it impossible to think of using exact techniques.

In Table 9.5 we show the results obtained with the crude, FB, SFB, and DSFB techniques, always in their balanced versions, using $\alpha = 0.7$, $\beta = 0.8$, $\beta_c = 0.2$, for a sample size $M = 10^5$. In this case, we divide variances and execution times by the corresponding metrics in the crude (standard) case. Since this is a complex case, the execution times are

Table 9.5. We consider here crude, FB, SFB, and DSFB methods for a very large system. We plot the ratio (in %) between the variance of the FB, SFB, and DSFB methods and the variance of the crude one, and the same with the execution times.

Method	RelVar	RelTime
FB	0.66%	684%
SFB	0.077%	419%
DSFB	0.035%	199%

larger than those observed for the previous examples; but even the slowest method, FB, takes less than 27 minutes to complete the estimation. In all cases, when using Importance Sampling techniques the variances obtained are between two and three orders of magnitude smaller than the variance of the crude simulation technique; this allows us to estimate γ with a higher precision with the same number of replications. The technique which a priori seems the more appropriate to this kind of system with failure propagations is DSFB; the experimental results confirm this, as DSFB not only has the best variance, but also the best execution time among the Importance Sampling techniques compared, and only twice the execution time of the crude technique. The numerical values in this example have been chosen such that with the relatively small number M of iterations considered, even the crude method allows us to obtain a confidence interval. At the same time, this underlines the importance of the concept of *efficiency*: even the FB method is more efficient than the crude technique, since we must take into account both the execution time and the obtained precision to evaluate a simulation method.

9.2.6 Conclusions

We discussed several methods designed to estimate the MTTF of a complex system modeled by a Markov chain, in the rare event context. The material in this part of the chapter has the following objectives: (i) first of all, to provide simulation techniques which are both efficient and very simple to implement, and at the same time, quite generic and covering a large number of the possible situations, (ii) to describe the idea of Importance Sampling in the context of Markovian models in dependability, and (iii) to introduce the reader to some general concepts to help understand the behavior of estimators in the area, namely the concepts of relative error of an estimation, of rarity parameterization, and the Bounded Relative Error property of an estimator. Another important idea in the section is that of balanced technique, specific to the family of Importance Sampling procedures described. The discussion should be helpful in choosing among the techniques described and in designing new variance reduction algorithms for the same or for other dependability measures.

9.3   Estimating the reliability of the system at time *t*

Here we want to estimate the reliability at t, $R(t)$, the probability that the system is operational during the whole interval $[0,t]$. If the system is highly dependable and if t is small, then $R(t)$ is close to one and we again face a rare event situation. The estimation of transient dependability metrics in the rare event case has received little attention so far. We thus complete the chapter with the description of a method to deal with this type of problem.

In this section we assume that there is only one component per class, that is, that $N_k = 1$ for all class k. This does not change the generality of the approach, since we allow failure rates to be the same for different components.

The failure rate of component k is λ_k, independently of the model's state. As in the previous section, the system can include an arbitrary repair subsystem, to bring failed components back to the operational state.

The next subsection presents a method which can be used to tackle our problem. The idea is to *force* the occurrence of some failures *before t*.

9.3.1   Pathset-based conditioning

In this context of multicomponent systems, a pathset is a set of components such that if all the members of the set are operational, then the system is also operational. In a coherent structure system, any super-set of a pathset is a pathset [20], [86]. A minimal pathset is a pathset not containing another pathset other than itself.

For instance, consider a communication network whose graph is the graph given in Figure 9.5. The nodes of the network are the vertices of the graph, and the links are represented by the graph's edges. Two nodes are marked (s and t in the figure, as *source* and *terminal*). The components of the system are the 30 network links. The network works if and only if there is at least one path between nodes s and t whose links are all working. In this example, a minimal pathset is simply a path between s and t that includes no cycle (that is, a path with no node appearing more than once). In the figure, such a pathset (one among the many possible pathsets in the graph) is indicated in bold.

We present here a simple conditioning method based on the pathsets of the system. The technique works with pathsets in general, but it will be obvious that it is better to use minimal ones only. We assume that there exists an efficient method to build pathsets of the system. This is true for many (while not all) cases of practical interest.

We denote by T_i the time of the first failure of component i. For the sake of clarity, we assume that T_i is an exponentially distributed random variable (with rate λ_i), even if this is not a necessary condition for the proposed approach to work efficiently (we can work with phase-type distributions as well).

Observe that we can build a Markov chain whose states are vectors with 30 binary components. In state $x = (x_1,\ldots,x_{30})$, $x_i = 1$ if component (edge) i is working, 0 otherwise. The set U of operational states is the set of states, x, such that the subset of components, i, for which $x_i = 1$ contains at least one pathset. Observe that this chain has 2^{30} states (slightly more than a billion (10^9)).

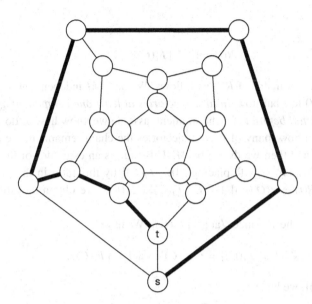

Figure 9.5 The "dodecahedron" topology with a source and a terminal (nodes s, t) and a minimal pathset in bold; for instance, another minimal pathset is composed of the single link between s and t

Given a pathset $C = \{i_1, \ldots, i_{|C|}\}$, we can define the event

$$O_t(C) = \{\forall i, 1 \leq i \leq |C|,\ T_i > t\}$$

and the complementary event

$$F_t(C) = \{\exists i, 1 \leq i \leq |C|,\ \text{such that } T_i \leq t\}.$$

The first event corresponds to trajectories of the chain X where no element of pathset C fails before t; by the definition of a pathset, this implies that the system is operational up to time t. The second event corresponds to trajectories where at least one element of pathset C fails; this is a necessary but not sufficient condition for a system failure.

As the r.v. T_i is exponentially distributed with rate λ_i, if we denote $\Lambda(C) = \sum_{i \in C} \lambda_i$, we have $\mathbb{P}\{O_t(C)\} = e^{-\Lambda(C)t}$ (thus, $\mathbb{P}\{F_t(C)\} = 1 - \mathbb{P}\{O_t(C)\} = 1 - e^{-\Lambda(C)t}$).

Let us now define the event $E_t = \{\forall s \in [0, t],\ X_s \in U\}$. The reliability of the system at time t is $R(t) = \mathbb{P}\{E_t\}$. Recall that $W_t = 1_{\{E_t\}}$.

Observe that

$$R(t) = \mathbb{P}\{E_t\} = \mathbb{P}\{E_t \mid O_t(C)\}\mathbb{P}\{O_t(C)\} + \mathbb{P}\{E_t \mid F_t(C)\}\mathbb{P}\{F_t(C)\}$$

$$= \mathbb{P}\{O_t(C)\} + \mathbb{P}\{E_t \mid F_t(C)\}\mathbb{P}\{F_t(C)\}. \tag{9.23}$$

Let us define now the random variable, Z_t, as follows:

$$Z_t = \mathbb{P}\{O_t(C)\} + \mathbb{P}\{F_t(C)\}\,W_t = e^{-\Lambda(C)t} + (1 - e^{-\Lambda(C)t})W_t. \tag{9.24}$$

We then have

$$R(t) = \mathbb{E}\{Z_t \mid F_t(C)\}. \tag{9.25}$$

This leads to a new estimator of $R(t)$ as follows. We build M independent trajectories of the chain from 0 to t but now, *in all trajectories at least one component of pathset C will be forced to fail before t* (for the moment, assume we know how to do this). As before, we count in how many of these trajectories the chain remains in the set U of operational states, and then we divide by M. This ratio is an estimator of the number $\mathbb{E}\{W_t \mid F_t(C)\} = \mathbb{P}\{E_t \mid F_t(C)\}$. Replacing $\mathbb{P}\{E_t \mid F_t(C)\}$ by this ratio in the previously given expression $R(t) = \mathbb{P}\{O_t(C)\} + \mathbb{P}\{E_t \mid F_t(C)\}\mathbb{P}\{F_t(C)\}$, we obtain an estimator of the reliability at t.

Let us look now at the variance $\mathrm{Var}\{Z_t \mid F_t(C)\}$. We have

$$\mathrm{Var}\{Z_t \mid F_t(C)\} = \mathbb{P}\{F_t(C)\}^2 \mathrm{Var}\{W_t \mid F_t(C)\}. \tag{9.26}$$

For $\mathrm{Var}\{W_t \mid F_t(C)\}$, we have

$$\mathrm{Var}\{W_t \mid F_t(C)\} = \mathbb{P}\{E_t \mid F_t(C)\}(1 - \mathbb{P}\{E_t \mid F_t(C)\}) \tag{9.27}$$

$$= \frac{R(t) - \mathbb{P}\{O_t(C)\}}{\mathbb{P}\{F_t(C)\}}\left(1 - \frac{R(t) - \mathbb{P}\{O_t(C)\}}{\mathbb{P}\{F_t(C)\}}\right) \tag{9.28}$$

and thus

$$\mathrm{Var}\{Z_t \mid F_t(C)\} = (R(t) - \mathbb{P}\{O_t(C)\})(1 - R(t)) \tag{9.29}$$

$$< R(t)(1 - R(t)) \tag{9.30}$$

$$= \mathrm{Var}\{W_t\}. \tag{9.31}$$

This guarantees a reduction in variance over the standard Monte Carlo estimator. It remains to describe the specific sampling procedure, that is, how to build trajectories of X conditional on $F_t(C)$. When sampling random variables T_i, with $i \in C$, instead of sampling from the original exponential distributions, we must sample from a modified distribution which takes into account this conditioning. More formally, we must first derive the distribution function of an exponential random variable T_1 conditional on the event that the minimum of the independent exponential random variables $T_1, T_2, \ldots, T_{|C|}$ with respective parameters $\lambda_1, \lambda_2, \ldots, \lambda_{|C|}$, is less than t, that is,

$$\mathbb{P}\{T_1 \leq x \mid \min(T_1, T_2, \ldots, T_{|C|}) < t\}.$$

This conditional distribution function can then be used, for example, with the inverse distribution function method, to generate samples for T_1. Then we must show how to proceed with T_2, etc.

For the sake of clarity, let us explain the conditional sampling procedure for the simple case of a pathset, C, composed of two links. If the pathset has $|C| > 2$ components, we can use the fact that $\min(T_1, T_2, \ldots, T_{|C|}) = \min(T_1, \min(T_2, \ldots, T_{|C|}))$, and that $\min(T_2,\ldots,T_{|C|})$ is an exponential random variable T_2' with rate $\lambda_2 + \cdots + \lambda_{|C|}$, leading again to the case of only two links in the pathset.

Let us denote by V the random variable $\min(T_1, T_2)$. Recall that $\Lambda(C) = \lambda_1 + \lambda_2$. If $x \leq t$, we have

$$\mathbb{P}\{T_1 \leq x \mid V < t\} = \frac{\mathbb{P}\{T_1 \leq x, V < t\}}{\mathbb{P}\{V < t\}}$$

$$= \frac{\mathbb{P}\{V < t \mid T_1 \leq x\}\,\mathbb{P}\{T_1 \leq x\}}{\mathbb{P}\{V < t\}}.$$

If $T_1 \leq x$, and $x \leq t$, then $T_1 \leq t$ and $\mathbb{P}\{V < t \mid T_1 \leq x\} = 1$. Thus, we have

$$\mathbb{P}\{T_1 \leq x \mid V < t\} = \frac{1 - e^{-\lambda_1 x}}{1 - e^{-\Lambda(C)t}}.$$

If $x > t$, we have

$$\mathbb{P}\{T_1 \leq x \mid V < t\} = \frac{\mathbb{P}\{T_1 \leq x, V < t\}}{\mathbb{P}\{V < t\}}$$

$$= \frac{\mathbb{P}\{T_1 \leq x, V < t, T_1 \leq t\} + \mathbb{P}\{T_1 \leq x, V < t, T_1 > t\}}{\mathbb{P}\{V < t\}}$$

$$= \frac{\mathbb{P}\{T_1 \leq t\} + \mathbb{P}\{t < T_1 \leq x, T_2 < t\}}{\mathbb{P}\{V < t\}}$$

$$= \frac{\mathbb{P}\{T_1 \leq t\} + \mathbb{P}\{t < T_1 \leq x\}\,\mathbb{P}\{T_2 < t\}}{\mathbb{P}\{V < t\}}$$

$$= \frac{1 - e^{-\lambda_1 t} + (e^{-\lambda_1 t} - e^{-\lambda_1 x})(1 - e^{-\lambda_2 t})}{1 - e^{-\Lambda(C)t}}.$$

This completes the derivation of the conditional distribution function for T_1.

Once the sample for T_1 has been generated, we need the distribution function for T_2 conditional on the generated value for T_1:

$$\mathbb{P}\{T_2 \leq x \mid V < t, T_1 = t_1\}.$$

If $t_1 \leq t$, we have

$$\mathbb{P}\{T_2 \leq x \mid V < t, T_1 = t_1 \leq t\} = \mathbb{P}\{T_2 \leq x \mid T_1 = t_1\}$$

$$= \mathbb{P}\{T_2 \leq x\}$$

$$= 1 - e^{-\lambda_2 x},$$

which is equal to the unconditional distribution of T_2. If $t_1 > t$, we have

$$\mathbb{P}\{T_2 \leq x \mid V < t, T_1 = t_1 > t\} = \mathbb{P}\{T_2 \leq x \mid T_2 \leq t\} = \frac{\mathbb{P}\{T_2 \leq x, T_2 \leq t\}}{\mathbb{P}\{T_2 \leq t\}}.$$

For $x \le t$, this gives

$$\mathbb{P}\{T_2 \le x \mid V < t, T_1 = t_1 > t\} = \frac{\mathbb{P}\{T_2 \le x\}}{\mathbb{P}\{T_2 \le t\}} = \frac{1 - e^{-\lambda_2 x}}{1 - e^{-\lambda_2 t}}. \tag{9.32}$$

Finally, if $x > t$, we have

$$\mathbb{P}\{T_2 \le x \mid V < t, T_1 = t_1 > t\} = \frac{\mathbb{P}\{T_2 \le t\}}{\mathbb{P}\{T_2 \le t\}} = 1. \tag{9.33}$$

This completes our derivation.

The preceding development concerns the case of a single pathset. When using K *disjoint* pathsets, whenever this is possible for some $K > 1$, the development is the same for each one. Since the pathsets are disjoint and the components independent, we simply apply the obtained distribution for the components of each one of the pathsets.

Assume now that we can build two disjoint pathsets, C_1 and C_2. Denote by $Z_t^{(1)}$ the process defined in (9.24) when the pathset C is C_1. Now, define

$$Z_t^{(2)} = \mathbb{P}\{O_t(C_1) \cup O_t(C_2)\} + \mathbb{P}\{F_t(C_1) \cap F_t(C_2)\} W_t$$
$$= e^{-(\Lambda(C_1) + \Lambda(C_2))t} + \left[1 - e^{-(\Lambda(C_1) + \Lambda(C_2))t}\right] W_t.$$

Similarly, we have as before,

$$R(t) = \mathbb{E}\{Z_t^{(2)} \mid F_t(C_1) \cap F_t(C_2)\}.$$

The procedure we propose consists of using this new process to build an estimator of $R(t)$ as described previously. All we need is to sample N trajectories of X conditioned on the event $F_t(C_1) \cap F_t(C_2)$ as we explained above. If we compute the variance of this new estimator, we obtain

$$\mathrm{Var}\left\{Z_t^{(2)} \mid F_t(C_1) \cap F_t(C_2)\right\} = [R(t) - \mathbb{P}\{O_t(C)\} - \mathbb{P}\{O_t(C_2)\}](1 - R(t))$$
$$< \mathrm{Var}\left\{Z_t^{(1)} \mid F_t(C_1)\right\}.$$

We see that the variance reduction is larger using two disjoint paths than only one. More generally, suppose that we can select K disjoint pathsets, $C_1, \ldots, C_K$, and let us define the corresponding process

$$Z_t^{(K)} = \mathbb{P}\left\{\bigcup_{k=1}^{K} O_t(C_k)\right\} + \mathbb{P}\left\{\bigcap_{k=1}^{K} F_t(C_k)\right\} W_t$$
$$= e^{-\left(\sum_{k=1}^{K} \Lambda(C_k)\right)t} + \left[1 - e^{-\left(\sum_{k=1}^{K} \Lambda(C_k)\right)t}\right] W_t.$$

We again have

$$R(t) = \mathbb{E}\{Z_t^{(K)} \mid \bigcap_{k=1}^{K} F_t(C_k)\}$$

and

$$\mathbb{V}\text{ar}\left\{Z_t^{(K)} \mid \bigcap_{k=1}^{K} F_t(C_k)\right\} = \left[R(t) - e^{-\left(\sum_{k=1}^{K} \Lambda(C_k)\right)t}\right][1 - R(t)].$$

The more disjoint paths we use, the better estimator we obtain (since the variance reduction is larger).

A final comment on the number of disjoint pathsets that can be built. In an example such as the previous one, where the definition of an operational system is based on the paths of a graph, the maximal number of disjoint pathsets that can be built between s and t is a well-known graph parameter called the graph's *breadth*. Its computation is polynomial in the size of the graph, and it can be obtained as a by-product of a minflow algorithm (see [86] and references therein).

9.3.2 Numerical results

In this section we illustrate the application of the proposed method for the evaluation of $R(t)$ in a specific example, and we compare its performance versus the standard Monte Carlo method.

The system under study comes from network reliability theory, and was used before to explain the vocabulary of this part of the chapter. We model a telecommunication network whose structure is represented by the graph of Figure 9.5. It has 30 components, the graph's edges correspond to the network's links, and each belongs to one of two classes. There are 16 components of type A and 14 of type B. Every component has the same failure rate, $\lambda = 1$ failure per hour; there are two repairmen, one for each type of component; the repair rate for the components of type A is $\mu_A = 100$ repairs per hour, and for the components of type B is $\mu_B = 1000$ repairs per hour. The structure function of the system [20], [86] is given by the graph in Figure 9.5: the system is up if there is an operational path (a set of operational links) between nodes s and t. Recall that the associated Markov model has exactly 2^{30} states.

We have implemented the method based on pathsets as described in Subsection 9.3.1, and a standard Monte Carlo method. For the numerical results given here, we used three disjoint minimal pathsets between s and t (the breadth of this graph is 3).

Table 9.6 shows the simulation results for the estimation of the unreliability $1 - R(t)$, computed for four different values of t: $t = 0.001, 0.01, 0.1$, and 1 hours. The sample size is 10^6 experiments for each different value of t indicated in the first column. The column $1 - \widehat{R}(t)$ presents the estimated values of the unreliability as computed by the proposed method. The third column contains an estimation of the accuracy improvement, given by the variances ratio (computed as the quotient of the variance of the standard Monte Carlo method $\mathbb{V}\text{ar}\{W_t\} = R(t)(1 - R(t))$, and the estimation $\widehat{Z}$ of the variance of the proposed estimator). The next columns present the ratio of the execution times (T_W for crude, T_Z for the new method), and the relative efficiency of the proposed variance reduction method relative to the standard Monte Carlo method (computed as the product of the accuracy improvement by the time reduction).

Table 9.6. Comparison between the proposed method and the crude Monte Carlo method

	$1 - \widehat{R}(t)$	$\dfrac{\mathrm{Var}\{W_t\}}{\widehat{Z}}$	$\dfrac{T_W}{T_Z}$	$\dfrac{\mathrm{Var}\{W_t \times T_W\}}{\widehat{Z} \times T_Z}$
$t = 0.001$	1.51×10^{-9}	0.99×10^7	2.65×10^{-1}	2.62×10^6
$t = 0.01$	5.63×10^{-7}	1.11×10^4	2.73×10^{-1}	3.03×10^3
$t = 0.1$	4.35×10^{-5}	2.53×10^1	3.17×10^{-1}	8.02×10^0
$t = 1$	5.01×10^{-4}	1.05×10^0	7.71×10^{-1}	8.10×10^{-1}

The proposed method always has a smaller variance than the standard Monte Carlo method, but the reduction depends on the value of t. For small values of t, the accuracy improvement is very important, then, it decreases as t increases, but at the same time, rarity also diminishes. Observe that the variance of the new estimator is bounded by

$$(R(t) - \mathbb{P}\{O_t(C)\})(1 - R(t))$$

with

$$\mathbb{P}\{O_t(C)\} = \prod_{i \in C} \mathbb{P}\{T_i > t\} = \prod_{i \in C} e^{-\lambda_i t}$$

which is a decreasing function of t; when t is small, the variance is also small, but when t grows, $\mathbb{P}\{O_t(C)\}$ goes to 0, and the bound approaches the value of the variance of the standard Monte Carlo method.

The execution time of the proposed method is a little bit larger than the standard Monte Carlo method, especially for the smallest values of t (as can be seen in Column 4); this can be explained by the necessity of finding the pathsets of the system and by the use of a more complex sampling procedure. The combined effect variance reduction-execution time (usually called relative efficiency) is presented in Column 5. We can see that in general the overall result is satisfactory, except for the case where $t = 1$, in which the execution time penalization is more important than the variance reduction obtained. For smaller values of t, the efficiency improvement is very important, and allows the evaluation in a reasonable time of systems which could not be evaluated by the standard Monte Carlo method.

9.3.3   Conclusions

In this chapter we have presented a recursive technique for the evaluation of transient reliability measures, based on the pathsets of the system. Its objective is to provide an efficient estimation tool in the context of rare events. The results obtained on a particular case show the interest of the method, presenting the accuracy improvements which can be obtained with a small computing time overhead. These improvements are especially important in the case of very reliable systems (i.e., very rare failure events), and where time windows are relatively small.

Notes

The general area of estimating metrics defined on Markov models and extensions has a rich literature associated with it. As stated before, the main difficulty when evaluating the dependability properties of a given complex system using simulation is rarity (second in the list comes the classical problem of efficiently building and validating models, not addressed in this book). In 2009, we published a book [99] devoted to this problem. It presents both the theoretical and the practical sides, and in both cases, the eleven chapters therein have a rich and up-to-date bibliography. The book also provides the mathematical background for the whole area. Moreover, one of its chapters [98] is devoted to Markovian models.

The main sources for this chapter are papers [12] and [13]. This means that the material here comes from joint works with Héctor Cancela and María Urquhart, from UDELAR, Uruguay, and Bruno Tuffin, from Inria, France. The two papers provide more material about the methods described here, while the chapter is more complete in presenting the area and some technical details of the simulation procedures in a unified way. For other papers related to the specific topics presented here, we must highlight [106] and [76] for the introduction of some useful concepts when designing or analyzing estimation procedures in the rare event case; the previously mentioned [98], especially for material concerning the first part of the chapter (Section 9.2); and [11], for the ideas used in Section 9.3.

In [99], the authors present the main approaches followed when analyzing stochastic processes using the Monte Carlo method, namely Importance Sampling and splitting procedures, plus some other ideas. Among them, we recommend the interested user to read about the quite recent "zero-variance" approach, that in the context of this chapter, can be described as follows. Consider the problem of estimating γ, as described in Subsection 9.1.1. Define $\gamma_y = \mathbb{P}\{\tau_D < \tau_0 \mid Y_0 = y\}$. It can be proved (see [54]) that the optimal change of measure of the Importance Sampling approach leading to a zero-variance estimator can be written as

$$P'_{x,y} = \frac{P_{x,y}\gamma_y}{\sum_z P_{x,z}\gamma_z}.$$

To apply this we need not only the answer to our main question, γ_0, but all the other γ_xs. At first glance, this seems useless in practice, but it appears to be a very productive idea, since it leads to effective changes of measures that replace the unknown elements in the previous expression by approximations. The literature on this approach is already important. Let us just mention the recent [8] and [55] for the case of static models.

10   Bounding techniques

One of the approaches that can be followed when dealing with very large models is to try to compute the bounds of the metrics of interest. This chapter describes bounding procedures that can be used to bound two important dependability metrics. We are concerned with highly dependable systems; the bounding techniques described here are designed to be efficient in that context.

First, we address the situation where the states of the model are weighted by real numbers, the model (the Markov chain) is irreducible and the metric is the asymptotic mean reward. The most important example of a metric of this type in dependability is the asymptotic availability, but many other metrics also fit this framework, for instance, the mean number of customers in a queuing system, or in part of a queuing network (in this case, the techniques must be considered in light traffic conditions, the equivalent to a highly dependable system in this queuing situation).

Second, we consider the evaluation of the Mean Time To Failure, that is, of the expectation of the delay from the initial instant to the occurrence of the first system's failure. The situation is that of a chain where all states but one are transient and the remaining state is absorbing; the MTTF corresponds to the mean absorption time of the chain.

This chapter basically follows [62] in the first part and [24] in the second one.

10.1   First part: Bounding the mean asymptotic reward

The system is represented by a continuous-time homogeneous and irreducible Markov chain X over the finite state space S with stationary distribution π (seen as a row vector). We denote $\pi_i = \mathbb{P}\{X_\infty = i\}$ where X_∞ is a stationary version of X. With each state, i, we associate a reward, $r_i \geq 0$; let r be the column vector of rewards. This chapter deals with the computation of the scalar product

$$R = \mathbb{E}\{r_{X_\infty}\} = \sum_{i \in S} r_i \pi_i = \pi r$$

which can be called the Mean Asymptotic Reward. Recall that in a dependability context, the states are in general called operational when they represent a system delivering its service as expected, and unoperational otherwise. If a reward equal to 1 is associated with the operational states and equal to 0 with the unoperational ones, then R is the asymptotic availability of the system. In a performance context, suppose that the model is a queuing network and that you are interested in the mean number of customers in queue q. If with each state, i, we associate a reward equal to the number of customers in station q when the model is in state i, the expectation, R, is equal to the desired mean number of customers in q. As a second example, if r_i is the number of active processors in some model of a fault-tolerant multiprocessor computer when its state is i, then R is a performability measure.

10.1.1   Related work

Before presenting the two methods described here for bounding R, let us briefly describe other related papers in the area. First note the approach of [110] which constructs, from the original model, two new models which respectively build a lower and an upper bound of metric R using particular job-local-balance equations. However, this technique doesn't give tight bounds and becomes more complicated to apply to complex systems. The starting works from which the technique described in this section is built are [22] and [23]. In [100], these results are improved, following the same research lines. A different improvement is in [69] and we follow this approach in the chapter. We also show how to improve it to obtain a more powerful bounding technique, following [62]. Briefly, in [100] the author derives a general iterative bounding technique having the following main differences with [69] and with our work: it can be applied without restrictions (while ours or the method of [69] needs some conditions to work) but is more expensive. The final form of the approach of [100] is quite different, however, because it is presented as an iterative bounding process in order to derive bounds on the total vector π and then on R. The key point in the complexity comparison is that [100] basically needs the inversion of a matrix to obtain each intermediate bound. The technique in [69] needs a strong condition to be applied, but it is much cheaper. The method in [62] is more expensive than the last one, but it needs much less restrictive conditions. This last feature means it can obtain tight bounds even for performance models on infinite state spaces, as shown in that paper.

To reduce the computational cost of the bounds obtained by the first algorithm of [69], a technique of "duplication of states" is proposed in [69], [70], and [71]. The main objective in [69] is to reduce the number of linear systems to solve. In [70], the authors propose a multi-step bounding algorithm to improve the obtained bounds without restarting the work from the beginning. But the spread between the bounds has a non-zero limiting value [71]. This problem is handled in [71] using a technique called bound spread reduction in order to reduce the error introduced at each step of the previous iterative procedure. A simple heuristic method to choose between the multi-step bounding algorithm and the bound spread reduction is developed and illustrated in [71]. The second method that we present in the next subsection can also use the same technique to improve its efficiency. We do not develop this point in the book.

The authors of [72] have chosen another approach to bound mean response times in heterogeneous multi-server queuing systems. From the original model, they construct two new models, one to obtain a lower bound of the considered measure and another one to obtain an upper bound. Another line of research is [15] which, improving previous work by the same author, gives better bounds when additional information, namely the distance to failure of each state (the minimal number of failures that put the system in an unoperational state) is available. In that paper the number of linear systems to be solved is also reduced.

10.1.2   The problem

In this chapter we denote by A the infinitesimal generator of X. We then have $\pi A = 0$. The goal is the evaluation of bounds of $R = \pi r$, *without computing vector* π. We denote by ϱ_1 and ϱ_2 a lower and an upper bound of the rewards, respectively, that is real numbers such that for all $i \in S$,

$$0 \le \varrho_1 \le r_i \le \varrho_2. \tag{10.1}$$

The state space S (by default, a finite set) is assumed to be decomposable into two disjoint sets, denoted by G and $\bar{G}$, such that "the chain spends most of its time" in the former. In many situations, the user can approximately identify a set with this informally described property. In a dependability context, a model corresponding to a repairable system with highly reliable components leads to the choice of G as the set of states having less than some fixed number of failed units. In a queuing model considered in a light traffic situation, we can associate G with the states where the network has less than some fixed number of customers.

The techniques discussed here attempt to give bounds of the asymptotic reward, R, by working with auxiliary Markov models obtained by replacing $\bar{G}$ (and the associated rates) with a "few" states. The situation we are looking for here is to have $|G| \ll |S|$.

Some notation associated with a subset of states
For any subset of states $C \subseteq S$, we denote

- π_C, the restriction of π to the set C (that is, a row vector with size $|C|$), r_C, the restriction of r to C, etc. (same notation as in Subsection 4.1.1);
- $\pi(C) = \mathbb{P}\{X_\infty \in C\} = \sum_{i \in C} \pi_i = \pi_C \mathbb{1}$, where $\mathbb{1}$ is a column vector having all its entries equal to 1, its dimension being defined by the context;
- $\hat{\pi}_C =$ distribution of X_∞ conditional on the event $\{X_\infty \in C\}$, row vector with size $|C|$; note that $\hat{\pi}_C = \pi_C / \pi(C)$;
- $\bar{C} = S \setminus C$, the complement of C;
- $in(C) = \{j \in C \mid \text{there exists } i \in \bar{C} \text{ with } A_{ij} > 0\}$; that is, $in(C)$ is the set of "entry" states of C;
- $out(C) = \{i \in C \mid \text{there exists } j \in \bar{C} \text{ with } A_{ij} > 0\}$; that is, $out(C)$ is the set of "exit" states of C;
- A_C, the block of A corresponding to the transitions inside subset C;
- $A_{C,C'}$, the block of A corresponding to the transitions from C to C'.

In Figure 10.1, we illustrate the notation.

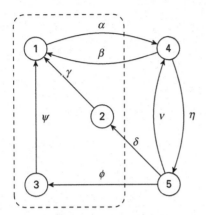

Figure 10.1 Example for illustrating some of the notation used; $G = \{1,2,3\}$, inside the area with the dashed border; $in(G) = G$, $out(G) = \{1\}$; $in(\bar{G}) = \{4\}$, $out(\bar{G}) = \bar{G}$; if $\pi = (\pi_1,\pi_2,\pi_3,\pi_4,\pi_5)$, then $\pi_G = (\pi_1,\pi_2,\pi_3)$, $\pi(G) = \pi_1 + \pi_2 + \pi_3$, and $\hat{\pi}_G = (\pi(G))^{-1}(\pi_1,\pi_2,\pi_3)$

Also, in the example of Figure 10.1, we have, for instance,

$$A_G = \begin{pmatrix} -\alpha & 0 & 0 \\ \gamma & -\gamma & 0 \\ \psi & 0 & -\psi \end{pmatrix}, \qquad A_{G,\bar{G}} = \begin{pmatrix} \alpha & 0 \\ 0 & 0 \\ 0 & 0 \end{pmatrix}.$$

Forcing the entries in G by a fixed state j

The main idea comes from the basic initial work by Courtois and Semal. It consists of building a family of Markov chains derived from X in the following way. For each state $j \in in(G)$, let us construct the new continuous-time homogeneous Markov chain $X^{(j)}$, by "forcing" the transitions from $\bar{G}$ into G to enter by state j. The infinitesimal generator of $X^{(j)}$ is denoted by $A^{(j)}$:

$$A^{(j)} = \begin{pmatrix} A_G & A_{G\bar{G}} \\ A_{\bar{G}G}^{(j)} & A_{\bar{G}} \end{pmatrix}.$$

The transition rate from any state $i \in \bar{G}$ to j is equal to $\sum_{l \in in(G)} A_{i,l}$. In other words, $A_{\bar{G}G}^{(j)} = A_{\bar{G}G}\mathbb{1}e_j$, where e_j is the jth row vector of the canonical base in $\mathbb{R}^{|G|}$. The other transition rates of $X^{(j)}$, that is inside G, from G to $\bar{G}$, and inside $\bar{G}$, are as in X.

If we call X the example given in Figure 10.1, we show in Figure 10.2 the chain $X^{(1)}$. First, we prove that $X^{(j)}$ has a unique stationary distribution by means of the following lemma.

LEMMA 10.1.1. *The Markov chain $X^{(j)}$ has a unique recurrent class, which includes state j, and no absorbing states.*

Proof. State j being fixed, we denote by S_j the class of states i such that i is reachable from j and j is reachable from i. The claim says that S_j can be reached from every state $k \in S$. Assume $S_j \neq S$ and let $k \in S \setminus S_j$. If $k \in out(\bar{G})$, by definition of $out(\bar{G})$, k is connected to j. If $k \in \bar{G} \setminus out(\bar{G})$ then there is necessarily a path from k to some $l \in out(\bar{G})$

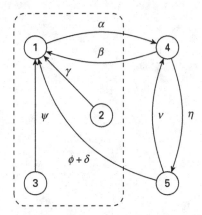

Figure 10.2   Calling X the chain depicted in Figure 10.1, this is $X^{(1)}$; see that X is irreducible, but now, in $X^{(1)}$ states 2 and 3 are transient; in this chain, $in(G) = out(G) = \{1\}$

(since X is irreducible), which is completely included in $\bar{G}$, and by definition of $X^{(j)}$, l is connected to j. There remains the case of a state $k \in G \setminus \{j\}$. If, in X, there is a path from k to j completely included in G, we are done. If not, since X is irreducible, there is at least a path from k to j passing through $\bar{G}$, thus entering $\bar{G}$ for the first time by some state l, and then we are in the first discussed case.   ∎

10.1.3   Auxiliary models

Since $X^{(j)}$ has a unique recurrent class, it has a unique stationary distribution, which we denote by $\pi^{(j)}$. Then, for each state $j \in in(G)$, we have the following result, which is a slight extension of results in [22]:

THEOREM 10.1.2.   *There exists a nonnegative vector β with support $in(G)$, that is, satisfying $\beta_j = 0$ if $j \notin in(G)$, such that $\beta \mathbb{1} = 1$ and*

$$\sum_{j \in in(G)} \beta_j \pi^{(j)} = \pi.$$

Proof.   This fundamental result is proved in [22]. Thanks to Lemma 10.1.1, the same proof shows that the only necessary assumption is the existence of a unique stationary distribution for $X^{(j)}$. In [62] it is shown that the result is still valid in the infinite state space case. We do not discuss this extension in this book, where we focus on finite state models.   ∎

From this theorem, the two following immediate results can be derived, put together as a corollary.

COROLLARY 10.1.3.   *If we denote $\pi^{(j)}(G) = \mathbb{P}\{X^{(j)}_\infty \in G\}$, then we have*

$$\min_{j \in in(G)} \pi^{(j)}(G) \leq \pi(G) \leq \max_{j \in in(G)} \pi^{(j)}(G), \tag{10.2}$$

and if $R^{(j)} = \pi^{(j)} r$, we have

$$\min_{j \in in(G)} R^{(j)} \leq R \leq \max_{j \in in(G)} R^{(j)}. \tag{10.3}$$

Proof. Let us denote by u the column vector defined by

$$u = \begin{pmatrix} \mathbb{1}_G \\ 0_{\bar{G}} \end{pmatrix},$$

where $\mathbb{1}_G$ (resp. $0_{\bar{G}}$) is the column vector having all its entries equal to 1 (resp. to 0), the dimension being equal to $|G|$ (resp. to $|\bar{G}|$). From Theorem 10.1.2, we have

$$\pi(G) = \pi u = \sum_{j \in in(G)} \beta_j \pi^{(j)} u = \sum_{j \in in(G)} \beta_j \pi^{(j)}(G).$$

This means that $\pi(G)$ is a weighted average of the $\pi^{(j)}(G)$, the weights being the β_js (recall that $\beta\mathbb{1} = 1$ and $\beta \geq 0$). As a consequence, $\pi(G)$ is somewhere between the smallest of the $\pi^{(j)}(G)$ and the largest:

$$\min_{j \in in(G)} \pi^{(j)}(G) \leq \sum_{j \in in(G)} \beta_j \pi^{(j)}(G) = \pi(G) \leq \max_{j \in in(G)} \pi^{(j)}(G).$$

In the same way, from $R = \pi r$, we have

$$\min_{j \in in(G)} \pi^{(j)} r \leq R = \sum_{j \in in(G)} \beta_j \pi^{(j)} r \leq \max_{j \in in(G)} \pi^{(j)} r.$$

This is basically the proof in [69], given in that paper for the particular case of the asymptotic availability measure. ∎

Corollary 10.1.3 gives the relationships that are used to derive the bounds of R. In the following, we develop a general approach to build a lower bound of $\min_{j \in in(G)} R^{(j)}$ and an upper bound of $\max_{j \in in(G)} R^{(j)}$.

10.1.4 Aggregation of states

To go further, we suppose that we are given a partition $\{C_I, I = 0, 1, \ldots, M\}$ of S and an integer, K, with $0 < K < M$, and that G is defined by

$$G = \bigcup_{I=0}^{K-1} C_I.$$

In a performance context, assuming that we are dealing with something like a queuing network or a stochastic Petri net, C_I can be, for instance, the set of states where the system, or some of its subsystems, has I customers or tokens. In a dependability case, if we work with a model of some fault-tolerant multi-component system, C_I can be, for instance, the set of states where there are I failed components. A good property for such a partition is that the higher the index I, the lower the probability that X_∞ belongs to C_I.

In what follows, we assume that transitions from class C_I to class C_J are not allowed if $J \leq I - 2$, that is, if the following condition holds:

Condition 1. *For any two integer indices, I and J, such that $J \leq I - 2$, for any states $i \in C_I$ and $j \in C_J$, we have $A_{i,j} = 0$.*

Observe that, given the irreducibility of X, this implies that for all index $I > 0$ there are at least two states $i \in C_I$ and $j \in C_{I-1}$ such that $A_{i,j} > 0$.

Following the examples used a few lines earlier, in the case of a queuing model or a Petri net, this means, for instance, that simultaneous departures are not allowed. In the dependability example, the condition says that simultaneous repairs are not allowed.

For each state $j \in in(G)$, we now consider the following aggregation of $X^{(j)}$. We define a continuous-time homogeneous Markov chain $X^{(j)\text{agg}}$, which is constructed from $X^{(j)}$ by collapsing the classes $C_K, C_{K+1}, \ldots, C_M$ of the partition into single states $c_K, c_{K+1}, \ldots, c_M$.

If we denote by $A^{(j)\text{agg}}$ the infinitesimal generator of $X^{(j)\text{agg}}$, recalling that $\pi^{(j)}(C_I)$ is the probability that $X^{(j)}_\infty$ belongs to C_I, the transition rates of $X^{(j)\text{agg}}$ are given by the following expressions:

- for all $h \in G$ (or $h \in out(G)$), and for all $I \geq K$,

$$A^{(j)\text{agg}}_{h,c_I} = \sum_{i \in C_I} A_{h,i},\tag{10.4}$$

- for any $I > K$,

$$A^{(j)\text{agg}}_{c_I,c_{I-1}} = \mu_I^{(j)} = \frac{\sum_{i \in C_I} \pi_i^{(j)} \sum_{l \in C_{I-1}} A_{i,l}}{\pi^{(j)}(C_I)},\tag{10.5}$$

$$A^{(j)\text{agg}}_{c_K,j} = \mu_K^{(j)} = \frac{\sum_{i \in C_K} \pi_i^{(j)} \sum_{l \in C_{K-1}} A_{i,l}}{\pi^{(j)}(C_K)},\tag{10.6}$$

- for any $I \geq K$ and $J > I$,

$$A^{(j)\text{agg}}_{c_I,c_J} = \lambda_{I,J}^{(j)} = \frac{\sum_{i \in C_I} \pi_i^{(j)} \sum_{l \in C_J} A_{i,l}}{\pi^{(j)}(C_I)}.\tag{10.7}$$

We denote by $S^{\text{agg}} = G \cup \{c_K, c_{K+1}, \ldots, c_M\}$ the state space of $X^{(j)\text{agg}}$. Since $X^{(j)}$ has a unique recurrent class (Lemma 10.1.1), it is immediate to see that $X^{(j)\text{agg}}$ also has a unique recurrent class containing j. Let us denote $\pi_i^{(j)\text{agg}} = \mathbb{P}\{X^{(j)\text{agg}}_\infty = i\}$, $i \in S^{\text{agg}}$, where $X^{(j)\text{agg}}_\infty$ denotes a stationary version of $X^{(j)\text{agg}}$. We then have the following well-known result on aggregation (see Lemma 5.3.1):

for any $g \in G$, $\pi_g^{(j)\text{agg}} = \pi_g^{(j)}$, and for any $I \geq K$, $\pi_{c_I}^{(j)\text{agg}} = \pi^{(j)}(C_I)$.\tag{10.8}

The chain $X^{(j)\text{agg}}$ is called in [22] the exact aggregation of $X^{(j)}$ with respect to the given partition. We adopt this terminology here (in Section 5.3, it is called the pseudo-aggregation of $X^{(j)}$ with respect to the given partition). Of course, to build it we need the stationary distribution $\pi^{(j)}$ of $X^{(j)}$, which is unknown. We define in the next subsection another chain which bounds, in a specific way, chain $X^{(j)\text{agg}}$, and from which the desired bounds of R are computed.

10.1.5 Bounding chain

This section is based on [69]. For each state $j \in in(G)$, let us define a homogeneous and irreducible Markov chain $Y^{(j)}$ over S^{agg} with the same topology as the aggregated chain $X^{(j)\text{agg}}$, in the following way. We keep the same transition rates inside the subset G and from G to the c_I, $I \geq K$, which are computable without knowledge of the stationary distribution of $X^{(j)}$. The transitions inside the set of states $\tilde{G} = \{c_K, c_{K+1}, \ldots, c_M\}$ and from c_K to j are changed as follows: we replace $\lambda_{I,J}^{(j)}$, the unknown exact aggregated rate from c_I to c_J, $I < J$, by some $\lambda_{I,J}^+$; we replace the unknown exact aggregated rate from c_I to c_{I-1}, $I > K$, by some μ_I^-, and the unknown exact aggregated rate from c_K to j by some μ_K^-. Assume for the moment that we can find these new rates such that

$$\text{for } K \leq I < J \leq M, \quad \lambda_{I,J}^+ \geq \lambda_{I,J}^{(j)}, \tag{10.9}$$

$$\text{for } I \geq K, \quad 0 < \mu_I^- \leq \mu_I^{(j)}. \tag{10.10}$$

Then, between the three chains, $X^{(j)}$, $X^{(j)\text{agg}}$, and $Y^{(j)}$, the following relation holds.

THEOREM 10.1.4. *If we denote by $y^{(j)}$ the stationary distribution of $Y^{(j)}$, we have*

$$\hat{\pi}_G^{(j)} = \hat{\pi}_G^{(j)\text{agg}} = \hat{y}_G^{(j)}. \tag{10.11}$$

Proof. The equality $\hat{\pi}_G^{(j)} = \hat{\pi}_G^{(j)\text{agg}}$ follows from the definition of pseudo-aggregation, since we have the stronger relation $\pi_G^{(j)} = \pi_G^{(j)\text{agg}}$ (see (10.8)).

Now, observe that again by the properties of exact aggregation, it is enough to consider the case of $\bar{G}$ composed of a single state, that is, by building the exact aggregation of both $\pi^{(j)\text{agg}}$ and $y^{(j)}$ where we collapse all $\bar{G}$ into a single state, which we denote as $\bar{g}$. So, consider that the state space of $X^{(j)\text{agg}}$ and of $Y^{(j)}$ is $G \cup \bar{g}$. Denote by μ the transition rate $A_{\bar{g},j}^{(j)}$ in $X^{(j)}$, and by ν the transition rate from $\bar{g}$ to j in $Y^{(j)}$. Last, assume that the states of $X^{(j)}$ and $Y^{(j)}$ are indexed in the order $\{\ldots, j, \bar{g}\}$.

As in Lemma 5.1.2, if we decompose the state space as $(G, \bar{G})$, vector $\pi_G^{(j)}$ satisfies

$$\pi_G^{(j)} \left(A_G - A_{G\bar{G}} A_{\bar{G}}^{-1} A_{\bar{G}G}^{(j)} \right) = 0,$$

where $\bar{G} = \{g\}$ in this proof. This relation defines $\pi_G^{(j)}$ up to a scalar factor. Now,

$$A_{G\bar{G}} = -A_G \mathbb{1}, \quad A_{\bar{G}}^{-1} = (-1/\mu) \quad \text{and} \quad A_{\bar{G}G}^{(j)} = (0_G \ \mu).$$

This gives

$$A_G - A_{G\bar{G}} A_{\bar{G}}^{-1} A_{\bar{G}G} = A_G - (-A_G \mathbb{1})\left(-\frac{1}{\mu}\right)(0_G \ \mu)$$

$$= A_G - A_G \mathbb{1}(0_G \ 1),$$

thus independent of μ. Since the only difference between the transition rates of $X^{(j)}$ and $Y^{(j)}$ is the rate from $\tilde{g}$ to j, the result is proved. ∎

Let $Y^{(j)}_\infty$ denote a stationary version of $Y^{(j)}$. Since the modification of $X^{(j)}$ leading to $Y^{(j)}$ consists of "pushing the process towards $\tilde{G}$", we have the intuitively clear following bounding result.

THEOREM 10.1.5. *The number* $\mathbb{P}\{Y^{(j)}_\infty \in G\}$, *denoted in this chapter by* $y^{(j)}(G)$, *is bounded above by* $\pi^{(j)\mathrm{agg}}(G) = \pi^{(j)}(G)$:

$$y^{(j)}(G) \leq \pi^{(j)}(G). \tag{10.12}$$

If one of the inequalities (10.9) or (10.10) is strict, then $y^{(j)}(G) < \pi^{(j)}(G)$.

Proof. See the Appendix in [69]. ∎

Define over S^{agg} the two column reward vectors r_1 and r_2, obtained by completing column vector r_G with rewards on the aggregated states $c_K, c_{K+1}, \ldots, c_M$ in $\tilde{G}$ equal to ϱ_1 in r_1 and equal to ϱ_2 in r_2.

THEOREM 10.1.6.

$$\min_{j \in in(G)} y^{(j)} r_1 \leq R, \tag{10.13}$$

$$\max_{j \in in(G)} y^{(j)} r_2 \geq R. \tag{10.14}$$

Proof. Consider a reward column vector, r', with $r'_i = r_i$ if $i \in G$, and $r'_i = \varrho$ if $i \in \tilde{G}$ for some $\varrho \geq 0$. Rewrite the scalar product $y^{(j)} r'$ as follows:

$$y^{(j)} r' = y^{(j)}_G r_G + y^{(j)}_{\tilde{G}} \varrho \mathbb{1}_{\tilde{G}}$$

$$= y^{(j)}(G) \hat{y}^{(j)}_G r_G + \varrho y^{(j)}(\tilde{G})$$

$$= y^{(j)}(G) \hat{\pi}^{(j)}_G r_G + \varrho(1 - y^{(j)}(G)) \tag{10.15}$$

$$= y^{(j)}(G)(\hat{\pi}^{(j)}_G r_G - \varrho) + \varrho, \tag{10.16}$$

where we used Theorem 10.1.4 in the third equality. If we use in the last equality $\varrho = \varrho_1$ where ϱ_1 satisfies (10.1), then $\hat{\pi}^{(j)}_G r_G - \varrho_1 \geq 0$, which means that if we replace $y^{(j)}(G)$ by $\pi^{(j)}_G(G)$, from Theorem 10.1.5 we obtain an upper bound of $y^{(j)} r'$. We have

$$y^{(j)} r_1 \leq \pi^{(j)}(G)(\hat{\pi}^{(j)}_G r_G - \varrho_1) + \varrho_1$$

$$= \pi^{(j)}_G r_G + \varrho_1(1 - \pi^{(j)}(G))$$

$$= \pi^{(j)}_G r_G + \sum_{i \in \tilde{G}} \varrho_1 \pi^{(j)}_i$$

$$\leq \pi^{(j)}_G r_G + \sum_{i \in \tilde{G}} r_i \pi^{(j)}_i \quad \text{by (10.1)}$$

$$= \pi^{(j)} r = R^{(j)}.$$

Recall that by Corollary 10.1.3, $\min_{j\,in(G)} R^{(j)}$ is a lower bound of R. So, the procedure consists of taking

$$R_{lb} = \min_{j\,in(G)} y^{(j)} r_1 \tag{10.17}$$

as the effective lower bound of the mean expected reward.

In the same way, consider again (10.16), but now set $\varrho = \varrho_2 \geq r_i$ for all $i \in S$. Now, $\hat{\pi}_G^{(j)} r_G - \varrho_2 \leq 0$. Then, if we replace as before $y^{(j)}(G)$ by $\pi_G^{(j)}(G)$, we obtain something smaller:

$$y^{(j)} r_2 \geq \pi^{(j)}(G)\left(\hat{\pi}_G^{(j)} r_G - \varrho_2\right) + \varrho_2$$

$$= \pi_G^{(j)} r_G + \varrho_2(1 - \pi^{(j)}(G))$$

$$= \pi_G^{(j)} r_G + \sum_{i\in\tilde{G}} \varrho_2 \pi_i^{(j)}$$

$$\geq \pi_G^{(j)} r_G + \sum_{i\in\tilde{G}} r_i \pi_i^{(j)} \quad \text{(by (10.1))}$$

$$= \pi^{(j)} r = R^{(j)}.$$

This leads to the following upper bound of R:

$$R_{ub} = \max_{j\,in(G)} y^{(j)} r_2, \tag{10.18}$$

which completes the proof. ∎

The bounds of R are obtained using (10.17) and (10.18). To do this, we must be able to build chain $Y^{(j)}$, that is, to build bounds $\lambda_{I,J}^+$ and μ_I^- of the corresponding unknown transition rates in $X^{(j)\text{agg}}$ as set in (10.9) and (10.10). We now describe a situation where this can be easily done [69]. The next section describes a more powerful technique, which has the same goal but doesn't need the strong assumption of [69].

First way of building bounds $\lambda_{I,J}^+$ and μ_I^-
The first bounds to try are simply

$$\forall I,J \text{ such that } K \leq I < J \leq M, \quad \lambda_{I,J}^+ = \max_{i\in C_I} \sum_{l\in C_J} A_{i,l}, \tag{10.19}$$

$$\forall I \text{ such that } K \leq I \leq M, \quad \mu_I^- = \min_{i\in C_I} \sum_{l\in C_{I-1}} A_{i,l}. \tag{10.20}$$

This is the idea followed in [69]. The use of Relation (10.19) immediately implies that $\lambda_{I,J}^+ > 0$. Let us examine now the bounds μ_I^-. In order to have $\mu_I^- > 0$ for all $I > K$, and for any value of $K < M$, we need a supplementary condition to be satisfied by X:

Condition 2. *For any index $I \neq 0$, for all states $i \in C_I$ there exists at least one state $j \in C_{I-1}$ such that $A_{i,j} > 0$.*

This can be quite restrictive as we will illustrate later, but the interest relies on the fact that it can be used to obtain the direct lower bounds of the $\mu_I^{(j)}$. In Subsection 10.1.6, we describe a different approach that does not need this assumption, so it can work with much more general models.

Illustrating the first way of building bounds $\lambda_{I,J}^+$ and μ_I^-

Let us illustrate this using a very simple model, the $E_2/M/1/H$ queuing system. The inter-arrivals are distributed according to the Erlang distribution with rank or order 2 and parameter η, that is, with density at $t \geq 0$ given by $\eta^2 t \exp(-\eta t)$. The service distribution is exponential with parameter μ. The queue can accept up to H customers. If N_t is the number of customers in the queue at time t ($0 \leq N_t \leq H$) and if S_t is the phase or stage of the inter-arrival distribution ($S_t \in \{1,2\}$, see for instance Chapter 1, page 16, for the first use of the phase concept in this book), then the process (N_t, S_t) is a CTMC with the state space $\{0, 1, \ldots, H\} \times \{1, 2\}$. We order the states as shown in Figure 10.3.

We naturally define $C_I = \{(I,1), (I,2)\}$, that is, C_I is the set of states corresponding to I customers in the system. Note that Condition 2 holds, and thus we can use $\lambda_{I,I+1}^+ = \eta$ and $\mu_I^- = \mu$.

In the bounding procedure, Figure 10.4 shows one of the $Y^{(j)}$, corresponding to the choice $K = 2$ and $j = (2,1)$.

To illustrate the procedure numerically, let us consider the metric $M = $ mean number of customers in equilibrium. This means that $r_{(h,s)} = h$, $h = 0, 1, \ldots, H$, $s = 1, 2$. In Table 10.1 we present some numerical results.

Observe that in the first case considered in Table 10.1 ($\eta = 3.14$, $\mu = 3$, $H = 500$, a light traffic situation), for $K = 35$ and using six significant digits, the lower and upper

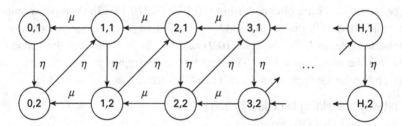

Figure 10.3   A Markovian representation of the $E_2/M/1/H$ queue

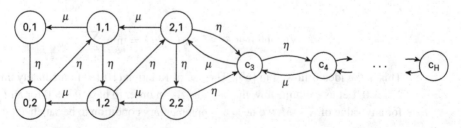

Figure 10.4   A bounding chain of the $E_2/M/1/H$ queue; here, $K = 2$ and entries to G are forced to be done by state $(2,1)$

Table 10.1. Bounding the mean number of customers in equilibrium, M, in the $E_2/M/1/H$ queue. The phase rate of the E_2 inter-arrival time distribution is η and the service rate is μ. The load is $\eta/(2\mu)$.

η	μ	load	H	K	M_{lb}	M	M_{ub}
3.14	3	0.5233	500	30	0.884832	0.884832	15.259886
3.14	3	0.5233	500	35	0.884832	0.884832	1.018087
3.14	3	0.5233	500	40	0.884832	0.884832	0.886032
1	1	0.5	1000	10	0.808961	0.809017	17.775462
1	1	0.5	1000	20	0.809017	0.809017	0.810146
1	1	0.5	1000	30	0.809017	0.809017	0.809017

bounds coincide. The same happens in the second case ($\eta = \mu = 1$, $H = 1000$, again a light traffic situation), now for $K = 30$. We can also see that the lower bound is more accurate than the upper bound, for the same computational effort. Even though this is a trivial example, for which the exact value is easily computable, it illustrates the power of the idea, when the chain "basically lives" in a small part of its state space.

10.1.6 A more general method

The goal in this subsection is to derive a method to avoid Condition 2 and still be able to bound the measure, R. We start by introducing the idea informally. Consider a birth and death process (over the nonnegative integers) and denote by λ_i (resp. μ_i) the birth rate (resp. death rate) associated with state $i \geq 0$ (resp. $i \geq 1$). The mean sojourn time in state $i \geq 1$ (or mean holding time) is $h_i = 1/(\lambda_i + \mu_i)$ and the probability that after visiting state i, the next visited state is $i - 1$, is $p_i = \mu_i/(\lambda_i + \mu_i)$. Observe then that

$$\mu_i = \frac{p_i}{h_i}. \tag{10.21}$$

The intuitive idea leading to our bounding technique is to write the aggregated rate in $X^{(j)}$ from class C_I from C_{I-1}, that is, the transition rate from c_I to c_{I-1} in $X^{(j)\text{agg}}$, $\mu_I^{(j)}$, in a similar form to (10.21). In the next subsection we write it as

$$\mu_I^{(j)} = \frac{p_I^{(j)}}{h_I^{(j)}},$$

and we derive expressions of $p_I^{(j)}$ and $h_I^{(j)}$ allowing us to obtain a bound of $\mu_I^{(j)}$.

Sojourn times and aggregation of states

We suggest that you read Chapter 5 before this point. Let us denote by $H_{I,n}^{(j)}$ the length of the nth sojourn of $X_\infty^{(j)}$ in class C_I. The first visited state of C_I during this sojourn is denoted by $V_{I,n}^{(j)}$ ($V_{I,n}^{(j)} \in in(C_I)$) and after leaving C_I, the next visited state is denoted by $W_{I,n}^{(j)}$ ($W_{I,n}^{(j)} \in in(\bar{C}_I)$).

The distribution of $V_{I,n}^{(j)}$, as a row vector $v_I^{(j)}(n)$ defined over C_I, can be written as follows (expression given in Theorem 5.2.1):

$$v_I^{(j)}(n) = v_I^{(j)}(1)(B_{C_I}^{(j)})^{n-1},$$

where $B_{C_I}^{(j)} = A_{C_I}^{-1}A_{C_I,\bar{C}_I}^{(j)}A_{\bar{C}_I}^{-1}A_{\bar{C}_I,C_I}^{(j)}$ and $v_I^{(j)}(1) = \pi_{C_I}^{(j)} - \pi_{\bar{C}_I}^{(j)}A_{\bar{C}_I}^{-1}A_{\bar{C}_I,C_I}^{(j)}$. Of course for all $n > 1$, $v_I^{(j)}(n)$ has non-zero entries only on states i belonging to $in(C_I)$.

The distribution of $H_{I,n}^{(j)}$ is given by the following expression (Theorem 5.2.1):

$$\mathbb{P}\{H_{I,n}^{(j)} > t\} = v_I^{(j)}(n)\exp(A_{C_I}t)\mathbb{1},$$

and its mean is

$$\mathbb{E}\{H_{I,n}^{(j)}\} = -v_I^{(j)}(n)A_{C_I}^{-1}\mathbb{1}.$$

In Corollary 5.1.7, it is in particular shown that vector

$$v_I^{(j)} = \frac{\pi_{C_I}^{(j)}A_{C_I}}{\pi_{C_I}^{(j)}A_{C_I}\mathbb{1}} \tag{10.22}$$

is the stationary distribution of the Markov chain $(V_{I,n}^{(j)})_n$, that is, if $v_I^{(j)}(1) = v_I^{(j)}$ then $v_I^{(j)}(n) = v_I^{(j)}$ for all $n \geq 1$. We denote by $v_{I,i}^{(j)}$ the component of $v_I^{(j)}$ corresponding to state $i \in C_I$. Observe that $v_G^{(j)} = e_j$, where e_j is the jth row vector of the canonical base in $\mathbb{R}^{|G|}$.

Let us consider now some relationships between chains $X^{(j)}$ and $X^{(j)\mathrm{agg}}$ from the sojourn time point of view. We denote by $h_I^{(j)}$ the mean holding time of $X^{(j)\mathrm{agg}}$ in state c_I, $K \leq I \leq M$, that is,

$$\text{for } I \geq K, \quad h_I^{(j)} = \frac{1}{\mu_I^{(j)} + \sum_{J>I}\lambda_{I,J}^{(j)}}. \tag{10.23}$$

In the following, we need the result given in the following lemma:

LEMMA 10.1.7.

$$\textit{For any } I \geq K, \quad h_I^{(j)} = -v_I^{(j)}A_{C_I}^{-1}\mathbb{1}. \tag{10.24}$$

Proof. The proof is straightforward (see Subsection 5.3.3). ∎

Lemma 10.1.7 says that the mean holding time of $X^{(j)\mathrm{agg}}$ in c_I is equal to the mean sojourn time of $X^{(j)}$ in C_I when it enters C_I by state i with probability $v_{I,i}^{(j)}$ (for instance, think of the first sojourn in C_I of a version of $X^{(j)}$ having as initial distribution the vector $\alpha^{(j)}$, such that $\alpha_{C_I}^{(j)} = v_I^{(j)}$).

Let us denote by $\hat{h}_{i,I}$ the mean sojourn time of $X^{(j)}$ in C_I conditional on the fact that the process enters the set C_I by state i. Observe that, for all $j \in in(G)$,

$$\hat{h}_{i,I} = \mathbb{E}\{H_{I,n}^{(j)} \mid V_{I,n}^{(j)} = i\} \text{ for all } n \geq 1. \tag{10.25}$$

From Relations (10.24) and (10.25), we can write

$$h_I^{(j)} = \sum_{i\in in(C_I)} v_{I,i}^{(j)} \hat{h}_{i,I}. \tag{10.26}$$

Now, for the purposes of this section, we have to consider the event "when the nth sojourn of $X^{(j)}$ in C_I ends, the next visited state belongs to C_{I-1}", that is, $\{W_{I,n}^{(j)} \in C_{I-1}\}$. It is straightforward to verify that the probability of this event is

$$\text{for } I > K, \quad \mathbb{P}\{W_{I,n}^{(j)} \in C_{I-1}\} = -v_I^{(j)}(n)A_{C_I}^{-1}A_{C_I,C_{I-1}}\mathbb{1}.$$

When $I = K$, we also have, for any $j \in in(G)$,

$$\mathbb{P}\{W_{K,n}^{(j)} = j\} = -v_K^{(j)}(n)A_{C_K}^{-1}A_{C_K,C_{K-1}}^{(j)}\mathbb{1},$$

where $A_{C_K,C_{K-1}}^{(j)}$ is the matrix equal to $A_{C_K,C_{K-1}}\mathbb{1}e_j$, and e_j is the jth row vector of the canonical base in $\mathbb{R}^{|C_{K-1}|}$.

The event similar to $\{W_{I,n}^{(j)} \in C_{I-1}\}$ in the aggregated chain $X^{(j)\text{agg}}$ is "when leaving state c_I, the chain jumps to c_{I-1}". Its probability is

$$p_I^{(j)} = \frac{\mu_I^{(j)}}{\mu_I^{(j)} + \sum_{J>I} \lambda_{I,J}^{(j)}}.$$

Observe that the transition rate $\mu_I^{(j)}$ from c_I to c_{I-1} in $X^{(j)\text{agg}}$ is

$$\mu_I^{(j)} = \frac{p_I^{(j)}}{h_I^{(j)}}. \tag{10.27}$$

The following result, similar to Lemma 10.1.7, holds.

LEMMA 10.1.8. *For any $I \geq K$, $p_I^{(j)} = -v_I^{(j)}A_{C_I}^{-1}A_{C_I,C_{I-1}}\mathbb{1}$.*

Proof. Let $I > K$. From (10.7) and from (10.5), we can write

$$\lambda_{I,J}^{(j)} = \frac{\pi_{C_I}^{(j)}A_{C_I,C_J}\mathbb{1}}{\pi_{C_I}^{(j)}\mathbb{1}}, \quad \mu_I^{(j)} = \frac{\pi_{C_I}^{(j)}A_{C_I,C_{I-1}}\mathbb{1}}{\pi_{C_I}^{(j)}\mathbb{1}}.$$

Then, since $A_{C_I,\bar{C}_I}\mathbb{1} = -A_{C_I}\mathbb{1}$,

$$\mu_I^{(j)} + \sum_{J>I}\lambda_{I,J}^{(j)} = \frac{\pi_{C_I}^{(j)}A_{C_I,\bar{C}_I}\mathbb{1}}{\pi_{C_I}^{(j)}\mathbb{1}}$$

$$= -\frac{\pi_{C_I}^{(j)}A_{C_I}\mathbb{1}}{\pi_{C_I}^{(j)}\mathbb{1}}.$$

This leads to

$$p_I^{(j)} = \frac{\mu_I^{(j)}}{\mu_I^{(j)} + \sum_{J>I} \lambda_{I,J}^{(j)}}$$

$$= \frac{\pi_{C_I}^{(j)} A_{C_I,C_{I-1}} \mathbb{1}}{\pi_{C_I}^{(j)} \mathbb{1}} \left(-\frac{\pi_{C_I}^{(j)} \mathbb{1}}{\pi_{C_I}^{(j)} A_{C_I} \mathbb{1}} \right)$$

$$= -\frac{\pi_{C_I}^{(j)} A_{C_I,C_{I-1}} \mathbb{1}}{\pi_{C_I}^{(j)} A_{C_I} \mathbb{1}}$$

$$= -v_I^{(j)} A_{C_I}^{-1} A_{C_I,C_{I-1}} \mathbb{1}.$$

The case of class C_K is proven in the same way. ∎

As for Lemma 10.1.7, Lemma 10.1.8 says that the probability that $X^{(j)\mathrm{agg}}$ will jump to c_{I-1} when leaving c_I is the same as the conditional probability for $X^{(j)}$ to jump to C_{I-1} when leaving C_I, given that $X^{(j)}$ enters C_I by state i with probability $v_{I,i}^{(j)}$. This implies that, if we denote by $\hat{p}_{i,I}$ the conditional probability that $X^{(j)}$ jumps to C_{I-1} when leaving C_I, given that the sojourn started in state $i \in C_I$, we have

$$p_I^{(j)} = \sum_{i \in in(C_I)} v_{I,i}^{(j)} \hat{p}_{i,I}. \tag{10.28}$$

The bounding algorithm

Let us assume that Condition 2 is not satisfied. The problem is the computation of the lower bounds of the $\mu_I^{(j)}$. To obtain the bounds of R given in (10.13) and (10.14), we proceed as follows.

First, let us consider a new subset of states of C_I, $in(C_I)^*$, which is the set of entry points i of C_I such that if $X^{(j)}$ enters C_I by i, there is a non-zero probability that the next visited class is C_{I-1}:

$$in(C_I)^* = \{i \in in(C_I) \mid \hat{p}_{i,I} > 0\}.$$

Our method then requires the two sets $in(C_I)^*$ and $in(C_I)$ to be equal. Let us put it explicitly.

Condition 3.   *For all $I \geq 1$ (or for all I large enough), and for all $i \in in(C_I)$, the probability to jump from C_I to C_{I-1} when the sojourn in C_I starts in i is non-zero, i.e., $in(C_I)^* = in(C_I)$.*

This condition is obviously much less restrictive than Condition 2. We did not find any realistic model where it does not hold. Observe that if Condition 3 holds for all $I \geq I_0$, for some I_0, then the bounding level, K, needs to satisfy $K \geq I_0$ as well.

Under Condition 3, the lower bounds of the $\mu_I^{(j)}$ are given in the following result.

THEOREM 10.1.9.   *For all $I \geq K$, for all $j \in in(G)$,*

$$\mu_I^* = \min_{i \in in(C_I)} \frac{\hat{p}_{i,I}}{\hat{h}_{i,I}} \leq \mu_I^{(j)}. \tag{10.29}$$

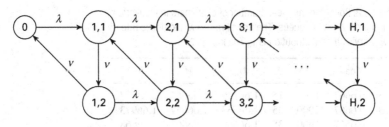

Figure 10.5   A Markovian representation of the $M/E_2/1/H$ queue

Proof. The result simply follows from Relations (10.26) and (10.28), writing that, for any $I \geq K$,

$$\mu_I^{(j)} = \frac{p_I^{(j)}}{h_I^{(j)}} = \frac{\sum_{i \in in(C_I)} v_{I,i}^{(j)} \hat{p}_{i,I}}{\sum_{i \in in(C_I)} v_{I,i}^{(j)} \hat{h}_{i,I}},$$

and then using the fact that $\sum_{i \in C_I} v_{I,i}^{(j)} = 1$.   ∎

Illustration using the $M/E_2/1/H$ queuing model

To illustrate this, let us go back to simple queues, but consider now the $M/E_2/1/H$ model, where the arrival rate is λ and the service time distribution has density equal to $v^2 t \exp(-vt)$ for any $t \geq 0$. As before with the $E_2/M/1/H$ system, we use the CTMC (N_t, S_t) on the state space $\{1, 2, \ldots, H\} \times \{1, 2\} \cup \{0\}$ (when the queue is empty, the service phase is not defined). We order the states as shown in Figure 10.5.

We define the same partition of the state space as before, for the $E_2/M/1/H$ queue: $C_0 = \{0\}$, $C_1 = \{(1,1), (1,2)\}$, $C_2 = \{(2,1), (2,2)\}, \ldots, C_H = \{(H,1), (H,2)\}$. See that Condition 2 does not hold, but Condition 3 does. To derive the bounds, consider first class C_I, $I \geq 2$. From state $(I,2)$, the probability of leaving C_I to go to the left, to class C_{I-1} (that is, to state $(I-1,1)$), is

$$\hat{p}_{(I,2),I} = \frac{v}{\lambda + v}.$$

The mean sojourn time in C_I when the process enters C_I by state $(I,2)$ is

$$\hat{h}_{(I,2),I} = \frac{1}{\lambda + v}.$$

When the process enters C_I by state $(I,1)$, we have

$$\hat{p}_{(I,1),I} = \frac{v}{\lambda + v} \qquad \hat{p}_{(I,2),I} = \left(\frac{v}{\lambda + v}\right)^2.$$

For the mean sojourn time, we have

$$\hat{h}_{(I,1),I} = \frac{1}{\lambda + v} + \frac{v}{\lambda + v} \hat{h}_{(I,2),I} = \frac{\lambda + 2v}{(\lambda + v)^2}.$$

This gives

$$\frac{\hat{p}_{(I,2),I}}{\hat{h}_{(I,2),I}} = v, \qquad \frac{\hat{p}_{(I,1),I}}{\hat{h}_{(I,1),I}} = \frac{v^2}{\lambda + 2v},$$

Table 10.2. Bounding the mean number of customers in equilibrium, M, in the $M/E_2/1/H$ queue. The arrival rate is λ and phase rate of the E_2 service time distribution is ν. The load is $2\lambda/\nu$.

λ	ν	load	H	K	M_{lb}	M	M_{ub}
1	3	0.6667	1000	15	1.66229	1.66667	2.27577
1	3	0.6667	1000	25	1.66663	1.66667	1.66973
1	3	0.6667	1000	35	1.66667	1.66667	1.66668
2	5	0.8	1000	15	2.88489	3.20000	79.33185
2	5	0.8	1000	30	3.19407	3.20000	3.69175
2	5	0.8	1000	45	3.19991	3.20000	3.20597

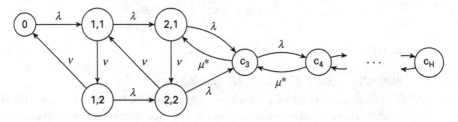

Figure 10.6   A bounding chain for the $M/E_2/1/H$ queue

and we have

$$\mu_I^* = \min\left(\nu, \frac{\nu^2}{\lambda+2\nu}\right) = \nu \min\left(1, \frac{\nu}{\lambda+2\nu}\right) = \frac{\nu^2}{\lambda+2\nu}.$$

Note that in this case, to enter $G = \bigcup_{I \leq K} C_I$ from $\bar{G}$ there is only one possibility, state $(K, 1)$. So, there is only one auxiliary process $Y^{(j)}$, once K has been chosen. Figure 10.6 illustrates this, for the choice $K = 2$. In the picture, we denote μ_I^* simply by μ^*.

As for the case of the $E_2/M/1/H$, this example illustrates the capability of these bounding approaches to evaluate standard metrics by working in a small part of the state space, if the chain spends "most of its life" there. Table 10.2 provides some numerical examples of the quality of the bounds.

Resuming the algorithm

Let us now resume the algorithm. The input data are the partition and a given level, K. The bounding technique consists of executing the following steps:

- Once the partition and the threshold, K, are fixed, compute the starred bounds given by (10.29). To do this,

  – for each class $C_I, I \geq K$, compute the $\hat{p}_{i,I}$ and the $\hat{h}_{i,I}$; alternatively, lower bounds of the $\hat{p}_{i,I}$ and upper bounds of the $\hat{h}_{i,I}$ can be used;
  – compute μ_I^* using (10.29).

- Generate G and then, for any $j \in in(G)$, find the stationary distribution, $y^{(j)}$, of chain $Y^{(j)}$ with the choice $\mu_I^- = \mu_I^*$. Possibly, use the techniques in [71] or [15] to reduce the number of linear systems to solve.
- Compute the lower and upper bounds of R using (10.17) and (10.18).

The main drawback of this algorithm is that the computation of the $\hat{p}_{i,I}$ and the $\hat{h}_{i,I}$ may be numerically intractable due to the possible size of class C_I. Even if they can be calculated, the induced cost may be too high for the user. A possible way to handle this problem is to try to obtain new bounds on these numbers. We do not explore this issue further in this book. In this section, we want only to illustrate the use of these approaches in cases where deriving the bounds can be done analytically. However, we can note that if Condition 2 holds, the bounds obtained by the new algorithm are in general better than those of [69], as stated in next result.

LEMMA 10.1.10. *If Condition 2 holds (which implies that Condition 3 holds as well), then for all $I \geq K$, $\mu_I^- \leq \mu_I^* \leq \mu_I$.*

Proof. From Lemmas 10.1.7 and 10.1.8, we have:

$$\hat{h}_{i,I} = -e_i A_{C_I}^{-1} \mathbb{1}, \tag{10.30}$$

$$\hat{p}_{i,I} = -e_i A_{C_I}^{-1} A_{C_I, C_{I-1}} \mathbb{1}. \tag{10.31}$$

Let us consider vector $A_{C_I, C_{I-1}} \mathbb{1}$. Each one of its components is greater than $\min_{i \in C_I} \sum_{j \in C_{I-1}} A_{i,j}$, that is, greater than μ_I^-. From the relations above, we have

$$\hat{p}_{i,I} \geq \mu_I^- \hat{h}_{i,I}$$

which ends the proof. ∎

Observe that in the previously discussed $E_2/M/1/H$ queue, the particular simple structure of the model deals with the fact that both methods are identical: if we analyze class C_I for the $E_2/M/1/H$ model, we obtain $\mu^* = \mu$.

10.1.7 Illustration

This section illustrates the efficiency of the bounding method proposed using a standard dependability model, a Machine Repairman Model (MRM), which leads to a large finite Markov chain that can not be handled by the technique published in [69]. In [62] there are more examples, in particular performance-oriented ones with infinite state spaces.

Bounding the asymptotic availability of an MRM

Our example is a standard multicomponent system subject to failures and repairs. There are two types of component. The number of components of type k is denoted by N_k and their time to failure is exponentially distributed with parameter λ_k, $k = 1, 2$. Think for instance of a communication network where the components are nodes or lines. In such

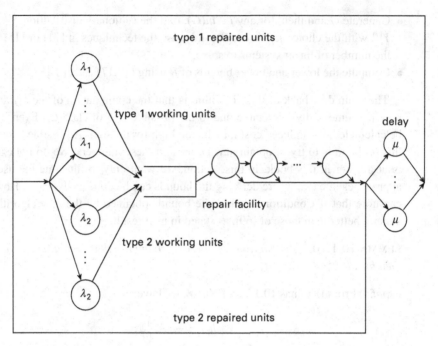

Figure 10.7   A Machine Repair Model

a system we can easily find a large number of components leading in turn to models with huge state spaces.

After a failure, the components enter a repair facility with one server and repair time distributed according to a d_k-phase Coxian distribution for type k machines, with mean m_k. Type 1 components are served with higher priority than type 2, and the priority is non-preemptive. We assume that type 2 units are immediately returned to operation when repaired, but that type 1 ones need a delay exponentially distributed with parameter μ to return to operation. Thus, in the model, type 1 customers go to a second infinite servers queue. This allows us to illustrate the method when the repair subsystem is more complex than a single queue. The example is pictured in Figure 10.7.

To define the state space, we use n_1 (resp. n_2) to represent the number of machines of type 1 (resp. 2) in the repair queue; we denote by k the type of the machine being repaired (with value 0 if the repair station is empty) and by d the phase of the current service with $1 \leq d \leq d_k$ ($d = 0$ if the repair station is empty). Denote by n_3 the number of machines of type 1 in transit (that is, in the delay station). We are interested in bounding the asymptotic availability of the system. Let us assume that the system is operational as soon as there are at least $nMin_1$ machines of type 1 and at least $nMin_2$ machines of type 2 operating.

On states $s = (n_1, n_2, d, k, n_3)$ we have a Markov chain on which we consider the subsets of states C_I defined by

$$C_I = \{s \mid n_1 + n_2 + n_3 = I\}$$

with $0 \leq I \leq N_1 + N_2$. They define a partition of the state space. Observe that Condition 2 is not satisfied and thus that the method in [69] can not be applied. On the contrary, Condition 3 holds.

Let us consider the following parameter values:

- $N_1 = 80$, $N_2 = 120$, $nMin_1 = 79$, $nMin_2 = 115$, $\lambda_1 = 0.00004$, $\lambda_2 = 0.00003$, $m_1 = m_2 = 1.0$, $d_1 = 6$, $d_2 = 5$, and $\mu = 3.0$. The size of the whole state space is $|S| = 4\,344\,921$. Using two small values of K, we obtain the following numerical results:

K	$|S^{\mathrm{agg}}|$	Lower bound	Upper bound
5	226	0.9997597121	0.9997597466
10	1826	0.9997597349	0.9997597349

- If we change the definition of the operational system to allow 77 type 1 units as the threshold, that is, if we change $nMin_1$ to $nMin_1 = 77$, we have again a state space S with cardinality $|S| = 4\,344\,921$ and for the same values of K we obtain

K	$|S^{\mathrm{agg}}|$	Lower bound	Upper bound
5	226	0.9999999698	0.9999999852
10	1826	0.9999999841	0.9999999841

As we see, the change in the meaning of the operational system leads to a significant change in the asymptotic availability metric.

As a technique to check the software used, let us consider the following situation. Let us keep the previous example with the values $N_1 = 80$ and $N_2 = 120$. We consider Coxian distributions for the repair times with two phases or stages, that is, $d_1 = d_2 = 2$, but we choose their parameters in such a way that they are equivalent to exponentially distributed service times: if $v_{k,d}$ is the parameter of the dth stage for a type k component and if l_d is the probability that phase d is the last one ($l_{d_k} = 1$), then for all phase d we have $v_{k,d}l_d = 1/m_k$ (technically, we put ourselves in a *strong lumpability* situation, see Chapter 4). Moreover, if the scheduling of the repair facility is changed to preemptive priorities, then type 2 units are invisible to type 1 ones, and with respect to type 1 components we have a product form queuing network. Standard algorithms can then be used to compute, for instance, the mean number of type 1 machines in the repair subsystem, which we denote by $\bar{N}_1$. Using the QNAP2 product of Simulog, we obtain for $\lambda_1 = 0.00004$, $m_1 = 0.2$, and $\mu = 3.0$, the value $\bar{N}_1 = 0.0006404$. Using our algorithm with $K = 3$, we obtain $0.0006403 < \bar{N}_1 < 0.0006413$.

10.1.8 Conclusions

This first part shows how to obtain upper and lower bounds of asymptotic performability measures. The asymptotic performability includes as particular cases the asymptotic availability in a dependability context, or standard asymptotic performance measures such as the mean number of customers, blocking probabilities, and loss probabilities, in performance evaluation. To be applied, the methods presented here need enough knowledge of the structure of the model to be able to derive analytically or to evaluate numerically certain values which are necessary to obtain the bounds. This is not always possible, but the methodologies described here illustrate what is possible using usually available information about the model. The next part of the chapter describes a completely different situation and procedure, with the same objective of bounding a dependability metric using a model too huge to be analyzed exactly.

10.2   Second part: Bounding the mean time to absorption

In this second part of the chapter, we explore a very different situation: how to bound the MTTF of a highly reliable system, which translates into the problem of bounding the mean time to absorption of an absorbing continuous-time Markov chain.

Bounding the MTTF when the model is large and the system is highly reliable is an almost unexplored problem. Here, the system is represented by a homogeneous continuous-time Markov chain, $X = \{X_t, t \geq 0\}$, with finite state space $\{1, \ldots, M, M+1\}$ and infinitesimal generator $(Q_{i,j})$. State $M+1$ is absorbing, and states $1, \ldots, M$ are transient. This means that from any state $i \leq M$ there is a path to state $M+1$, that is, a sequence of states $(i, i_1, i_2, \ldots, i_k, M+1)$ (reduced to $(i, M+1)$ if $k = 0$) with $Q_{i,i_1} Q_{i_1,i_2} \cdots Q_{i_k,M+1} > 0$. With these assumptions, the chain is absorbed (with probability 1) after a random finite time T, called the absorption time. Formally, $T = \inf\{t \geq 0 \mid X_t = M+1\}$ and $\mathbb{P}(T < \infty) = 1$. We denote by $q_i = \sum_{j \neq i} Q_{i,j}$ the output or departure rate of state i. It is useful to give a name to the set of transient states. We set $\Omega = \{1, 2, \ldots, M\}$.

Associated with each state $i \leq M$, there is a reward, r_i, which is a real number interpreted as the rate at which the system generates a benefit while in state i (or at which the system cumulates a cost). Setting $r_{M+1} = 0$, the cumulated reward up to absorption is the random variable

$$R_\infty = \int_0^\infty r_{X_t} \, dt.$$

When for all $i \leq M$ we set $r_i = 1$, we have $R_\infty = T$. When X represents the state of a system subject to failures and repairs, if states $1, \ldots, M$ represent the system working as specified and $M+1$ represents a failed system, $\mathbb{E}\{T\}$ is also called the Mean Time To Failure of the system (MTTF).

The goal of the analysis developed here is the evaluation of the expectation of the random variable R_∞, called the Expected Cumulated Reward up to Absorption (ECRA):

$$\text{ECRA} = \mathbb{E}\{R_\infty\}.$$

10.2.1   Exact analysis of the ECRA metric

Let us denote by P the restriction to Ω of the transition probability matrix of the discrete Markov chain canonically embedded at the transition epochs of X (that is, if Q is the restriction of $(Q_{i,j})$ to Ω and if Λ is the diagonal matrix $\Lambda = \mathrm{diag}(q_i,\ i \leq M)$, then $P = I + \Lambda^{-1}Q$). Matrix P is strictly sub-stochastic (that is, the sum of the elements of at least one of its rows is strictly less than 1), otherwise state $M+1$ would not be reachable from Ω.

Let r be the column vector indexed by Ω whose ith entry is r_i/q_i. Let $\varrho_i = \mathbb{E}_i\{R_\infty\}$, where $\mathbb{E}_i$ denotes the expectation operator conditioned on the fact that the initial state is state $i \in \Omega$. Denote by ϱ the column vector indexed on Ω whose ith entry is ϱ_i. Since all states in Ω are transient, matrix $I - P$ is not singular and $\varrho = (I - P)^{-1}r$, that is, ϱ is the solution to the linear system $(I - P)\varrho = r$, with size $|\Omega|$. If $\alpha_i = \mathbb{P}\{X_0 = i\}$, with $\alpha_{M+1} = 0$, and if α is the vector indexed by Ω whose ith entry is α_i, then ECRA $= \sum_{i=1}^{M} \alpha_i \varrho_i$.

So, computing the ECRA metric is a linear problem. This section describes a scheme allowing the computation of the lower and upper bounds of this expectation by working with linear systems much smaller than $|\Omega|$ in size. The method is designed for the case of $|\Omega| \gg 1$ and when the chain is *stiff* (this typically happens when X models a highly dependable system).

10.2.2 Fast and slow transitions and states

We say that the transition (i,j) is a *fast* one if $Q_{i,j} \geq \theta > 0$ where θ is a given positive threshold. It is a *slow* one if $0 < Q_{i,j} < \theta$. A state $i \neq M+1$ is a *fast state* if at least one transition out of it is fast; otherwise, i is a *slow state* [7]. This defines a partition of the state space of X into three subsets: F, the set of fast states, S, the set of slow ones, and $\{M+1\}$. We assume that $F \neq \emptyset$ and $S \neq \emptyset$. Also recall that the distribution of X_0 has its support in S.

Let the transition rate matrix, $(Q_{i,j})$, be reordered so that the states belonging to S are numbered from 1 to $|S|$ and the states belonging to F from $|S|+1$ to $|S|+|F| = M$. Its submatrix, Q, can then be partitioned according to (S,F), as well as any other matrix indexed on Ω, and the same for vectors. For example,

$$Q = \begin{pmatrix} Q_S & Q_{SF} \\ Q_{FS} & Q_F \end{pmatrix} \quad \text{and} \quad r = \begin{pmatrix} r_S \\ r_F \end{pmatrix}.$$

If, for instance, we look at the model depicted in Figure 10.10 where ϕ is a failure rate and $\gamma \gg \phi$ is a repair rate, we naturally choose a threshold, θ, such that we have $S = \{7, 6, 5\}$ and $F = \{6', 5', 4, 3\}$.

Assumption: We assume that in the graph of chain X there is no circuit composed only of fast transitions. This is the usual situation in dependability models, where fast transitions are associated with repairs. In some models, however, the choice of the θ threshold can make this assumption invalid. See [7] for an approximation used to handle this problem.

10.2.3   Uniformization

Let us denote by $Y = \{Y(k), k \geq 0\}$ the discrete-time Markov chain associated with X, obtained by uniformization of the latter with respect to the uniformization rate $\eta \geq \max_{1 \leq i \leq M}\{q_i\}$. We denote by $U = (U_{i,j})$ the restriction of the transition probability matrix of Y to Ω (that is, $U = I + Q/\eta$). As for matrix Q, we similarly decompose U into the corresponding blocks using the analogous notation U_S, U_{SF}, U_{FS}, and U_F.

The number ECRA defined on X can be computed by dividing the corresponding metric on Y by η. Let us write a superscript Y to explicitly indicate that we are working on Y. Then,

$$R_\infty^Y = \sum_{k=0}^\infty r_{Y(k)},$$

$$\mathbb{E}\{R_\infty^Y\} = \alpha(I - U)^{-1}r$$

and

$$\text{ECRA} = \frac{1}{\eta}\mathbb{E}\{R_\infty^Y\} = \frac{1}{\eta}\alpha(I - U)^{-1}r.$$

All this means that we can work with chain Y to derive the ECRA metric defined on X. From this point, our goal is to derive bounds of $\mathbb{E}\{R_\infty^Y\}$.

10.2.4   Decomposing the cumulated reward

Let us write

$$R_\infty^{Y,S} = \sum_{k=0}^\infty r_{Y(k)}1_{\{Y(k)\in S\}}, \qquad R_\infty^{Y,F} = \sum_{k=0}^\infty r_{Y(k)}1_{\{Y(k)\in F\}}.$$

That is, $R_\infty^{Y,S}$ (resp. $R_\infty^{Y,F}$) is the reward cumulated on the slow states (resp. on the fast ones) by Y until absorption. Then obviously $R_\infty^Y = R_\infty^{Y,S} + R_\infty^{Y,F}$ and the same relation holds for the respective expectations.

Denoting

$$\varrho_i^{Y,S} = \mathbb{E}\{R_\infty^{Y,S} \mid Y(0) = i\}, \qquad \varrho_i^{Y,F} = \mathbb{E}\{R_\infty^{Y,F} \mid Y(0) = i\},$$

and

$$\varrho_i^Y = \mathbb{E}\{R_\infty^Y \mid Y(0) = i\},$$

we also have $\varrho_i^Y = \varrho_i^{Y,S} + \varrho_i^{Y,F}$. As stated before, we assume that the initial state is a slow one (the natural situation). Correspondingly, we define the column vectors ϱ^Y, $\varrho^{Y,S}$, and $\varrho^{Y,F}$ indexed on S, with entry i equal to ϱ_i^Y, $\varrho_i^{Y,S}$ and $\varrho_i^{Y,F}$, respectively. Again, $\varrho^Y = \varrho^{Y,S} + \varrho^{Y,F}$.

10.2.5   Stochastic complement

Let us denote by $\tilde{Y}$ the chain obtained by stochastically complementing [65] the subset of states $S \cup \{M + 1\}$ in Y. This means considering the chain obtained by copying Y

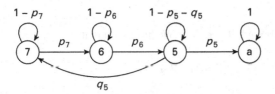

Figure 10.8 Consider the chain obtained by uniformizing the model given in Figure 10.10 with respect to some rate, $\eta \geq \max(7\phi, 6\phi + \gamma)$. The picture represents the stochastic complement of that chain with respect to $S \cup \{a\}$, where a denotes the absorbing state, and $S = \{7,6,5\}$. We have $p_7 = 7\phi/\eta$ and $p_6 = 6\phi/\eta$; p_5 and q_5 are much more involved (see 10.32).

while it lives inside $S \cup \{M+1\}$ and by freezing it in state $i \in S$ if, after visiting i, Y goes to some state in F; in this case, if Y enters again $S \cup \{M+1\}$ by state k, $\tilde{Y}$ jumps from i to k. Process $\tilde{Y}$ is also called the *reduction* of Y with respect to $S \cup \{M+1\}$. Its transition probability matrix, indexed on $S \cup \{M+1\}$, is

$$\begin{pmatrix} \tilde{U} & \tilde{u} \\ 0 & 1 \end{pmatrix}$$

where

$$\tilde{U} = U_S + U_{SF}(I - U_F)^{-1}U_{FS} \tag{10.32}$$

and $\tilde{u}$ is the vector $\tilde{u} = 1 - \tilde{U}1$.

For instance, consider the model illustrated in Figure 10.10, where ϕ is any component's failure rate and $\gamma \gg \phi$ is its repair rate (for the description of the system it models, see Subsection 10.2.7 below). Once uniformized with respect to some rate, η, we illustrate in Figure 10.8 the stochastic complement of the uniformized chain, with respect to the set $\{7,6,5,a\}$, where a denotes the absorbing state of the chain.

As for Y, we define the cumulated reward until absorption $R_\infty^{\tilde{Y}}$ and the vector $\tilde{\varrho}$ indexed on S whose ith entry is the conditional expectation

$$\tilde{\varrho}_i = \mathbb{E}\{R_\infty^{\tilde{Y}} \mid \tilde{Y}(0) = i\}.$$

A key observation now is that, by definition of the stochastic complement,

$$\tilde{\varrho} = \varrho^{Y,S}. \tag{10.33}$$

10.2.6 Bounding procedure

In this section we describe our bounding procedure. We first derive a lower bound of $\varrho^Y = \varrho^{Y,S} + \varrho^{Y,F}$, by computing a lower bound of $\tilde{\varrho} = \varrho^{Y,S}$. Then, an upper bound of ϱ^Y is computed in two parts: first, an upper bound of $\tilde{\varrho} = \varrho^{Y,S}$ is obtained; then, an upper bound of $\varrho^{Y,F}$ ends the bounding scheme.

In the following, all the inequalities between vectors or between matrices mean that the inequality holds componentwise: for instance, if $a = (a_i)$ and $b = (b_i)$ are vectors with the same dimension, then $a \leq b \Longleftrightarrow a_i \leq b_i$ for all i.

Lower bound of ECRA

We provide here a lower bound of vector $\tilde{\varrho}$. We have seen that

$$\tilde{\varrho} = (I - \tilde{U})^{-1} r_S. \tag{10.34}$$

The problem is that, in general, matrix $\tilde{U}$ is unknown. In some specific models, however, matrix $\tilde{U}$ can be exactly computed with a moderate effort, in which case we can evaluate $\tilde{\varrho}$ exactly and then use it as our lower bound of ϱ^Y (see the numerical examples).

Assume here that we have no access to $\tilde{U}$. In this case, we can use the following approach. Fix some (small) integer $N \geq 2$. For all states $i, j \in S$, consider the set $\mathcal{P}_N$ of all the paths having the form $\pi = (i, f_1, f_2, \ldots, f_{N-1}, j)$, where $f_1, f_2, \ldots, f_{N-1} \in F$. The probability of such a path is

$$p(\pi) = U_{i, f_1} U_{f_1, f_2} \cdots U_{f_{N-1}, j};$$

then,

$$\sum_{\pi \in \mathcal{P}_N} p(\pi) = (U_{SF} U_F^{N-2} U_{FS})_{i,j}.$$

Define matrix U' by

$$U' = U_S + U_{SF} \left(I + U_F + U_F^2 + \cdots + U_F^{N-2} \right) U_{FS}. \tag{10.35}$$

Clearly, $U' \leq \tilde{U}$. Assume that we use a value of N such that $U_F^{N-1} \neq 0$, so that U' is strictly sub-stochastic. If ϱ' is the solution to the system

$$(I - U')\varrho' = r_S, \tag{10.36}$$

then

$$\varrho' \leq \tilde{\varrho} = \varrho^{Y,S} \leq \varrho^Y.$$

Vector ϱ', solution to linear system (10.36) with size $|S|$, is our lower bound of ϱ^Y. To build the matrix of the system, we need to explore the set of fast states up to a distance $N - 2$ from S. As can be seen in the examples, we could obtain accurate bounds in all the explored cases with quite small values of N.

Upper bound

For the upper bound of ϱ^Y, we first find an upper bound of $\varrho^{Y,S}$, then an upper bound of $\varrho^{Y,F}$.

Upper bound of $\varrho^{Y,S}$

Define $U'' = \tilde{U} - U' \geq 0$, another sub-stochastic matrix. We have

$$U'' = U_{SF} \left(U_F^{N-1} + U_F^N + \cdots \right) U_{FS}$$

$$= U_{SF} U_F^{N-1} (I - U_F)^{-1} U_{FS}.$$

Define

$$V = (I - U_F)^{-1}.$$

The number $V_{i,j}$ is the mean number of visits of Y to $j \in F$ starting from $i \in F$, before leaving F. The main idea now is to build an upper bound of matrix U''.

For this purpose, we use the following result, adapted from [80].

LEMMA 10.5. *Since there is no circuit in the graph of X composed of fast transitions only, we can partition the states of F into D classes $C_1, \ldots, C_D$, where C_1 is the set of states directly connected to S by a fast transition, C_2 is the set of states directly connected to C_1 by a fast transition, etc. Integer D is the maximal distance from any state in F to set S. Then, if we define*

$$\lambda = \max_{f \in F} \sum_{g \in F, (f,g) \, slow} Q_{f,g}$$

and

$$\mu = \min_{f \in F} \sum_{g \in F, (f,g) \, fast} Q_{f,g},$$

and if we assume that $\lambda < \mu$ (this is the natural situation), any sojourn of Y in F has a mean duration less than σ, where

$$\sigma = \frac{\eta}{\mu} \frac{1 - (1 - \psi)^D}{\psi (1 - \psi)^D}, \tag{10.37}$$

and $\psi = \lambda/\mu < 1$.

Proof. We assume here that $Y(0) \in F$. Let us transform $S \cup \{M + 1\}$ into a single absorbing state 0. Let us denote by $I(k)$ the index of the class where $Y(k)$ is, $I(k) = d \iff Y(k) \in C_d$, $d = 0, 1, \ldots, D$, with the convention $I(k) = 0$ if $Y(k) = 0$.

We use an auxiliary discrete-time Markov chain, $Z = \{Z(k), k \geq 0\}$, defined on the set of indexes $\{0, 1, 2, \ldots, D\}$. State 0 is absorbing and the remaining states are transient. From any state $d > 0$, transitions are possible to $d - 1$ with probability $1 - \psi$ and to D with probability ψ. Figure 10.9 shows this chain in the case where $D = 4$.

The proof consists of showing first that "Y visits fewer classes than Z visits states". This comes from the fact that for any $k \geq 0$, the random variable, $I(k)$, is stochastically

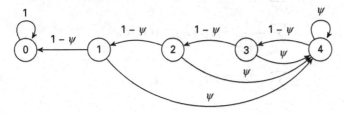

Figure 10.9 Auxiliary chain Z for $D = 4$

smaller than $Z(k)$, which is obtained by recurrence on k as follows. The claim is obviously true for $k = 0$, since, by construction, $I(0) = Z(0)$. The recurrence assumption is that $I(k) \leq_{\text{st}} Z(k)$. We must prove that $\mathbb{P}\{I(k+1) \geq l\} \leq \mathbb{P}\{Z(k+1) \geq l\}$ for all possible l. Write

$$\mathbb{P}\{I(k+1) \geq l\} = \sum_{d=0}^{D} p(l,d)\mathbb{P}\{I(k) = d\},$$

where $p(l,d) = \mathbb{P}\{I(k+1) \geq l \mid I(k) = d\}$, and similarly,

$$\mathbb{P}\{Z(k+1) \geq l\} = \sum_{d=0}^{D} q(l,d)\mathbb{P}\{Z(k) = d\},$$

where $q(l,d) = \mathbb{P}\{Z(k+1) \geq l \mid Z(k) = d\}$. For $h = 0,1,\ldots,d-1$, we have $q(h,d) = 1$, and for $h = d, d+1, \ldots, D$, we have $q(h,d) = \psi$.

We see immediately that $p(l,d) \leq q(l,d) = 1$ for $l = 0,1,\ldots,d-1$. For $l \geq d$,

$$p(l,d) = \sum_{f \in C_d} \mathbb{P}\{I(k+1) \geq l \mid Y(k) = f\}\mathbb{P}\{Y(k) = f \mid Y(k) \in C_d\}.$$

The conditional probability, $\mathbb{P}\{I(k+1) \geq l \mid Y(k) = f\}$, can be written

$$\mathbb{P}\{I(k+1) \geq l \mid Y(k) = f\} = \frac{u_f}{u_f + v_f} = \frac{1}{1 + \dfrac{v_f}{u_f}},$$

where

$$u_f = \frac{1}{\eta} \sum_{g \in C_l \cup C_{l+1} \cup \cdots \cup C_D} Q_{f,g}, \qquad v_f = \frac{1}{\eta} \sum_{g \in C_{d-1} \cup C_d \cup \cdots \cup C_{l-1}} Q_{f,g}.$$

Observe that u_f is a "small" probability and v_f is a "large" one, since the former comes from transition rates of slow transitions and the latter includes the rates of fast ones.

By definition of λ and μ,

$$\frac{1}{1 + \dfrac{v_f}{u_f}} \leq \frac{1}{1 + \dfrac{\mu/\eta}{\lambda/\eta}} = \frac{\psi}{1 + \psi}.$$

So, for $l \geq d$,

$$p(l,d) = \sum_{f \in C_d} \mathbb{P}\{I(k+1) \geq l \mid Y(k) = f\}\mathbb{P}\{Y(k) = f \mid Y(k) \in C_d\}$$

$$= \sum_{f \in C_d} \frac{1}{1 + \dfrac{v_f}{u_f}} \mathbb{P}\{Y(k) = f \mid Y(k) \in C_d\}$$

$$\leq \sum_{f \in C_d} \frac{\psi}{1 + \psi} \mathbb{P}\{Y(k) = f \mid Y(k) \in C_d\}$$

$$= \frac{\psi}{1 + \psi}.$$

Resuming, we have in all cases $p(l,d) \leq q(l,d)$.

Now, fix l and look at $p(l,d)$ as a function of d only: $p(l,d) = g(d)$, then,

$$\mathbb{P}\{I(k+1) \geq l\} \leq \sum_{d=0}^{D} g(d)\mathbb{P}\{I(k) = d\}$$

$$= \mathbb{E}\{g(I(k))\}$$

$$\leq \mathbb{E}\{g(Z(k))\}$$

$$\leq \mathbb{P}\{Z(k+1) \geq l\}.$$

The last inequalities in the derivation come from the recurrence assumption and the fact that g is an increasing function. We have thus proved that $I(k) \leq_{st} Z(k)$ for all $k \geq 0$. This means that the number of states visited by Z before absorption is stochastically larger than the number of classes visited by Y.

In the second step of the proof, we look at the time Y spends in each of the classes. We prove here that, on average, this time is bounded above by the ratio μ/η. Denote by β_f the probability of starting a sojourn in C_d by state $f \in C_d$, and by β the row vector whose fth entry is β_f, indexed on C_d. Let U_d be the restriction to C_d of matrix U, also indexed on that class of states. If L is the length of the sojourn time of Y in C_d when the first visited state is chosen according to distribution β, then we have $\mathbb{P}\{L > k\} = \beta(U_d)^k \mathbb{1}$ (see Chapter 5). The expected value of L is

$$\mathbb{E}\{L\} = \sum_{k \geq 0} \beta(U_d)^k \mathbb{1} = \beta(I - U_d)^{-1}\mathbb{1}.$$

Write $U_d\mathbb{1} = u$. Entry f of u is $u_f = 1 - \sum_{g \notin C_d}(U_d)_{f,g}$. For any f in any one of the D classes,

$$u_f \leq 1 - q, \qquad q = \frac{\mu}{\eta} \leq 1,$$

which follows by definition of μ and η. So, we have $u \leq (1-q)\mathbb{1}$. In the same way, $(U_d)^2\mathbb{1} = U_d u \leq (1-q)U_d\mathbb{1} \leq (1-q)^2\mathbb{1}$. By recurrence, it is then immediate to check that $(U_d)^k\mathbb{1} \leq (1-q)^k\mathbb{1}, k \geq 0$. Choosing now $\eta > \mu$, we obtain

$$\mathbb{E}\{L\} = \sum_{k \geq 0} \beta(U_d)^k \mathbb{1} \leq \sum_{k \geq 0} \beta(1-q)^k\mathbb{1} = \sum_{k \geq 0}(1-q)^k = \frac{1}{q} = \frac{\eta}{\mu}.$$

The final step is to prove that the expected absorption time of Z, when the initial state is state D, is given by

$$\frac{1 - (1-\psi)^D}{\psi(1-\psi)^D},$$

which can easily be done by recurrence on D. This concludes the proof. ∎

The preceding Lemma means that for any $i,j \in F$, $V_{i,j} \leq \sigma$, and thus that

$$U'' \leq \sigma U_{SF}U_F^{N-1}U_{FS}.$$

Denote by B_N this last matrix:

$$B_N = \sigma\, U_{SF} U_F^{N-1} U_{FS}. \tag{10.38}$$

Then, if $\sum_{n=0}^{\infty} (U' + B_N)^n < \infty$,

$$\varrho^{Y,S} = \tilde{\varrho}$$

$$= \sum_{n=0}^{\infty} \tilde{U}^n r_S$$

$$= \sum_{n=0}^{\infty} (U' + U'')^n r_S$$

$$\leq \sum_{n=0}^{\infty} (U' + B_N)^n r_S.$$

Denote this last vector by φ:

$$\varphi = \sum_{n=0}^{\infty} (U' + B_N)^n r_S.$$

Now, since the states of F are transient, we have $\lim_{N \to \infty} B_N = 0$. This means that, for N large enough, matrix $U' + B_N$ is strictly sub-stochastic, and thus that vector φ is well defined. Observe that φ is obtained as the solution to the linear system

$$(I - U' - B_N)\varphi = r_S. \tag{10.39}$$

Upper bound of $\varrho^{Y,F}$

For $i, j \in S$, denote by $W_{i,j}$ the mean number of visits that Y makes to state j, starting from i, and by v_i the mean number of visits that Y makes to the whole subset F, again conditional on $Y(0) = i$. Observe that

$$v_i = \sum_{j \in S} W_{i,j} U_{j,F}$$

where $U_{j,F} = \sum_{f \in F} U_{j,f}$. Also observe that this number, v_i, is computable on the reduced process (on the stochastic complement $\tilde{Y}$): if $W = (W_{i,j})$, we have

$$W = (I - \tilde{U})^{-1}.$$

The ith entry of vector $\varrho^{Y,F}$ can be bounded above by $\sigma\, v_i r^*$ where

$$r^* = \max\{r_f, f \in F\}.$$

If we define vectors u_F and v, indexed on S, by $(u_F)_i = U_{i,F}$ and $v = (v_i)$, then

$$v = W u_F. \tag{10.40}$$

Now,

$$v = W u_F = \sum_{n=0}^{\infty} \tilde{U}^n u_F \leq \sum_{n=0}^{\infty} (U' + B_N)^n u_F.$$

Denote by τ this last vector:

$$\tau = \sum_{n=0}^{\infty} (U' + B_N)^n u_F.$$

Observe that τ is the solution to the linear system

$$(I - U' - B_N)\tau = u_F. \tag{10.41}$$

So, the upper bound of $\varrho^{Y,F}$ is

$$\varrho^{Y,F} \leq \sigma r^* \tau.$$

Finally, the upper bound of ϱ is:

$$\varrho \leq \sigma r^* \tau + \varphi.$$

The case when the stochastic complement is known

In some cases, it is possible to compute the exact stochastic complement (without any matrix inversion). This happens when the system has some special structure, such as a single return state to S coming from F, a single exit state from F, or in the cases where all non-zero rows in U_{FS} are proportional to each other. For instance, consider the special case where there is a single return state to S, that is, a single state $s \in S$ such that for some $f \in F$, $U_{f,s} > 0$, and that for all $g \in F$ and $z \in S, z \neq s$, $U_{g,z} = 0$. Note that this simply means that in matrix U_{FS} all the columns are composed of 0 except one. In these cases, matrix $(I - U_F)^{-1} U_{FS}$ can be evaluated easily, with small computational complexity. For more details, see [65] and also [93].

Assume then that $\tilde{U}$ is known. In this case, the lower bound of ϱ is simply given by $\tilde{\varrho}$, after solving the linear system

$$(I - \tilde{U})\tilde{\varrho} = r_S. \tag{10.42}$$

The obtained bound is clearly better than the lower bound given by ϱ'. The upper bound is obtained by summing $\sigma r^* v$ and $\tilde{\varrho}$. The first term is computed after solving the linear system

$$(I - \tilde{U})v = u_F. \tag{10.43}$$

Again, this bound is better than in the general case.

This situation is illustrated in the example described in Subsection 10.2.7.

Summary

Let us summarize the proposed technique in algorithmic form: one for the general case, where $\tilde{U}$ is unknown, and the second one for the case where the exact computation of the stochastic complement is possible.

ECRA bounding technique – general case

1. Choose θ
2. Choose η
3. Choose N
4. Build U' (Equation (10.35))
5. Compute ϱ' (linear system (10.36))
6. Compute σ (Equation (10.37))
7. Compute B (Equation (10.38))
8. Compute φ (linear system (10.39))
9. Compute τ (linear system (10.41))
10. Bounds: $\varrho' \leq \varrho \leq \sigma r^* \tau + \varphi$

Remark: After step 7 we must check whether matrix $U' + B$ is strictly sub-stochastic. If it is not, a larger value of N must be tried, repeating steps 3 to 7. In our examples we always used fairly small values of N and we never found this situation.

ECRA bounding technique – exact stochastic complement case

1. Choose θ
2. Choose η
3. Compute $\tilde{U}$ (Equation (10.32))
4. Compute $\tilde{y}$ (linear system (10.34))
5. Compute σ (Equation (10.37))
6. Compute v (linear system (10.43))
7. Bounds: $\tilde{\varrho} \leq \varrho \leq \sigma r^* v + \tilde{\varrho}$

Computational complexity

Concerning the cost of the procedure, observe that in the general case, the cost is dominated by the resolution of three linear systems having the size of S, usually much less than the size of the whole state space. Moreover, some supplementary gain can be obtained using the fact that two of the linear systems to solve correspond to the same matrix. The rest of the cost is concentrated on the computation of matrices U' and B (the latter is "almost" a sub-product of the former). In most of the examples we have considered here, $N \ll |S|$, which explains why the linear systems' resolution dominates the global cost. Since N will always be small or moderate in applications, the case of $|S| \leq N$ is not a very important one, since in that case, S will be small and the global procedure will run very quickly (or we will use a superset S' of S, and $|S'| \gg N$).

When the stochastic complement is available, the bounds are better, as explained before, and also less expensive to find since we now have only two linear systems to solve, both with the same matrix, and no path-based part.

10.2.7   Examples

This section illustrates the behavior of the method through a few examples. In the first one we present the evaluation of a multicomponent system subject to failures and repairs. The second example studies the behavior of a system with a set of identical components. The third one presents a grid system based on [34], used for studying the case where the original slow set has cardinality equal to 1 and a superset of S is used to calculate a more efficient approximated solution, but keeping a reduced total computational cost. The last example presents a database model, where we are interested in studying the economic loss due to failures. An important feature of the illustrations is that they are built in such a way that the exact solution is always available (even when the state spaces are huge); this allows the interested reader to verify the results and also other exact values for different related quantities.

In the examples, the results are presented in two types of table: the first one gives the set of parameters characterizing the example and the numerical results: the exact solution (ECRA), the lower and upper bounds (L.B. and U.B., respectively), and the relative error (R.E.) defined by $|(L.B.+U.B.)/2-ECRA|/ECRA$. The second table type gives information related to the main computational gains, which includes the reduction of the state space size and the parameters related to the path-based approach. For the first three examples, the ECRA measure is a basic reliability measure: the MTTF, that is, the reward of a state is 1 if the system is operational when in that state, 0 otherwise. The last example presents a model with more general rewards.

In the tables, we use the following notation: $2.0e6$ means 2.0×10^6 and $3.0e{-}4$ means 3.0×10^{-4}.

A multicomponent system with deferred repairs

Consider a system composed of C identical components subject to failures and repairs. The repair facility starts working only when there are H failed components and it has only one server. Once started, the repair facility remains active until there is no failed unit in the system (hysteresis phenomenon). The system starts with all its C components up. The time to failure of any component is exponentially distributed with rate ϕ. The system is considered operational if there are at least K components up. We are interested in the MTTF measure, that is, the expected operational time until the system has only $K - 1$ components up. Observe that for this description to make sense, we need $C \geq H + K$. Finally, the components' lifetimes and repair times are independent of each other.

All states that have a repair exit rate are fast states here. This means that the total number of slow states is equal to H.

In the first set of results the repair time is exponentially distributed with rate γ. For a total number of components $C = 1000$ (think of a network) we performed

Table 10.3. Numerical results for the system with $C = 1000$ components. We use $\gamma = 1$ for all cases. L.B.: Lower Bound, U.B.: Upper Bound, R.E.: Relative Error.

H	K	ϕ	L.B.	ECRA	U.B.	R.E.
3	1	$1e{-}4$	$2.5378e16$	$2.5466e16$	$2.6268e16$	$1.4e{-}2$
100	10	$1e{-}5$	$2.7186e16$	$2.7187e16$	$2.7189e16$	$1.8e{-}5$
100	10	$1e{-}4$	$2.5381e16$	$2.5471e16$	$2.5484e16$	$1.5e{-}3$

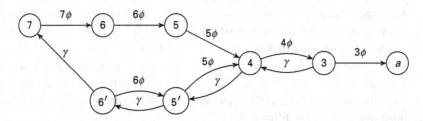

Figure 10.10   The multicomponent model with deferred repairs, for $C = 7$ (the initial number of components in the system), $H = 3$ (number of failed components necessary to activate the repairing facilities), $K = 3$ (minimal number of operational components necessary for the system to operate). The component failure rate is ϕ, the component repair rate is γ and we assume $\phi \ll \gamma$. The set of slow states is $S = \{7,6,5\}$ and the set of fast states is $F = \{3,4,5',6'\}$. The absorbing state is denoted here by a.

some experiments varying the values of the failure rate as well as the parameters H and K.

The space state is composed of states $i = C, C - 1, \ldots, C - H, \ldots, K, \ldots, 1, 0$, plus the "clones" $(C - 1)', (C - 2)', \ldots, (C - H + 1)'$ necessary because of the hysteresis phenomenon. We can then collapse states $K - 1, \ldots, 1, 0$ into a single absorbing state, say $0'$. This new state space has $C + H - K + 1$ states, with H slow states and $C - K$ fast states. It is illustrated in Figure 10.10.

In this particular model, there is only one exit state from S and there is only one entering state into F. Obviously, this simple topology allows a numerical analysis of the example for virtually any values of its parameters.

Table 10.3 shows the main parameters and numerical results. Table 10.4 shows the computational gains and the path-based approach parameters (following the order of the results presented in Table 10.3). For the cases analyzed, the bounding method is efficient even when a small number of states is effectively generated.

Now, let us consider a system with the same characteristics and a total number of components $C = 5000$. The varied parameter was H. Table 10.5 shows the numerical results and Table 10.6 presents the additional information about the sizes of the corresponding state spaces.

The parameter N used in all cases, in this example, is at least equal to H. The choice of this number is based on the particular model structure and the idea of generating at

Table 10.4. State space reduction and parameter N for the system with $C = 1000$ components

Orig. state space size	New state space size	Reduction rate (ratio)	N
1003	3	334	15
1100	100	11	106
1100	100	11	106

Table 10.5. Numerical results for the system with $C = 5000$ components. In these cases, we use $\gamma = 1$, $K = 1$ and $\phi = 10^{-4}$. L.B.: Lower Bound, U.B.: Upper Bound, R.E.: Relative Error.

H	L.B.	ECRA	U.B.	R.E.
20	$2.5032e16$	$2.5471e16$	$2.7480e16$	$3.0e{-}2$
50	$2.5447e16$	$2.5472e16$	$2.5777e16$	$5.5e{-}3$
100	$2.5458e16$	$2.5472e16$	$2.5629e16$	$2.8e{-}3$
500	$2.5441e16$	$2.5472e16$	$2.5478e16$	$4.9e{-}4$

Table 10.6. State space reduction and parameter N for the system with $C = 5000$ components

Orig. state space size	New state space size	Reduction rate (ratio)	N
5020	20	251	20
5050	50	101	60
5100	100	51	108
5500	500	11	516

least one path into the set of fast states. This special topology means that the fact of generating at least one path into F does not significantly increase the computational cost and better bounds can be obtained. It is worth noting that the idea of generating at least one path into F is not a necessary assumption for the method to work, given that all the bounds are computed based on having a lower bound for the exact stochastic complement. This is a choice that can be made during the dependability analysis and is based on the tradeoff between better bounds and computational cost.

Let us introduce a special way of increasing the original state space used for the bounding method and at the same time keeping a very small one to obtain the exact MTTF solution. For the exact solution, the repair rate remains exponentially distributed with rate γ. However, for applying the bounding method let us consider that after a failure, the components enter a repair facility with one server and that the repair time

Table 10.7. Numerical results for the system with $C = 10\,000$ components, using a Coxian repair time distribution. The values for the parameters are: $\gamma = 1$, $K = 1$ and $\phi = 10^{-5}$. L.B.: Lower Bound, U.B.: Upper Bound, R.E.: Relative Error.

H	s	L.B.	ECRA	U.B.	R.E.
10	10	2.7170e16	2.7187e16	2.7227e16	4.2e−4
10	100	2.7186e16	2.7187e16	3.3415e16	1.5e−1

Table 10.8. State space reduction and parameter N for the system with $C = 10\,000$ components. In the first row, the Coxian distribution has $s = 10$ phases; in the second, $s = 100$.

Orig. state space size	New state space size	Reduction rate (ratio)	N
100 100	10	10 010	20
1 001 000	10	100 100	30

is distributed according to a Coxian distribution with mean $1/\gamma$. The parameters of the Coxian distribution are then chosen in such a way that the distribution is equivalent to an exponential one. If v_s is the parameter of the sth stage and if l_s is the probability that phase s is the last one, then, for all phases s, we have $v_s l_s = 1/\gamma$. Technically, we put ourselves in a *strong lumpability* situation (see Chapter 4), which allows us to obtain the exact solution from the previous case, and to test our bounding technique in the case of a large state space.

Consider the system with $C = 10000$ components, for which we know the exact MTTF value. Using the Coxian distribution version, the bounding method is performed for the cases where the number of phases is $s = 10$ and $s = 100$. The state space cardinality increases 10 and 100 times, respectively, with respect to the case of exponentially distributed repair times. Table 10.7 shows the numerical results, as well as the model parameters. Table 10.8 shows the second set of the numbers related to the bounding method.

The numbers in Table 10.7 deserve a special comment related to the upper bound results. For the case where $s = 10$, the bounds are tighter than for the case where $s = 100$. This can be explained by the way the time spent on the fast set is bounded. One of the parameters is D, the largest distance between any state in the fast set to the slow set. In this very special case, D is the cardinality of the fast set. The other point is that ϕ is not much smaller than γ. However, even for this extreme case, the approximation is good enough with a high reduction rate and with a small number of states generated.

Table 10.9. Numerical results for the system with several identical components, using the symmetries in the Markov chain (strong lumpability). The values for the parameters are $\phi c = 10^{-6}$, $\phi(1 - c) = 10^{-7}$, and $\gamma = 1$. L.B.: Lower Bound, U.B.: Upper Bound, R.E.: Relative Error.

C	L.B.	ECRA	U.B.	R.E.
10	$2.9285e7$	$2.9285e7$	$2.9285e7$	0
30	$3.9932e7$	$3.9933e7$	$3.9933e7$	$1.2e{-}5$
40	$4.2765e7$	$4.2765e7$	$4.2767e7$	$2.3e{-}5$
50	$4.4959e7$	$4.4959e7$	$4.4961e7$	$2.2e{-}5$

Table 10.10. State space reduction for the system with many identical components, using the symmetries in the Markov chain (strong lumpability).

Orig. state space size	New state space size	Reduction rate (ratio)
66	10	6
496	30	16
861	40	21
1326	50	26

Table 10.11. Numerical results for the system with many identical components, without using the symmetries in the Markov chain. The exact values of the MTTF are obtained by exploiting the strong lumpability property. The values for the parameters are $\phi c = 10^{-6}$, $\phi(1 - c) = 10^{-7}$, and $\gamma = 1$. L.B.: Lower Bound, U.B.: Upper Bound, R.E.: Relative Error.

C	L.B.	ECRA	U.B.	R.E.
11	$3.0116e7$	$3.0177e7$	$3.0206e7$	$5.3e{-}4$
12	$3.0925e7$	$3.1026e7$	$3.1031e7$	$1.5e{-}3$
18	$3.4943e7$	$3.4951e7$	$3.4997e7$	$5.4e{-}4$
20	$3.5977e7$	$3.5977e7$	$3.5990e7$	$1.8e{-}4$

System with many identical components

This example is based on [80], where a system is composed of C identical and independent components. The evolution of any of them is modeled by the Markov chain shown in Figure 10.11. The failure rate is ϕ, the *coverage factor* is c, and the repair rate is γ. We will calculate the MTTF measure.

Table 10.12. State space reduction and parameter *N* for the system with many components without using the strong lumpability property

C	Orig. state space size	New state space size	Reduction rate (ratio)	N
11	177 147	2048	86	6
12	531 441	4096	130	6
18	382 742 098	262 144	1460	5
20	3 486 784 401	1 048 576	3325	5

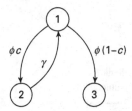

Figure 10.11 Evolution of any of the components of the system. The failure rate is ϕ, the coverage factor is c and the repair rate is γ. State 1 means that the component is up; in state 2, it is down and being repaired; in state 3, it is down and not repairable.

For a first group of systems, with a total number of components, C, in the set $\{10, 30, 40, 50\}$, we explore how the method performs when the exact stochastic complement is used (this is possible in this case). All the systems were modeled exploring the symmetries in the Markov chain which results in a state space reduction (due to the *strong lumpability* property). A state is classified as fast if it has at least one exit transition rate greater than or equal to γ (that is, using previous notation for the threshold θ, we have here $\theta = \gamma$). Table 10.9 shows the numerical results. Table 10.10 shows the reduction rate reached by applying the method.

Let us now model all the systems *without* exploiting the strong lumpability property. The idea is to generate models with much larger state spaces and to apply the bounding method to them. This allows us to work on large models and still have access to the exact value of the MTTF. Table 10.11 shows the numerical results and Table 10.12 shows the reduction rate and the used value of N. We can see that tight bounds can be provided even when large reduction rates occur.

As a last case study, consider that the system has $C = 11$ components. Here, we apply the bounding technique for two different values of N: $N = 3$ and $N = 6$. Table 10.13 shows the numerical results. As expected, increasing the total number of possible explored paths yields a smaller relative error measure. However, even for a case where the maximum size path is $N = 3$, a good approximation is reached.

Grid: an example of a further refinement of the method

Consider the example presented in [34] used for studying the sensitivity of system reliability/availability with respect to the coverage parameter, c, and look at it as a grid

Table 10.13. Numerical results for the system with many identical components, without using the symmetries in the Markov chain. There are $C = 11$ components here. The exact values of the MTTF are obtained exploiting the strong lumpability property. The values for the parameters are $\phi c = 10^{-4}$, $\phi(1 - c) = 10^{-5}$, and $\gamma = 1$. We vary N in this illustration. L.B.: Lower Bound, U.B.: Upper Bound, R.E.: Relative Error.

N	L.B.	ECRA	U.B.	R.E.
3	$3.0199e5$	$3.0202e5$	$3.0566e5$	$6.0e{-}3$
6	$3.0191e5$	$3.0202e5$	$3.0558e5$	$5.7e{-}3$

Table 10.14. Numerical results for the multiprocessor system with $C = 5000$ processors. L.B.: Lower Bound, U.B.: Upper Bound, R.E.: Relative Error.

L.B.	ECRA	U.B.	R.E.
$1.9984e11$	$2.0000e11$	$2.0034e11$	$4.5e{-}4$

Table 10.15. Numerical results for the grid with $C = 5000$ processors, using a superset of S in the bounds computation.

L.B.	ECRA	U.B.	R.E.
$1.9999e11$	$2.0000e11$	$2.0000e11$	$2.5e{-}5$

composed of C processors. We assume that only one node up is enough for considering the system as operational. Each processor fails at rate $\phi = 10^{-10}$ and is repaired at rate $\gamma = 1$, and we consider a large number of connected elements in the grid: $C = 5000$. Any failure is repairable with probability c. That is, for any state $i = 1, \ldots, C$ there is a transition to the absorbing state, with rate $i\phi(1 - c)$. The state transition diagram is shown in Figure 10.12.

Applying the slow state set definition introduced in Subsection 10.2.2, using any reasonable threshold value, this system has only one slow state, the initial one, where all processors are working. The numerical results are shown in Table 10.14. Even for the case where we have just one slow state, the relative error is small and the computational savings are obviously very significant.

Let us now illustrate the idea of redefining the slow states set, considering some states that belong to the fast states as slow ones. Or, saying this in a different way, let

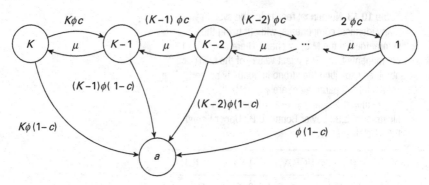

Figure 10.12 Markov model of the grid. The absorbing state is here denoted by a.

us perform the stochastic complement (and all the analysis) using S' instead of S, where $S \subset S'$. We do this by "adding 1000 states to S", that is, by using a set S' where $|S'| = 1001$, in order to reach a better approximation. Table 10.15 shows the new numerical results. In this case, increasing the total number of states in the slow set results not only in a better lower bound, as expected, but also in a better upper bound. This is because of the way we obtain the upper bound of $\varrho^{Y,F}$, given that we consider the worst case for each visit in F.

Database system

In this last example, we study a database system composed of CPUs, hard disks, and memory units. Each component in the system has a similar behavior to the component presented in Figure 10.11.

First, suppose that we are interested in the CPU and hard disk unit performance. Let us study a system with three CPUs and five hard disks, and let us calculate the MTTF measure. In order to increase the original state space and at the same time to be able to compute the exact solution, we apply the same approach as used in Subsection 10.2.7. Each component has a repair time distributed according to an *s-stage* Coxian distribution. Table 10.16 shows the numerical results for this first database system. Table 10.17 shows the reduction rate reached by applying the method. For the exact solution, we model the same system using the *strong lumpability* property.

Now, let us also consider the memory performance. In this last case study, our system has four CPUs, five disks, and three memory units. Suppose that for each component failure without repair, we have an economic loss $L = \$500$ for the CPU component, $L = \$5000$ for the disk component, and $L = \$1000$ for the memory component, and for each failure with repair, we have an economic loss $L = \$100$ for the three types of components. We now calculate the total amount of money lost until system breakdown (state where all components are failed). This scenario can represent, for instance, a commercial website, where the loss of performance affects the total amount of capital cumulated. Table 10.18 shows the numerical results for this database system. The values are negative due to the fact that the system is losing capital.

Table 10.16. Numerical results for the database system. We are interested in the CPU and hard disk units' behavior. The exact values of the MTTF are obtained by exploiting the strong lumpability property. The values for the parameters are: CPU, $\phi c = 10^{-5}$, $\phi(1 - c) = 10^{-6}$, and $\gamma = 1$; Disk, $\phi c = 10^{-3}$, $\phi(1 - c) = 10^{-4}$, and $\gamma = 5$. L.B.: Lower Bound, U.B.: Upper Bound, R.E.: Relative Error.

s	L.B.	ECRA	U.B.	R.E.
3	$1.8139e6$	$1.8333e6$	$1.8354e6$	$4.7e{-}3$
10	$1.8150e6$	$1.8333e6$	$1.8870e6$	$9.6e{-}3$

Table 10.17. State space reduction and parameter N for the database system composed of CPU and hard disk units

C	Orig. state space size	New state space size	Reduction rate (ratio)	N
3	$390\,625$	256	$1\,525$	7
10	$429\,998\,169$	256	$1\,679\,680$	7

Table 10.18. Numerical results for the database system, considering the memory units' behavior. The exact value of the expected cumulated economical loss is obtained by exploiting the strong lumpability property. The values for the parameters are: CPU: $\phi c = 2e^{-6}$, $\phi(1 - c) = 1e^{-5}$; Disk: $\phi c = 1e^{-5}$, $\phi(1 - c) = 1e^{-5}$; Memory: $\phi c = 1e^{-5}$, $\phi(1 - c) = 1e^{-4}$, and $\gamma = 1$ for all components.

L.B.	ECRA	U.B.	R.E.
$-4.4753e10$	$-4.4741e10$	$-4.4740e10$	$1.2e{-}4$

10.2.8 Conclusions

This second part of the chapter describes a method for determining the bounds of the expected cumulated reward metric. The results are based on a classification of the states and transitions as slow and fast and on the reduction of the state space using stochastic complementation, plus the use of a path-based approach. We showed that by reducing the original stochastic process that represents the system under study with respect to the state classification presented and by using the path-based approach, tight bounds can be obtained with a much lower computational cost than is necessary to solve the original problem.

The contents of this chapter come from joint research work done with Stephanie Mahévas, researcher with Ifremer, France, and Ana Paula Couto da Silva, professor at the Federal University of Minas Gerais, Brazil.

References

[1] A. M. Abdel-Moneim and F. W. Leysieffer, Weak lumpability in finite Markov chains, *J. of Appl. Probability*, 19:685–691, 1982.

[2] S. Asmussen, *Applied Probability and Queues*, 2nd ed. Springer-Verlag, 2003.

[3] A. Avizienis, J. C. Laprie, B. Randell, and C. Landwehr, Basic concepts and taxonomy of dependable and secure computing, *IEEE Trans. on Dependable and Secure Computing*, 1:11–33, January 2004.

[4] J. Banks, J. S. Carson, B. L. Nelson, and D. M. Nicol, *Discrete-Event System Simulation*, 5th ed. Prentice-Hall international series in industrial and systems engineering. Prentice Hall, July 2009.

[5] R. E. Barlow and L. C. Hunter, Reliability analysis of a one unit system, *Operations Research*, 9(2):200–208, 1961.

[6] M. Bladt, B. Meini, M. Neuts, and B. Sericola, Distribution of reward functions on continuous-time Markov chains, in *Proc. of the 4th Int. Conf. on Matrix Analytic Methods in Stochastic Models (MAM4)*, Adelaide, Australia, July 2002.

[7] A. Bobbio and K. S. Trivedi, An aggregation technique for the transient analysis of stiff Markov chains, *IEEE Trans. on Computers*, C-35(9):803–814, September 1986.

[8] Z. I. Botev, P. L'Ecuyer, and B. Tuffin, Markov chain importance sampling with applications to rare event probability estimation, *Statistics and Computing*, 23(2):271–285, 2013.

[9] P. N. Bowerman, R. G. Nolty, and E. M. Scheuer, Calculation of the Poisson cumulative distribution function, *IEEE Trans. on Reliability*, 39:158–161, 1990.

[10] P. Brémaud, *Markov Chains: Gibbs Fields, Monte Carlo Simulation and Queues*. Springer, 1998.

[11] H. Cancela, M. El Khadiri, and G. Rubino, Rare events analysis by Monte Carlo techniques in static models, in G. Rubino and B. Tuffin, Eds, *Rare Event Simulation Using Monte Carlo Methods*, chapter 7, pages 145–170. Chichester: J. Wiley, 2009.

[12] H. Cancela, G. Rubino, and B. Tuffin, MTTF Estimation by Monte Carlo Methods using Markov models, *Monte Carlo Methods and Applications*, 8(4):312–341, 2002.

[13] H. Cancela, G. Rubino, and M. Urquhart, Pathset-based conditioning for transient simulation of highly dependable networks, in *Proc. of the 7th IFAC Symp. on Cost Oriented Automation*, Ottawa, Canada, 2004. (Also presented in "5th International Conference on Monte Carlo and Quasi-Monte Carlo Methods", Juan les Pins, France, June 2004.)

[14] J. A. Carrasco, Failure distance based on simulation of repairable fault tolerant systems, in *Proc. of the 5th Int. Conf. on Modelling Techniques and Tools for Computer Performance Evaluation*, pages 351–365, 1991.

[15] J. A. Carrasco, Improving availability bounds using the failure distance concept, in *4th Int. Conf. on Dependable Computing for Critical Applications (DCCA'4)*, San Diego, USA, January 1994.

[16] J. A. Carrasco, H. Nabli, B. Sericola, and V. Sunié, Comment on "Performability analysis: A new algorithm", *IEEE Trans. on Computers*, C-59:137–138, January 2010.

[17] F. Castella, G. Dujardin, and B. Sericola, Moments' analysis in homogeneous Markov reward models, *Methodology and Computing in Applied Probability*, 11(4):583–601, 2009.

[18] G. Ciardo, R. Marie, B. Sericola, and K. S. Trivedi, Performability analysis using semi-Markov reward processes, *IEEE Trans. on Comput.*, C-39:1251–1264, October 1990.

[19] B. Ciciani and V. Grassi, Performability evaluation of fault-tolerant satellite systems, *IEEE Trans. on Commun.*, COM-35:403–409, April 1987.

[20] C. J. Colbourn, *The Combinatorics of Network Reliability*. New York: Oxford University Press, 1987.

[21] A. E. Conway and A. Goyal, Monte Carlo simulation of computer system availability/reliability models, in *Proc. of the 17th Symp. on Fault-Tolerant Computing*, pages 230–235, Pittsburg, USA, July 1987.

[22] P.-J. Courtois and P. Semal, Bounds for the positive eigenvectors of nonnegative matrices and for their approximations by decomposition, *J. of the Assoc. for Computing Machinery*, 31(4):804–825, October 1984.

[23] P.-J. Courtois and P. Semal, Computable bounds for conditional steady-state probabilities in large Markov chains and queueing models, *IEEE Trans. on Selected Areas in Commun.*, 4(6):926–936, September 1986.

[24] A. P. Couto da Silva and G. Rubino, Bounding the mean cumulated reward up to absorption, in A. Langville and W. J. Stewart, Eds, *Markov Anniversary Meeting*, chapter 22, pages 169–188. Boson Books, 2006.

[25] A. Csenki, *Dependability for Systems with a Partitioned State Space. Markov and semi-Markov theory and computational implementation*, volume 90 of *Lecture Notes in Statistics*. Springer-Verlag, 1994.

[26] J. N. Darroch and E. Seneta, On quasi-stationary distributions in absorbing discrete-time finite Markov chains, *J. of Appl. Probability*, 2:88–100, 1965.

[27] H. A. David, *Order Statistics*. New York – London – Sidney – Toronto: John Wiley & Sons, Inc., 1981.

[28] E. de Souza e Silva and H. R. Gail, Calculating cumulative operational time distributions of repairable computer systems, *IEEE Trans. on Comput.*, C-35:322–332, April 1986.

[29] E. de Souza e Silva and H. R. Gail, Calculating availability and performability measures of repairable computer systems using randomization, *J. of the Assoc. for Computing Machinery*, 36:171–193, January 1989.

[30] E. de Souza e Silva and H. R. Gail, An algorithm to calculate transient distributions of cumulative rate and impulse based reward, *Commun. in Stat. – Stochastic Models*, 14(3), 1998.

[31] P. Diaconis, The Markov chain Monte Carlo revolution, *Bulletin of the American Math. Soc.*, 46(2):179–205, November 2008.

[32] L. Donatiello and V. Grassi, On evaluating the cumulative performance distribution of fault-tolerant computer systems, *IEEE Trans. on Comput.*, 40:1301–1307, November 1991.

[33] L. Donatiello and B. R. Iyer, Analysis of a composite performance measure for fault-tolerant systems, *J. of the Assoc. for Computing Machinery*, 34:179–189, January 1987.

[34] J. B. Dugan and K. S. Trivedi, Coverage modeling for dependability analysis of fault-tolerant systems, *IEEE Trans. on Comput.*, 38(6):775–787, June 1989.

[35] G. S. Fishman, *Discrete-Event Simulation: Modeling, Programming, and Analysis*. Springer Series in Operations Research and Financial Engineering. Springer, 2012.

[36] D. Furchtgott and J. F. Meyer, A performability solution method for degradable nonre-pairable systems, *IEEE Trans. on Comput.*, C-33:550–554, July 1984.

[37] P. Glasserman, *Monte Carlo Methods in Financial Engineering*. Springer, 2003.

[38] P. W. Glynn and D. L. Iglehart, Simulation methods for queues: An overview, *Queueing Systems*, 3(3):221–255, 1988.

[39] B. V. Gnedenko, Y. K. Belyayev, and A. D. Solovyev, *Mathematical Methods of Reliability*. New York: Academic Press, 1969.

[40] H. Gould, J. Tobochnik, and C. Wolfgang, *An Introduction to Computer Simulation Methods: Applications to Physical Systems*, 3rd ed. Boston, MA, USA: Addison-Wesley Longman Publishing Co., Inc., 2005.

[41] A. Goyal, P. Shahabuddin, P. Heidelberger, V. F. Nicola, and P. W. Glynn, A unified frame-work for simulating Markovian models of highly dependable systems, *IEEE Trans. on Comput.*, 41(1):36–51, January 1992.

[42] A. Goyal and A. N. Tantawi, Evaluation of performability for degradable computer systems, *IEEE Trans. on Comput.*, C-36:738–744, June 1987.

[43] V. Grassi, L. Donatiello, and G. Iazeolla, Performability evaluation of multi-component fault-tolerant systems, *IEEE Trans. on Rel.*, 37(2):216–222, June 1988.

[44] W. K. Grassmann, M. I. Taksar, and D. P. Heyman, Regenerative analysis and steady-state distributions for Markov chains, *Operations Research*, 33(5):1107–1116, 1985.

[45] P. Heidelberger, Fast simulation of rare events in queueing and reliability models, *ACM Trans. on Modeling and Comput. Simulations*, 54(1):43–85, January 1995.

[46] B. R. Iyer, L. Donatiello, and P. Heidelberger, Analysis of performability for stochastic models of fault-tolerant systems, *IEEE Trans. on Comput.*, C-35:902–907, October 1986.

[47] P. Joubert, G. Rubino, and B. Sericola, Performability analysis of two approaches to fault tol-erance. Technical Report 3009, INRIA, Campus de Beaulieu, 35042 Rennes Cedex, France, July 1996.

[48] S. Karlin and H. W. Taylor, *A First Course in Stochastic Processes*. New York – San Francisco – London: Academic Press, 1975.

[49] J. Katzmann, System architecture for non-stop computing, in *14th IEEE Comp. Soc. Int. Conf*, pages 77–80, 1977.

[50] J. G. Kemeny and J. L. Snell, *Finite Markov Chains*. New York Heidelberg Berlin: Springer-Verlag, 1976.

[51] V. G. Kulkarni, V. F. Nicola, R. M. Smith, and K. S. Trivedi, Numerical evaluation of per-formability and job completion time in repairable fault-tolerant systems, in *Proc. IEEE 16th Fault-Tolerant Computing Symp., (FTCS'16)*, pages 252–257, Vienna, Austria, July 1986.

[52] S. S. Lavenberg, *Computer Performance Modeling Handbook*. Notes and reports in com-puter science and applied mathematics. Academic Press, 1983.

[53] G. Le Lann, Algorithms for distributed data-sharing systems which use tickets, in *Proc. of the 3rd Berkeley Workshop on Distributed Data Base and Comput. Networks*, Berkeley, USA, 1978.

[54] P. L'Ecuyer, M. Mandjes, and B. Tuffin, Importance sampling in rare event simulation, in G. Rubino and B. Tuffin, Eds, *Rare Event Simulation using Monte Carlo Methods*, chapter 2, pages 17–38. Chichester: J. Wiley, 2009.

[55] P. L'Ecuyer, G. Rubino, S. Saggadi, and B. Tuffin, Approximate zero-variance importance sampling for static network reliability estimation, *IEEE Trans. on Reliability*, 60(3):590–604, 2011.

[56] J. Ledoux, On weak lumpability of denumerable Markov chains. *Stat. and Probability Lett.*, 25:329–339, 1995.

[57] J. Ledoux, A geometric invariant in weak lumpability of finite Markov chains, *J. of Appl. Probability*, 34(4):847–858, 1997.

[58] J. Ledoux, G. Rubino, and B. Sericola, Exact aggregation of absorbing Markov processes using the quasi-stationary distribution, *J. of Appl. Probability*, 31:626–634, 1994.

[59] J. Ledoux and B. Sericola, On the probability distribution of additive functionals of jump Markov processes, in *Proc. of the Int. Workshop on Appl. Probability (IWAP'08)*, Compiègne, France, July 2008.

[60] E. E. Lewis and F. Böhm, Monte Carlo simulation of Markov unreliability models, *Nucl. Eng. and Design*, 77:49–62, 1984.

[61] C. Liceaga and D. Siewiorek, Automatic specification of reliability models for fault tolerant computers. Technical Report 3301, NASA, July 1993.

[62] S. Mahévas and G. Rubino, Bound computation of dependability and performance measures, *IEEE Trans. on Comput.*, 50(5):399–413, May 2001.

[63] M. Malhotra, J. K. Muppala, and K. S. Trivedi, Stiffness-tolerant methods for transient analysis of stiff Markov chains, *Microelectronics and Rel.*, 34(11):1825–1841, 1994.

[64] T. H. Matheiss and D. S. Rubin, A survey and comparison of methods for finding all vertices of convex polyhedral sets, *Math. of Operations Research*, 5(2):167–185, 1980.

[65] C. D. Meyer, Stochastic complementation, uncoupling Markov chains and the theory of nearly reducible systems, *Siam Review*, 31(2):240–272, 1989.

[66] J. F. Meyer, On evaluating the performability of degradable computing systems. *IEEE Trans. on Comput.*, C-29:720–731, August 1980.

[67] J. F. Meyer, Closed-form solutions for performability, *IEEE Trans. on Comput.*, C-31:648–657, July 1982.

[68] J. Misra, Detecting termination of distributed computations using markers, in *Proc. of the 2nd Annual ACM Symp. on Principles of Distributed Computing*, Montreal, Canada, 1983.

[69] R. R. Muntz, E. de Souza e Silva, and A. Goyal, Bounding availability of repairable computer systems, *IEEE Trans. on Comput.*, 38(12):1714–1723, December 1989.

[70] R. R. Muntz and J. C. S. Lui, *Evaluating Bounds on Steady-State Availability of Repairable Systems from Markov Models*, pages 435–454. Marcel Dekker, Inc., 1991.

[71] R. R. Muntz and J. C. S. Lui, Computing bounds on steady state availability of repairable computer systems, *J. of Assoc. for Computing Machinery*, 41(4):676–707, July 1994.

[72] R. R. Muntz and J. C. S. Lui, Bounding the response time of a minimum expected delay routing system, *IEEE Trans. on Comput.*, 44(5):1371–1382, December 1995.

[73] J. K. Muppala and K. S. Trivedi, *Numerical Transient Solution of Finite Markovian Queueing Systems*, chapter 13, pages 262–284. Oxford University Press, 1992.

[74] H. Nabli and B. Sericola, Performability analysis: a new algorithm, *IEEE Trans. on Comput.*, C-45:491–494, April 1996.

[75] H. Nabli and B. Sericola, Performability analysis for degradable computer systems, *Comput. and Math. with Applicat.*, 39(3-4), 217–234, 2000.

[76] M. K. Nakayama, General conditions for bounded relative error in simulations of highly reliable Markovian systems, *Advances in Appl. Probability*, 28:687–727, 1996.

[77] V. F. Nicola, M. K. Nakayama, P. Heidelberger, and A. Goyal, Fast simulation of highly dependable systems with general failure and repair processes, *IEEE Trans. on Comput.*, 42(12):1440–1452, December 1993.

[78] K. R. Pattipati, Y. Li, and H. A. P. Blom, A unified framework for the preformability evaluation of fault-tolerant computer systems, *IEEE Trans. on Comput.*, 42:312–326, March 1993.

[79] I. G. Petrovsky, *Lectures on Partial Differential Equations*. New York: Interscience Publishers, 1962.

[80] O. Pourret, *The Slow-Fast Approximation for Markov Processes in Reliability (text in French)*. PhD thesis, University of Paris-Sud Orsay, 1998.

[81] J. G. Propp and D. B. Wilson, Coupling from the past, in D. A. Levin, Y. Peres, and E. L. Wilmer, Eds, *Markov Chains and Mixing Times*. Oxford: American Mathematical Society, 2008.

[82] M. Raynal and G. Rubino, An algorithm to detect token loss on a logical ring and to regenerate lost tokens, in *Proc. of the Int. Conf. on Parallel Process. and Applicat.*, L'Aquila, Italy, 1987.

[83] C. P. Robert and G. Casella, *Monte Carlo Statistical Methods (Springer Texts in Statistics)*. Secaucus, NJ, USA: Springer-Verlag New York, Inc., 2005.

[84] R. T. Rockafellar, *Convex Analysis*. Princeton, New Jersey: Princeton University Press, 1970.

[85] S. Ross, *Stochastic Processes*, 2nd ed. Wiley, April 1996.

[86] G. Rubino, Network reliability evaluation, in K. Bagchi and J. Walrand, Eds, *The State-of-the art in Performance Modeling and Simulation*. Gordon and Breach Books, 1998.

[87] G. Rubino and B. Sericola, Accumulated reward over the n first operational periods in fault-tolerant computing systems. Technical Report 1028, INRIA, Campus de Beaulieu, 35042 Rennes Cedex, France, May 1989.

[88] G. Rubino and B. Sericola, Distribution of operational times in fault-tolerant systems modeled by semi-Markov reward processes, in *Proc. of the 12th Int. Conf. on Fault-Tolerant Systems and Diagnostics (FTSD'12)*, Prague, Czech Republic, September 1989.

[89] G. Rubino and B. Sericola, On weak lumpability in Markov chains, *J. of Appl. Probability*, 26:446–457, 1989.

[90] G. Rubino and B. Sericola, Sojourn times in Markov processes, *J. of Appl. Probability*, 27:744–756, 1989.

[91] G. Rubino and B. Sericola, A finite characterization of weak lumpable Markov processes. Part I: The discrete time case, *Stochastic Processes and their Applications*, 38:195–204, 1991.

[92] G. Rubino and B. Sericola, Successive operational periods as dependability measures, in A. Avizienis and J. C. Laprie, Eds, *Dependable Computing and Fault-Tolerant Systems, Vol. 4*, pages 239–254. Springer-Verlag, 1991. (Also presented in "1st International Conference on Dependable Computing for Critical Applications", Santa Barbara, USA, August 1989.)

[93] G. Rubino and B. Sericola, Interval availability analysis using operational periods, *Performance Evaluation*, 14(3-4):257–272, 1992.

[94] G. Rubino and B. Sericola, A finite characterization of weak lumpable Markov processes. Part II: The continuous time case, *Stochastic Processes and their Applications*, 45:115–125, 1993.

[95] G. Rubino and B. Sericola, Interval availability distribution computation, in *Proc. of the 23rd IEEE Int. Symp. on Fault Tolerant Computing (FTCS'23)*, Toulouse, France, June 1993.

[96] G. Rubino and B. Sericola, Sojourn times in semi-Markov reward processes. Application to fault-tolerant systems modelling, *Reliability Engineering and Systems Safety*, 41, 1993.

[97] G. Rubino and B. Sericola, Interval availability analysis using denumerable Markov processes. Application to multiprocessor subject to breakdowns and repair, *IEEE Trans. on Comput. (Special Issue on Fault-Tolerant Computing)*, 44(2), February 1995.

[98] G. Rubino and B. Tuffin, Markovian models for dependability analysis, in G. Rubino and B. Tuffin, Eds, *Rare Event Simulation using Monte Carlo Methods*, chapter 6, pages 125–144. Chichester: J. Wiley, 2009.

[99] G. Rubino and B. Tuffin, Eds. *Rare Event Simulation using Monte Carlo Methods*. Chichester: J. Wiley, 2009.

[100] P. Semal, Refinable bounds for large Markov chains, *IEEE Trans. on Comput.*, 44(10):1216–1222, October 1995.

[101] E. Seneta, *Non-Negative Matrices and Markov Chains*, Springer Verlag, 1981.

[102] B. Sericola, Closed-form solution for the distribution of the total time spent in a subset of states of a homogeneous Markov process during a finite observation period, *J. of Appl. Probability*, 27, 1990.

[103] B. Sericola, Interval availability of 2-states systems with exponential failures and phase-type repairs, *IEEE Trans. on Rel.*, 43(2), June 1994.

[104] B. Sericola, Occupation times in Markov processes, *Communications in Statistics – Stochastic Models*, 16(5), 2000.

[105] B. Sericola, *Markov Chains: Theory, Algorithms and Applications*. Iste Series. Wiley, 2013.

[106] P. Shahabuddin, Importance sampling for the simulation of highly reliable Markovian systems, *Management Science*, 40(3):333–352, March 1994.

[107] T. J. Sheskin, A Markov partitioning algorithm for computing steady-state probabilities, *Operations Research*, 33(1):228–235, 1985.

[108] R. M. Smith, K. S. Trivedi, and A. V. Ramesh, Performability analysis: measures, an algorithm, and a case study, *IEEE Trans. on Comput.*, C-37:406–417, April 1988.

[109] W. J. Stewart, *Probability, Markov Chains, Queues and Simulation: The Mathematical Basis of Performance Modeling*. Princeton, NJ, USA: Princeton University Press, 2009.

[110] N. M. van Dijk, A simple bounding methodology for non-product form finite capacity queueing systems, in *Proc. of the 1st Int. Workshop on Queueing Networks with Blocking*, Raleigh, USA, May 1988.

[111] D. Williams, *Probability with Martingales*. Cambridge University Press, 1991.

Index